Disasters and Society – From Hazard Assessment to Risk Reduction

Proceedings of the
International Conference

Universität Karlsruhe (TH)
Germany

July 26 - 27, 2004

Editors: Dörthe Malzahn, Tina Plapp

Organised by

**Graduiertenkolleg "Naturkatastrophen", Universität Karlsruhe (TH)
Sonderforschungsbereich "Starkbeben", Universität Karlsruhe (TH)
Center for Disaster Management and Risk Reduction Technology,
Universität Karlsruhe (TH) and GeoForschungsZentrum Potsdam**

Bibliografische Information Der Deutschen Bibliothek

Die Deutsche Bibliothek verzeichnet diese Publikation in der Deutschen Nationalbibliografie; detaillierte bibliografische Daten sind im Internet über http://dnb.ddb.de abrufbar.

ISBN 3-8325-0585-7

Logos Verlag Berlin
Comeniushof, Gubener Str. 47,
10243 Berlin
Tel.: +49 030 42 85 10 90
Fax: +49 030 42 85 10 92
INTERNET: http://www.logos-verlag.de

Table of Contents

Understanding and Modelling of Hazards

Disaster Management

Risk Reduction in Industrialised Societies

Panel Discussion

Editorial Notes

Some authors of accepted conference contributions did not submit a paper for the proceedings. The decision about paper acceptance was made by the scientific committee.

Neither the author(s), nor the editors and the publisher make any warranty, expressed or implied, that the statements, calculation methods, procedures and programmes included in this proceedings volume are free from error. The use of any information from this book is at the users own risk, and the author(s), editors and publisher disclaim any liability for damage, whether direct, incidental or consequential, arising from the information, methods or procedures of this book.

The conference organisers, editors and publisher are not responsible for statements or opinions made in the papers. The papers have been reproduced from the authors' original typescript to minimize delay.

Preface

The damage caused by disasters and their impact on societies have increased over the past decades. Extreme events cannot always be prevented. However, we can minimize the amount of damage which they inflict upon society. It is therefore important to gain a profound understanding of potentially hazardous processes, reliable forecasting methods, effective systems for warning and disaster management as well as strategies for disaster mitigation and risk reduction. Vulnerability and disaster impact on society are influenced by many factors. It is therefore necessary to combine methods and knowledge from various academic disciplines and from all relevant sectors of society.

This book compiles oral and poster presentations given at the international conference "Disasters and Society – From Hazard Assessment to Risk Reduction". The book provides insights into the state of disaster research in different *academic disciplines* as well as insights into the *practice* of risk and disaster management. The contributions cover the entire functional chain associated with disasters in industrial societies. They are written by national and international specialists from universities, administration, civil defence, private companies and re-insurers. The contributions reflect the diversity and the interdisciplinary character of disaster research. They report work on different natural hazards such as severe storms, heavy precipitation, flood, earthquakes, and mass movements (e.g. land slides) as well as research results in the area of human-made hazards. The applied methodologies come from many different areas such as the natural sciences, engineering, informatics, geography, sociology, actuarial statistics and mathematics. The book is structured according to the six conference sessions:

- Understanding and modelling of hazards (Chair: C. Kottmeier)
- Hazard and risk assessment (Chair: F. Nestmann)
- Forecasting and early warning (Chair: J. Zschau)
- Information and communication (Chair: F. Wenzel)
- Disaster management (Chair: F. Gehbauer)
- Risk reduction in industrialized societies (Chair: U. Werner)

As a particular strength of the book, each of the six parts given above contains work addressing different types of hazards as well as specialists from a variety of areas of expertise.

The international conference "Disasters and Society – From Hazard Assessment to Risk Reduction" took place in July 2004 at the Universität Karlsruhe (TH) in the city of Karlsruhe, Germany. The conference was organised by three interdisciplinary research institutions which are situated at the Universität Karlsruhe (TH), collaborate and have developed Karlsruhe into a centre for disaster research in Germany. The three institutions are:

- The Postgraduate Programme "Natural Disasters" (Graduiertenkolleg 450 "Naturkatastrophen")
- The Collaborative Research Center "Strong Earthquakes: A Challenge for Geoscience and Civil Engineering" (Sonderforschungsbereich 461 "Starkbeben: Von geowissenschaftlichen Grundlagen zu Ingenieurmaßnahmen").
- The Center for Disaster Management and Risk Reduction Technology (CEDIM)

The research programmes of the three institutions are depicted in the next sections.

The conference "Disasters and Society – From Hazard Assessment to Risk Reduction" was funded by the Deutsche Forschungsgemeinschaft (DFG). The conference proceedings were produced with the financial support of CEDIM.

The editors would like to thank all authors for their contributions and their efforts to enable a timely publishing of the book.

Karlsruhe, August 2004

Dörthe Malzahn
Tina Plapp

The Postgraduate Programme "Natural Disasters" (GRK 450)

All natural disasters have in common that they are complex phenomena. An interdisciplinary cooperation is required to improve the understanding and prediction of the basic mechanisms which cause natural hazards, to quantify risk, to understand the impacts of disasters on society and to develop measures and tools for damage reduction. The Universität Karlsruhe (TH) presents unique conditions for such an interdisciplinary programme. The Postgraduate Programme "Natural Disasters" (GRK 450) was established in October 1998 and is supported by a total of 15 institutes of the Universität Karlsruhe (TH) from the department of civil engineering, environmental and geosciences, the department of computer science, the department of physics, the department of mathematics, and the department of economics and business engineering. The Postgraduate Programme "Natural Disasters" is structured in three periods of three years. The third period will start in October 2004. The programme is funded by the Deutsche Forschungsgemeinschaft (DFG) and the federal state of Baden-Württemberg.

The Postgraduate Programme "Natural Disasters" was established with the aim of developing adequate modelling methods to meet the rising demand for problem-oriented know-how and solutions in the field of natural disaster research. The programme promotes doctoral students by providing funds and the opportunity to work and prepare their PhD theses in a comprehensive research context. The focus of the programme is to endow the participants with the ability to understand and evaluate the relevant relationships of natural disasters and with the ability to propose and implement adequate solutions needed for an optimal disaster management. Scientists trained in this manner are increasingly demanded not only by research but also by government institutions, by the insurance industry and by businesses offering commercial disaster management advice.

The Postgraduate Programme pursues the following key objectives:
- *Investigation of the entire chain of effects*, which reaches from vulnerability assessment and risk prediction to measures of damage reduction and to the study of the risk culture maintained by the respective societies. This goal is realized by a network of up to 14 PhD and 2 postdoc projects.
- *Analysis and characterisation of different types of natural hazards*, which differ with respect to physical model complexity, data base precision and the possibility of predictability and damage reduction.
- *Development, application and validation of modern methods of mathematics and computer science.* This concerns the handling of fuzzy and imprecise information as well as the prediction of the behaviour of complex systems of which intrinsic mechanisms are known only vaguely. A further aim is to provide tools for rapid information processing and information exchange to enable an effective disaster management during disasters.

- *Improvement of rescue and recovery machinery* which are decisive for the quality of initiated disaster response measures.

Karlsruhe, August 2004

Fritz Gehbauer
Graduiertenkolleg „Naturkatastrophen"

Collaborative Research Center "Strong Earthquakes: A Challenge for Geosciences and Civil Engineering" (CRC 461)

Strong earthquakes in the Romanian Vrancea area have caused a high toll of casualties and extensive damage over the last centuries. The average recurrence rate makes another strong event within the next two decades highly probable and provides a challenge to mitigate its impact. Romanian and German scientists from various fields (geology, seismology, civil engineering, operation research) organized themselves in the Collaborative Research Center "Strong Earthquakes: A Challenge for Geosciences and Civil Engineering" (Germany) and the Romanian Group for "Strong Vrancea Earthquakes" in a multidisciplinary attempt towards earthquake mitigation. The Collaborative Research Center is funded as strategic research project by the Deutsche Forschungsgemeinschaft (DFG). Funding commenced in 1996, the third project period will end in 2004, and another period until 2008 is foreseen.

Key objectives are:

- Understanding of the tectonic processes that are responsible for the strong intermediate depth seismicity beneath Vrancea;
- Developing realistic models and predictions of ground motion;
- Prognosis of potential damage in case of a strong earthquake;
- Risk reduction by appropriate civil engineering concepts.

Highlights of recent accomplishments are: Installation of a state-of-the-art accelerometer network for monitoring of strong ground motion; installation of a Global Positioning System network for three-dimensional deformation analysis; realization of an international seismological tomographic experiment with 120 stations in and around the Vrancea region; instrumentation of a multidisciplinary test site including a test building and other buildings in Bucharest; development of a soil model for the Romanian capital Bucharest; installation of an earthquake early warning system with 30 seconds lead time; development of the damage simulation tool EQSIM; conceptualization of a Disastermanagement-Tool that integrates modeling tools, decision support systems, and communication tools and connects those components with a dynamic databank. Geosciences developed an unprecedented understanding of the intermediate depth seismicity beneath Vrancea as ongoing detachment of a subducted slab. This model allows (a) to understand stress transfer processes in the seismogenic volume and to scrutinize non-Poisson earthquake statistics, (b) to quantify the non-radial shape of isoseismic patterns and the associated non-radial attenuation properties, and (c) to constrain the minimum depth of strong upper mantle seismicity.

Forthcoming objectives are:

- Quantitative thermo-mechanical modeling of deep mantle processes responsible for the seismicity;
- Probabilistic site-specific hazard assessment on the best possible data base;

- Development of an urban and regional shakemap;
- Development of a near real-time earthquake information system;
- Finalization and application of the Disastermanagement-Tool

Karlsruhe, August 2004

Friedemann Wenzel
Sonderforschungsbereich 461 "Starkbeben"

The Center for Disaster Management and Risk Reduction Technology (CEDIM)

The frequency of catastrophes - especially natural disasters - and their resulting damage have increased in the last decades. This trend is documented in statistical information for example from reinsurance companies. Risks are growing since more and more population is concentrated in dangerous regions due to the fact that these areas are very often favoured living spaces. Additionally the rapid growth of cities to megacities or conurbations contributes to that phenomenon.

The goal of the "Center for Disaster Management and Risk Reduction Technology" (CEDIM) – founded in December 2002 by the Universität Karlsruhe (TH) and the GeoForschungsZentrum Potsdam - is to understand the risks, detect them early and cope with the consequences at a better level. For that purpose, the research of catastrophes requires collaboration from different scientific disciplines. Only by collectively enhancing the scientific basis the current damage due to catastrophes can be reduced significantly. CEDIM unites experts from different scientific disciplines like geosciences, engineering sciences and economic sciences. Researchers are specialised in fields such as meteorology, water management, civil engineering, geophysics, economics, insurance and risk management, social sciences and geo-informatics.

CEDIM researchers are working in different projects which are linked to each other:

Project Risk Map Germany: The aim of the project is to provide information about several natural hazards and man-made hazards. In addition, the vision is to map the *risk* which is defined through hazard and vulnerability. The work is therefore also concentrated on modelling and estimating monetary losses due to catastrophes and their consequences on human beings, infrastructure and nature. The challenge is to integrate methodologies of the involved disciplines and to produce discipline spanning results that are comparable across these disciplines, thus creating a common basis for further work on risk analysis, risk management and risk mitigation. One main and new aspect is the assessment and – if possible –prediction of direct and indirect economic losses and macroeconomic effects of disastrous events.

Project Flood Risk Information and Modelling Tool: The project aims at developing information and modelling tools for the quantification and visualization of flood risk in big catchment areas such as the Elbe catchment area. Another goal is to implement the tools. An internet capable *Flood Risk Atlas* as well as a *River Basin Flood Modelling Tool* are core results.

Project Megacities: Growing cities and therefore growing risk potential will become a major feature of human development of the next 30 years. Rapid city development in disaster prone regions cannot be sustainable without significant efforts in disaster mitigation as the vulnerability of megacities grows continuously. Because of the speed of change in large cities, the main challenge in risk reduction research is a dynamic approach that takes temporal changes of hazards, vulnerability and exposure appropriately into account. Megacities must be understood as entities where everything is highly variable in time and everything is interacting in a complex non-linear way. As urban growth will determine human development of the next decades understanding and managing risks in megacities will become a key challenge in developing the world in a sustainable way.

Karlsruhe, August 2004

Lothar Stempniewski

Center for Disaster Management and Risk Reduction Technology CEDIM

Understanding and Modelling of Hazards

Tropical Cyclones: A Natural Hazard in the Tropics and the Midlatitudes

Sarah C. Jones

Meteorological Institute, University of Munich, Theresienstr. 37, D-80333 München, Germany

Abstract

Tropical cyclones represent one of the most destructive natural hazards. Due to their strong winds and heavy precipitation they are responsible for significant loss of life and damage to property and constitute a hazard to aviation and shipping. Around 40% of all tropical cyclones move into the midlatitudes and undergo extratropical transition, transforming into extratropical storms that may still produce intense rainfall, very large waves and hurricane force winds. In order to mitigate the threat posed by tropical cyclones it is necessary to provide sufficient warning time to allow for adequate preparation and evacuation. However, providing accurate forecasts of tropical cyclone track and intensity poses a particular challenge for numerical weather forecast models. The basic processes that determine the structure, intensity and motion of a tropical cyclone are described and the changes a tropical cyclone experiences as it moves polewards and interacts with the midlatitude flow discussed. The challenges associated with modelling and forecasting tropical cyclones are explored.

Keywords: tropical cyclones, extratropical transition.

Figure 1: Satellite image of Hurricane Mitch on 26 October 1998. Courtesy of NOAA.

1 Tropical Cyclones: A Natural Hazard in the Tropics

Tropical cyclones represent one of the most destructive natural hazards. Tropical cyclones with maximum sustained surface wind speeds above 17 m s^{-1} are generally called tropical storms and given an individual name. When the maximum sustained surface wind speed is at least 33 m s^{-1} the system is variously called a hurricane, a typhoon, a severe tropical cyclone, a severe cyclonic storm or a tropical cyclone, depending on the geographical location (Neumann, 1993). The strongest tropical cyclones have wind speeds in excess of 70 m s^{-1}. Tropical cyclones produce copious rainfall amounts with precipitation totals that can exceed 1000 mm in 24 h. Consequences of the strong wind and heavy precipitation are storm surge, serious flooding, landslides and heavy seas. Thus tropical cyclones are responsible for significant loss of life and damage to property and constitute a hazard to aviation and shipping.

A number of factors determine the impact of a tropical cyclone. One of the most serious tropical cyclone disasters occurred in Bangladesh in 1970, when the storm surge of between 6 and 9 metres associated with a tropical cyclone making landfall close to high tide was responsible for an estimated 300 000 deaths. Hurricane Mitch (1998) caused the loss of over 9 000 lives in Central America after tremendous rainfall caused flooding and landslides. The severity of the rainfall can be attributed to both the slow motion of the storm and the interaction between the circulation of Mitch and the mountainous terrain. The economic damage of US$5.5 billion caused by Mitch (Munich Re, 2000) includes an estimated 50% loss of agricultural crops in Honduras and the damage and destruction of over 70 000 homes (Guiney and Lawrence, 1998). The tropical cyclone that caused the highest monetary damage to date was Hurricane Andrew (1992). The strong winds, storm surge and heavy seas led to US$26.5 billion economic losses with US$17 billion insured losses, the highest insured losses of any natural disaster (Munich Re, 2000). Despite the threat posed by tropical cyclones they provide essential rainfall to many of the countries they cross.

2 Tropical Cyclones: Structure, Intensity and Motion

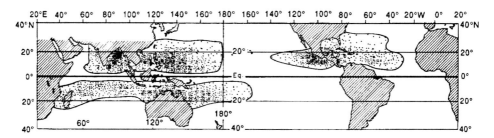

Figure 2: Regions of tropical cyclone formation. Taken from Gray (1975).

The regions in which tropical cyclones develop (Fig. 2) are determined by a number of conditions (e.g. Gray, 1968) including: ocean temperatures of at least 26.5 °C through a depth of at least 60 m, little variation in horizontal wind with height, a pre-existing disturbance containing abundant deep convection, and a distance of at least 500 km from the equator.

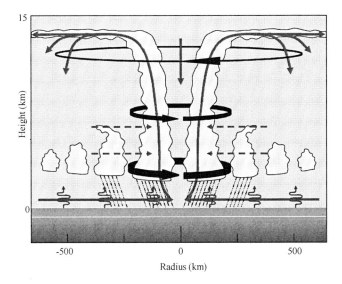

Figure 3: Schematic vertical cross section through a tropical cyclone. Modified from Emanuel (1988).

The most familiar view of a tropical cyclone is that from space. Satellite imagery, such as that of Hurricane Mitch (1998) in Fig. 1, shows us the largely circular pattern of clouds with a cloud-free region, the eye, at the centre. A schematic vertical cross-section through the centre of a tropical cyclone illustrates its main structural features (Fig. 3). The damaging winds are part of a primary circulation depicted by the dark arrows. This circulation is cyclonic in the inner region and strongest near the surface. In an intense tropical cyclone the cyclonic circulation extends up to the tropopause. In the upper troposphere at larger radii the circulation is anticyclonic.

The main energy source of a tropical cyclone is the flux of latent and sensible heat from the ocean surface (the wavy arrows). The so-called secondary circulation of a tropical cyclone is indicated by the lighter-coloured arrows. Although the secondary circulation is much weaker than the primary circulation, it is of vital importance for the maintenance and intensification of a tropical cyclone. In the secondary circulation the air flows inwards in the boundary layer and is warmed and moistened by the surface fluxes. The air rises in the deep, precipitating convection in the eyewall, flows outwards in the upper troposphere, and descends slowly at large radius. Radiative cooling offsets part of the adiabatic warming in the outer descent region. There is descent in the eye. The role of the deep convection in the eyewall is to redistribute vertically the energy gained from the ocean, resulting in the air ascending near the center of the storm being warmer than the air in the environment of the tropical cyclone. This temperature difference drives the secondary circulation (Emanuel, 1988). For a tropical cyclone to intensify there must be inflow in the middle troposphere (indicated by the dashed arrows). This inflow brings rings of rotating fluid towards the centre. Conservation of angular momentum requires that this fluid spins faster, leading to stronger tangential flow.

The motion of a tropical cyclone (Fig. 4) can be attributed to the large-scale flow across its centre (Elsberry, 1995). A large part of the motion of a tropical cyclone is associated with the large-scale environmental flow in the tropics or with transient weather systems in the environment of the tropical cyclone (e.g. an upper-level trough). A second important component arises from the modification of the environment by the circulation of the tropical cyclone, which leads to the formation of large-scale asymmetric flow.

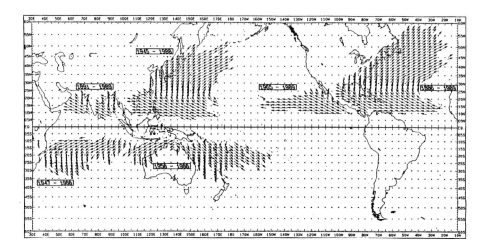

Figure 4: Annual average of tropical cyclone motion during the indicated period for various regions. Taken from Neumann (1993).

3 Tropical Cyclones: A Natural Hazard in Midlatitudes

The threat to life and property from a tropical cyclone described above is widely appreciated. It is perhaps less well known that a tropical cyclone can evolve into a fast-moving and occasionally rapidly developing midlatitude storm that contains intense rainfall and strong winds (Jones et al., 2003). Over 40% of all tropical cyclones undergo this process, known as extratropical transition or ET (Hart and Evans, 2001; Klein et al. 2000). The acceleration of a tropical cyclone into the midlatitudes leads to an increased asymmetry in the wind field and rapid growth in surface wave-heights. The structure of the tropical cyclone changes from a nearly symmetric distribution of clouds and circular enclosed eye to the asymmetric structure typical of an extratropical cyclone, as seen for Hurricane Mitch in Fig. 5. Typically, the movement of a tropical cyclone into the midlatitudes is accompanied by a decrease in intensity. However, the interaction with an extratropical system during ET may result in rapid intensification of the ex-tropical cyclone. During ET the interaction with the large-scale midlatitude flow influences the formation of precipitation. Thus, heavy precipitation extends over a larger area poleward of the tropical cyclone center so that heavy precipitation can fall over land without the tropical cyclone making landfall. If the heavy precipitation associated with the central region of the tropical cyclone then falls in the same region as the pre-storm precipitation, the potential for flooding is increased. Furthermore, the increased asymmetry in precipitation and wind fields can lead to the strongest winds being associated with dry air; over locations such as southwest Australia this enhances the bush fire hazard.

The societal impact of ET can be substantial. In North America loss of life occurred because of severe flooding associated with the ET of Tropical Storm Agnes (1972). In his study of the ET of Hurricane Hazel (1954), Palmén (1958) was the first to describe the rapid intensification and high amounts of precipitations that are associated with an ET over land. This case resulted in 83 deaths in the Toronto area of southern Ontario.

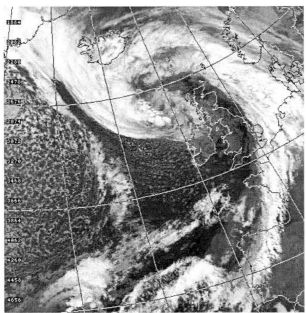

Figure 5: Infrared satellite image of ex-Hurricane Mitch on 9 November 1998. Courtesy of Dundee Satellite Receiving Station.

In the northwest Pacific, severe flooding and landslides have occurred in association with ET. An example is the ET of Tropical Storm Janis (1995) over Korea, in which at least 45 people died and 22,000 people were left homeless. In one southwest Pacific ET event (Cyclone Bola) over 900 mm of rain fell over northern New Zealand. Another event brought winds gusting to 75 ms^{-1} to New Zealand's capital city, Wellington, resulting in the loss of 51 lives when a ferry capsized. Extratropical transition has produced a number of weather related disasters in eastern Australia, due to severe flooding, strong winds and heavy seas (e.g. Cyclone Wanda in 1974). In Western Australia there can be an increased fire hazard associated with ET. Tropical systems which re-intensify after ET in the North Atlantic constitute a hazard for Canada and for northwest Europe. The extratropical system which developed from Hurricane Lili (1996) was responsible for seven deaths and substantial economic losses in the United Kingdom, France and Germany.

More difficult to quantify is the indirect effect of tropical cyclones on the development of severe weather in the midlatitudes. The interaction between a tropical cyclone and the midlatitude flow can create favourable conditions for the development of severe weather well to the east of the tropical cyclone itself. One of the key features responsible for the Great Storm of October 1987 could be traced back to the air flowing out of Hurricane Floyd (Hoskins and Berrisford, 1988). This storm was one of the most costly European winter storms with the loss of 17 lives, economic losses of US$3.7 billion and insured losses of US$3.1 billion.

4 Forecast Challenges

In order to mitigate the threat posed by tropical cyclones it is necessary to provide sufficient warning time to allow for adequate preparation and evacuation. However, since tropical cyclones develop over the tropical oceans where there is a lack of in-situ data and due to the complex physical processes responsible for their development and intensification, providing accurate forecasts of their track and intensity poses a particular challenge for operational forecasters. As the motion of a tropical cyclone is determined to a large extent by the large-scale steering flow, numerical weather forecast models provide useful forecast guidance. Track forecast errors continue to decrease steadily so that the average 3 day track error is now of the order of 350 km; this is comparable to a 2 day track error 20 years ago. Tropical cyclone intensification depends on the complex interplay between surface fluxes, deep convection, vortex dynamics, and the oceanic response. Many of the crucial factors are still poorly understood so that their representation in numerical weather prediction models is incomplete. Thus, intensity forecasts do not exhibit the same steady improvement as track forecasts. Average 72 h forecast errors are around 20 knots and individual forecast errors can be greatly in excess of this figure.

During ET the increased forward speed results in a decreased warning time for hazardous weather and, if the timing is misjudged, in large errors in the forecast track. The major challenge of forecasting reintensification as an extratropical system is complicated by the fact that the interpretation of satellite imagery appropriate for a tropical cyclone would suggest that the system is weakening. Due to the change in the mechanisms responsible for the development of precipitation, current operational techniques for quantitative precipitation forecasting are often "stressed" during ET. The numerical prediction of ET is hindered by the necessity of representing both the tropical cyclone and the midlatitude circulation into which it is moving. Problems relate to the difficulty in accurately initializing both the tropical cyclone structure and the midlatitude flow as well as to the current inability of operational models to adequately represent the inner-core of a tropical cyclone whilst covering a spatial scale sufficient to forecast the evolution of the midlatitude flow. Furthermore, physical processes (convection, boundary-layer processes, air-sea interaction) play a crucial role in ET and may not be represented adequately in current operational models.

References

Elsberry, R. L.: "Tropical cyclone motion" - Chapter 4 in *Global Perspectives on Tropical Cyclones,* WMO/TD-No. 693, Report No. TCP-38, World Meteorological Organization, Geneva, Switzerland.

Emanuel, K. A.: Towards a general theory of hurricanes. *American Scientist,* 76, 371-379, 1988.

Gray, W. M.: Global view of the origin of tropical disturbances and storms. *Mon. Wea. Rev.,* 96, 669-700, 1968.

Gray, W. M.: Tropical cyclone genesis. Dept. of Atmos. Sci. Paper No. 323, Colorado State University, Ft. Collins, CO 80523, 121 pp., 1975.

Guiney, J. L. and Lawrence, M. B.: Preliminary report on Hurricane Mitch. Available at http://www.nhc.noaa.gov, 1998.

Hart, R. E. and Evans, J. L.: A climatology of extratropical transition of Atlantic tropical cyclones. *J. Clim.,* 14, 546-564, 2001.

Jones, Sarah C., Harr, Patrick A., Abraham, Jim, Bosart, Lance F., Bowyer, Peter J., Evans, Jenni L., Hanley, Deborah E., Hanstrum, Barry N., Hart, Robert E., Lalaurette, François, Sinclair, Mark R., Smith, Roger K., Thorncroft, Chris: "The Extratropical Transition of Tropical Cyclones: Forecast Challenges, Current Understanding, and Future Directions," *Weather and Forecasting,* 18, 1052-1092, 2003.

Klein, P. M., Harr, P. A., and Elsberry, R. L.: Extratropical transition of western North Pacific tropical cyclones: Midlatitude and tropical cyclone contributions to reintensification. *Mon. Wea. Rev.,* 130, 2240-2259, 2001.

Munich Re Group: *World of Natural Hazards,* 2000.

Neumann, C. J., "Global Overview" - Chapter 1 in *Global Guide to Tropical Cyclone Forecasting,* WMO/TC-No. 560, Report No. TCP-31, World Meteorological Organization, Geneva, Switzerland, 1993.

Palmén, E.: Vertical circulation and release of kinetic energy during the development of hurricane Hazel into an extratropical storm. *Tellus,* 10, 1-23, 1958

Impact of Long-term Correlations on the Distribution of Extreme Events

Dörthe Malzahn

Institute for Mathematical Stochastics, University of Karlsruhe, D-76218 Karlsruhe, Germany

Abstract

Using simple models, I study how the probability density $p(T)$ of the recurrence intervals T between subsequent threshold exceedances changes for statistically dependent data. I show qualitatively and quantitatively that persistence and a periodic trend yield different patterns for the clustering of threshold exceedances.

Keywords: distribution of recurrence intervals, clustering of threshold exceedances

1 Introduction

Standard models for the estimation of the probability of occurrence of extreme events and threshold exceedances from a finite sample of observations $Z(N) = \{z_1, \ldots, z_N\}$ are based on the assumption that the observations z_i are statistically independent. In contrast, the existence of long-term correlations has been shown for a variety of naturally occurring time series' such as river runoffs (Koscielny-Bunde et al., 2003). These studies consider the observed time series $\mathcal{Z}(t)$ as realization of a stochastic process with trend (average value) $\bar{\mathcal{Z}}(t)$ and show that the time series of the fluctuations $\Delta \mathcal{Z}(t) \doteq \mathcal{Z}(t) - \bar{\mathcal{Z}}(t)$ with respect to the trend exhibits correlations $C(t_2, t_1) \sim |t_2 - t_1|^{-\gamma}$ which decay algebraically with the time difference $|t_2 - t_1|$ for values $|t_2 - t_1| \geq \frac{1}{2}$ year. Consequently, one is lead to conclude that the assumption of statistical independence is violated. Moreover, processes with a slowly decaying auto-correlation function are persistent and therefore likely to generate a clustering of threshold exceedances.

Statistical dependence may have only a minor effect on the asymptotic distribution of the maximum. If the data is statistically dependent but satisfies a certain set of conditions, the *type* of the asymptotic distribution of the maximum is preserved (see Hsing et al., 1988, O'Brien, 1987).[1] These conditions are moderate enough to allow a clustering of threshold exceedances. In this paper, I study the *probability density of recurrence intervals* between subsequent exceedances of a given threshold. This density will change in *type* due to statistical dependence in the data.

The paper is structured as follows: I briefly inspect the probability density $p(T)$ of the size of the recurrence intervals for threshold exceedances of independent identically distributed data and continue with a study of the impact of persistence using the Ornstein-Uhlenbeck process as a simple model (Subsection 2.1). The Ornstein-Uhlenbeck process is a Markov process with no trend and an exponentially decaying auto-correlation function. It shows persistent behavior. Although a trend and persistent behavior appear similar in data, they are of a different statistical nature and result in different patterns for the clustering of threshold exceedances. To show this, I examine the probability density of the size of the recurrence intervals for threshold exceedances of data which have a periodic trend and statistically independent fluctuations (Subsection 2.2). I summarize my conclusions in Section 3.

---

[1] Note however that the higher order statistics changes.

2 Distribution of Recurrence Intervals between Subsequent Extreme Events

I will assume a series of ordered observations $\{z_i\}$ of a continuous stochastic process $\mathcal{Z}(t)$ which are taken at discrete times $z_i = z(t_i)$ with constant time step Δt between subsequent observations. For example, the values z_i may represent extremal or average values of a process of interest in the time interval $\left[\Delta t\left(i - \frac{1}{2}\right), \Delta t\left(i + \frac{1}{2}\right)\right]$.

I am interested in the probability density $p(n)$ of the size of the recurrence interval $T = n\Delta t$ between subsequent exceedances of a given threshold u. Due to the discreteness of the observations, the recurrence interval $T = n\Delta t$ will be measured in integer multiples of the time step Δt. The probability density $p(n)$ is exponential for independent identically distributed observations (see e.g. Embrechts et al., 1997, p.305)

$$p(n) = (1 - q_u)^n q_u \tag{1}$$

where $n \geq 0$ is a natural number [2] and q_u denotes the probability to find a threshold exceedance $z_i \geq u$. $p(0)$ is the probability to find a set of neighboring observations z_i, z_{i+1} which both exceed the threshold. The interpretation of such instances is ambiguous and can correspond to both a tight clustering of high-threshold exceedances or instances where a threshold exceedance lasts longer than time step Δt.

When dealing with extreme events, many works use the return period $T(u)$ which is the average (mean) size of the recurrence interval [3]

$$T(u) = \Delta t \sum_{n=0}^{\infty} p(n)\, n = \frac{\Delta t}{q_u}(1 - q_u) \approx \frac{\Delta t}{q_u} \;. \tag{2}$$

Note however, that the exponential distribution Eq. (1) is not well characterized by its mean Eq. (2). Specifically, we can calculate the standard deviation $\sigma_{T(u)}$ of the return period and find that it is of the same size as the return period itself. From Eq. (1) follows

$$\sigma_{T(u)} = \frac{T(u)}{\sqrt{(1 - q_u)}} \approx T(u) \;. \tag{3}$$

To give an illustration, the dashed curve in Figures 1 and 2 (right panel) show an example of Eq. (1) for $q_u = 0.01$. For events with smaller probability q_u, the density Eq. (1) becomes very flat in the sense that a broad range of values n has almost equal probability. The most probable value is always $p(0)$.

The probability density of the recurrence interval is sensitive to the statistical nature of the process and can supply useful additional information. To demonstrate this, I have considered simple statistical models for generation of the data $\{z_i\}$. The models share the common property that they have a bounded trend and that the probability density of the data $\{z_i\}$ has an exponentially decaying tail. For each model, I have selected a constant model specific threshold u such that the probability of a threshold exceedance $z_i \geq u$ equals the same value $q_u = 0.01$. I will show that the probability distribution of the recurrence interval for subsequent threshold exceedances differs strongly between the models despite the fact that the return period (mean recurrence time) is the same.

2.1 Example 1: Data With No Trend and Statistically Dependent Fluctuations

This example was inspired by the finding of Koscielny-Bunde et al. (2003) that the fluctuations of natural time series' such as river runoffs have algebraically decaying auto-correlation

---

[2] I include the value $p(0)$ and normalize $\sum_{n=0}^{\infty} p(n) = 1$.
[3] The factor $(1 - q_u) \approx 1$ in Eq. (2) is due to the normalization of $p(n)$, see above.

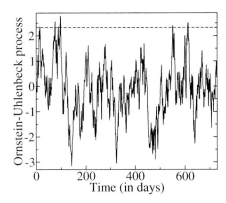 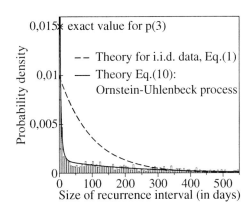

Figure 1: *Left:* Time series of the Ornstein-Uhlenbeck process for $D = 2k$ and $k = 1/10$. The process has no trend and shows persistence due to the statistically dependent fluctuations. The threshold u (dashed line) was set such that $q_u = 0.01$. *Right:* Corresponding probability density of the size of the recurrence intervals $p(n)$ between subsequent threshold exceedances. Simulation: Histogram. Theory: Bold line. The dashed line shows the density $p(n)$ for i.i.d. data (Eq. (1) with $q_u = 0.01$).

functions $C(t_2, t_1) \sim |t_2 - t_1|^{-\gamma}$ for time differences $|t_2 - t_1| \geq \frac{1}{2}$ year. The latter leads to persistent behavior of the fluctuations which resembles a local trend, but is of a different statistical nature. Persistence is expected to lead to a clustering of threshold exceedances.

I will study this effect using the Ornstein-Uhlenbeck process (Uhlenbeck and Ornstein, 1930, Wang and Uhlenbeck, 1945) as statistical model for the observations $\{z_i\}$. The Ornstein-Uhlenbeck process is stationary with limit density

$$G_0(z) = \sqrt{\frac{k}{\pi D}} \exp\left(-\frac{z^2}{D/k}\right) \qquad (4)$$

The process has the Markov property, that is, it is fully described by the limit density Eq. (4) and the transition probability

$$G(z(t)|z(t_0)) = \frac{1}{\sqrt{2\pi\sigma^2(t - t_0)}} \exp\left(-\frac{(z(t) - \mu(t, t_0))^2}{2\sigma^2(t - t_0)}\right) \qquad (5)$$

with time-dependent variance $\sigma^2(t - t_0) = (2k)^{-1}D\left(1 - e^{-2k(t-t_0)}\right)$ and mean $\mu(t, t_0) = z(t_0)e^{-k(t-t_0)}$. On average, the process has no trend as can be seen from Eq. (4) and I will set the variance to one, corresponding to $D = 2k$. The free parameter k can be interpreted as inverse correlation length in time. The left panel of Figure 1 shows a realization of the Ornstein-Uhlenbeck process for $k = 1/10$.

Equation (4) yields the probability for a threshold exceedance

$$q_u = 1 - \Phi(u) \qquad (6)$$

with error function $\Phi(u) = (2\pi)^{-1/2} \int_{-\infty}^{u} dx\, e^{-x^2/2}$. I set $q_u = 0.01$ which corresponds to a threshold value $u = 2.326$. For the remainder of the paper, I will interpret the unit time Δt as one day. That is, an event $z_i \geq u$ occurs on average roughly three times a year. The right panel of Figure 1 shows the resulting empirical result (histogram) for the probability $p(n)$ to find a recurrence interval of size $n\Delta t$ between subsequent threshold exceedances.

The Ornstein-Uhlenbeck process is simple enough to allow an approximate analytical calculation of the probability density $p(n)$. The exact calculation of $p(n)$ involves integration over a path of $(n+2)$ variables at times $t_i = t_0 + i\Delta t$ where $i = 0, 1, \ldots, n+1$

$$p(n) = \frac{1}{q_u} \int_u^\infty dz_{n+1} \, G(z_{n+1}|z_n) \int_u^\infty dz_0 \, G_0(z_0) \prod_{i=1}^n \left\{ \int_{-\infty}^u dz_i \, G(z_i|z_{i-1}) \right\} \tag{7}$$

The exact evaluation of Eq. (7) becomes tedious for larger values of n. Instead, I propose the following approximate treatment which is based on an iterative scheme

$$\hat{G}_{l+1}(z) \doteq \int_{V(l)} dz' \, G(z|z') \, \hat{G}_l(z') \tag{8}$$

where $G(z|z')$ denotes the transition probability Eq. (5), $\hat{G}_0(z')$ is given by the limit distribution Eq. (4) and the integration interval is $V(0) = [u, \infty]$ for $l = 0$ and $V(l) = [-\infty, u]$ for all $l \geq 1$. In each step of the iteration $l \to (l+1)$, I approximate the expression Eq. (8) by a Gaussian

$$\hat{G}_{l+1}(z) \approx \frac{1}{\sqrt{2\pi\sigma_{l+1}^2}} \exp\left(-\frac{(z - \mu_{l+1})^2}{2\sigma_{l+1}^2} \right) \prod_{i=0}^l N_i \tag{9}$$

The parameters $\{\mu_{l+1}, \sigma_{l+1}^2\}$ are obtained by matching the first two moments of Eq. (8) with Eq. (9) and $N_i = (2\pi\sigma_i^2)^{-1/2} \int_{V(i)} dz \, \exp\left(-\frac{1}{2\sigma_i^2}(z - \mu_i)^2\right)$. Detailed results are given in Appendix A. The probability density $p(n)$ is obtained from Eq. (9) as

$$p(n) \approx (1 - N_{n+1}) \prod_{i=1}^n N_i \quad \text{for } n \geq 1 \tag{10}$$

$$p(0) \approx (1 - N_1)$$

Note that the approximate result Eq. (10) can be understood as a generalization of Eq. (1) to a Markov process. The factors N_i are obtained as a function of the time step Δt and of the inverse correlation length k (see Appendix A, Eq. (16)-(21)). The approximate result for $p(n)$ is displayed in the right panel of Figure 1 by a bold line. It agrees well with the empirical result for $p(n)$ (histogram). In comparison, the dashed line shows the probability density $p(n)$ for independent identically distributed (i.i.d.) data as given by Eq. (1). Persistence boosts the probability of extremely small recurrence intervals. In the present example, I obtain particularly large probabilities for recurrence intervals of size $T \leq 4\Delta t$ where $p(0) = 0.552$, $p(1) = 0.059$, $p(2) = 0.026$ and $p(3) = 0.015$ (cross in Figure 1, right panel). These exact values were obtained from Eq. (7). The probability $p(n)$ to find recurrence intervals of size $n\Delta t$ between threshold exceedances of the Ornstein-Uhlenbeck process is diminished for a range of intermediate values n and shows a slightly heavier tail for $n \to \infty$ in comparison to i.i.d. data. By construction, the return period $T(u)$ coincides with the return period of the i.i.d. data. However, the standard deviation $\sigma_{T(u)}$ of the return period is significantly larger in comparison to i.i.d. data corresponding to the fact that $p(n)$ is more homogeneous (flatter) for most values of n. This implies, that the standard deviation $\sigma_{T(u)}$ of the return period of the persistent process is *larger* than the return period $T(u)$ itself.

I expect similar results for processes which have a slower decaying auto-correlation function than the Ornstein-Uhlenbeck process.

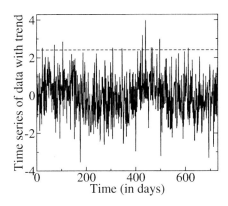 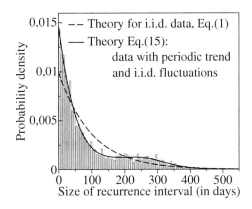

Figure 2: *Left:* Time series with periodic trend and statistically independent fluctuations. The threshold u (dashed line) was set such that $q_u = 0.01$. *Right:* Corresponding probability distribution of the recurrence intervals $p(n)$ between subsequent threshold exceedances. Simulation: Histogram. Theory: Bold line. The dashed line shows the density $p(n)$ for i.i.d. data (Eq. (1) with $q_u = 0.01$).

2.2 Example 2: Data With Periodic Trend and Statistically Independent Fluctuations

In comparison to persistence, I demonstrate in this subsection that a periodic trend in the data has a different impact on the distribution $p(n)$ of the size of the recurrence intervals between subsequent threshold exceedances. I consider data which are composed of a periodic trend $f(t)$ with period t_b and statistically independent fluctuations (Gaussian white noise), that is, the data $\{z_i\}$ are generated from the time-dependent density

$$p(z,t) = \frac{1}{\sqrt{2\pi}} \exp\left(-\frac{1}{2}\big(z(t) - f(t)\big)^2\right) \tag{11}$$

The average probability that the process Eq. (11) exceeds a given constant threshold u is

$$q_u = \frac{1}{t_b} \int_0^{t_b} dt\, p\big(z(t) \geq u|t\big) = \frac{1}{t_b} \int_0^{t_b} dt\, \big(1 - \Phi(u - f(t))\big) \tag{12}$$

Assume that one has found a threshold exceedance $z(t) \geq u$. The probability that this event occurs at time t is

$$p\big(t|z(t) \geq u\big) = \frac{1}{q_u t_b}\big(1 - \Phi(u - f(t))\big) \tag{13}$$

It changes periodically with time t due to the trend. The probability to find a recurrence interval of size n is

$$p(n|t) = \big(1 - \Phi\big(u - f(t + \Delta t + n\Delta t)\big)\big) \prod_{i=1}^{n} \Phi\big(u - f(t + i\Delta t)\big) \tag{14}$$

$$p(0|t) = 1 - \Phi\big(u - f(t + \Delta t)\big)$$

Note the similarities between Eq. (1) and Eq. (14). Combining Eq. (13) with Eq. (14), yields the average probability density for the size of the recurrence intervals

$$p(n) = \int_0^{t_b} dt\, p(n|t)\, p\big(t|z(t) \geq u\big) \tag{15}$$

Figure 2 (left panel) shows an example. The process has the trend $f(t) = A \sin(2\pi t (t_b)^{-1})$ with amplitude $A = 0.4$. The time step Δt represents one day and I have chosen the period $t_b = 365.258\Delta t$. The average probability for a threshold exceedance is set to $q_u = 0.01$ which corresponds to a threshold $u = 2.415$. The right panel of Figure 2 shows the resulting empirical density (histogram) of the size of the recurrence intervals between threshold exceedances. For comparison, I have displayed the theory Eq. (15) (bold line) as well as the theory for i.i.d. data, Eq. (1) (dashed line). The density Eq. (15) is formed in a characteristic way which resembles a damped oscillation around Eq. (1). For several time intervals, I find an increased probability and for the complementary time intervals a decreased probability. This result is expected due to the periodic trend. The standard deviation $\sigma_{T(u)}$ of the return period $T(u)$ is only slightly increased in comparison to i.i.d. data.

Besides periodicity, we find in nature other processes which have an intrinsic time scale. One example are forest fires which are characterized by a time scale for sufficient forest regeneration such that a new non-localized fire may develop. Phillips (2002) studied fire return intervals in the Dinkey Creek watershed in California. The area has a particular high rate of lightning-caused fires with a rate of up to one every 1.36 years. The reported empirical density of the fire return intervals shows a clear suppression of small recurrence intervals ($p(1 \text{ year}) < 5\%$) with a peak at $p(3 \text{ years}) \approx 27\%$ and a steady decrease to values of a view percent for $T \geq 9$ years. Processes of this type are sufficiently well characterized by the return period although it should be noted that the most probable value differs from the mean size of the recurrence interval.

3 Discussion and Conclusions

In this paper, I have studied the probability density $p(T)$ for recurrence intervals of given size T between succeeding exceedances of a threshold u. The first point of the paper was to stress that the probability density $p(T)$ is typically *not* well characterized by its average or mean value $T(u) = \int dt \, p(T) \, T$ which is commonly referred to as *return period*. Therefor, the return period should only be understood as a different way to visualize the probability q_u for an extreme event in the time span Δt. The value of $T(u)$ itself is typically an rather unreliable estimate which has a standard deviation of the same order as $T(u)$.

The probability density $p(T)$ of the recurrence intervals is sensitive to the statistical nature of the process and can supply useful additional information. When studying the probability density $p(T)$ to learn more about a process of interest, the choice of the threshold value u is arbitrary and can be set low enough to exploit the data sufficiently well. Moreover, the probability density $p(T)$ itself is an interesting object for applications (such as insurance) where the time-distribution of extreme events is of particular importance.

I have considered simple examples of processes which generate a series of statistically *dependent* observations and studied the resulting impact on the probability density $p(T)$. Persistence has the effect that tight clusters with extremely small recurrence intervals have a strongly increased probability. The remainder of $p(T)$ becomes extremely flat. A periodic trend imprints a characteristic pattern on $p(T)$. The probability of certain ranges of values T is increased while the probability is decreased in all complementary regions.

Acknowledgements

This work was supported by the Deutsche Forschungsgemeinschaft (DFG) through the Graduiertenkolleg 450 "Naturkatastrophen". I thank R. Friedrichs for fruitful discussions.

Appendix A Distribution of Recurrence Intervals for the Ornstein-Uhlenbeck Process

A good approximation of Eq. (7) is obtained by using Eq. (8) iteratively in combination with the ansatz Eq. (9). The parameters $\{\mu_{l+1},\ \sigma_{l+1}^2\}$ are determined by the first and second moment of Eq. (8) and by the volume $N_i = (2\pi\sigma_i^2)^{-1/2}\int_{V(i)} dz\ \exp\left(-\frac{1}{2\sigma_i^2}(z-\mu_i)^2\right)$.

The approximation yields an iterative scheme $\{\mu_l, \sigma_l^2, N_l\} \to \{\mu_{l+1}, \sigma_{l+1}^2, N_{l+1}\}$ with start values

$$\mu_1 = \frac{\exp(-k\Delta t)}{(1-\Phi(u))}\frac{\exp\left(-\frac{1}{2}u^2\right)}{\sqrt{2\pi}} \tag{16}$$

$$\sigma_1^2 = 1 + \frac{\exp(-2k\Delta t)}{(1-\Phi(u))}\frac{u\exp\left(-\frac{1}{2}u^2\right)}{\sqrt{2\pi}} - (\mu_1)^2 \tag{17}$$

$$N_1 = \Phi\left(\frac{u-\mu_1}{\sigma_1}\right) \tag{18}$$

and

$$\mu_{l+1} = \exp(-k\Delta t)\left(\mu_l - \frac{\sigma_l}{N_l\sqrt{2\pi}}\exp\left(-\frac{1}{2}\frac{(u-\mu_l)^2}{\sigma_l^2}\right)\right) \tag{19}$$

$$\sigma_{l+1}^2 = 1 - \exp(-2k\Delta t)\left(1 - \sigma_l^2 - \frac{\sigma_l^2}{N_l^2 2\pi}\exp\left(-\frac{(u-\mu_l)^2}{\sigma_l^2}\right)\right) \tag{20}$$

$$\quad - \exp(-2k\Delta t)\frac{(u-\mu_l)\sigma_l}{N_l\sqrt{2\pi}}\exp\left(-\frac{1}{2}\frac{(u-\mu_l)^2}{\sigma_l^2}\right)$$

$$N_{l+1} = \Phi\left(\frac{u-\mu_{l+1}}{\sigma_{l+1}}\right) \tag{21}$$

References

Embrechts, Paul, Klüppelberg, Claudia, and Mikosch, Thomas: *Modelling Extremal Events.* Springer-Verlag, Berlin Heidelberg, 1997.

Hsing, T., Hüsler, J., and Leadbetter, M. R.: On the exceedance point process for a stationary sequence. *Probab. Theory and Related Fields* 78: 97-112, 1988.

Koscielny-Bunde, Eva, Kantelhardt, Jan W., Braun, Peter, Bunde, Armin, and Havlin, Shlomo: Long-term persistence and multifractality of river runoff records: Detrended fluctuation studies. In *Proceedings of the Conference Hydrofractals 2003*, Journ. of Hydrology, Elsevier. eprint arXiv:physics/0305078.

O'Brien, George L.: Extreme values for stationary and Markov sequences. *Ann. Probab.* 15: 281-291, 1987.

Phillips, Catherine: Fire-return intervals in mixed-conifer forests of the Kings river sustainable forest ecosystems project area. General Technical Report PSW-GTR-183, USDA Forest Service, Sonora, CA 95370, 2002.

Uhlenbeck, G. E., and Ornstein, L. S.: On the theory of Brownian motion. *Phys. Rev.* 36: 823-841, 1930.

Wang, M. C., and Uhlenbeck, G. E.: On the theory of Brownian motion II. *Rev. Modern Phys.* 17: 323-342, 1945.

Storm Damage Caused by Winter Storm "Lothar" in the Black Forest – Orographic and Soil Influences

Julia Schmoeckel[1,2], Christoph Kottmeier[1]
[1] *Institute for Meteorology and Climate Research, University of Karlsruhe (TH and Forschungszentrum Karlsruhe), D-76128 Karlsruhe, Germany;*
[2] *Postgraduate College <Natural Disasters>, University of Karlsruhe (TH), D-76128 Karlsruhe, Germany*

Abstract

In the black forest (Schwarzwald) the extraordinarily strong storm "Lothar" on December 26, 1999 caused large widespread damage. An empirical analysis of the windthrow pattern for a region of 5000 km^2 and affected sites of >1 ha is performed to derive the influence of orographic factors (slope magnitude and orientation, landscape curvature) and soil characteristics. Slopes with an inclination of 15° to 25° were significantly more affected than steeper slopes or flat terrain, also forest on north-westerly and south-easterly hills was damaged more than forest on hills with other orientation. Trees on wet soils were affected more than trees on dry or moist soils, especially in convex areas such as in valleys. The windthrow pattern is partly explainable by the effect of wind modification over complex orography causing higher wind speeds at the flanks of isolated mountains and in saddles.

Keywords: orography, winter storm, forest damage, Lothar.

1 Introduction

On December 26 1999, winter storm LOTHAR caused large damage in the forests of France, Germany and Switzerland. In Germany, especially the federal state Baden-Württemberg was concerned. In this region, the amount of damaged wood added to 300% of the mean annual harvest (figure 1).

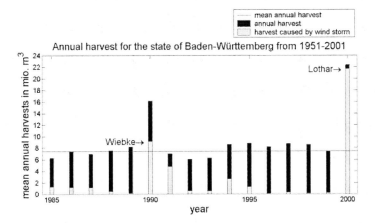

Figure 1: Annual harvest for the state of Baden-Württemberg from 1951 – 2001 (Federal Statistical Office).

Depending on the topographic parameters (ground level, slope, exposure) and meteorological conditions (atmospheric stratification, wind in the free atmosphere), the wind flow generally tends to pour around or across mountains. As a result, increasing velocity appears on hilltops, in valleys or saddles. These processes can be simulated by three dimensional atmospheric flow models. However, there are no data of real damage patterns being available so far that are suitable for comparisons with the results of such numerical models. The idea of this project is, to derive information about the influence of the orography on the wind field by an analysis of the damage pattern in the black forest.

2 Mapping of Windthrow Areas

The storm damage caused by windstorm LOTHAR are analysed empirically in the forests of the western Black Forest (about 5000 km²). The damage pattern was recorded by an airborne survey with a Color Line Scanner (CLS) in high resolution (2 m x 2 m on the ground) (Kottmeier et al. 2002). With the CCD sensors used in the CLS, the intensity of the reflected radiation from the surface is measured in three different channels: green (500 – 570 nm), red (580 – 680 nm) and near infrared (720 – 830 nm) (see also Bochert et al. 2000). The intensity of the reflected radiation of vegetation shows a steep slope in the range between 690 nm and 740 nm (figure 2). The higher the chlorophyll content of the plants, the more this slope is shifted to longer wavelengths.

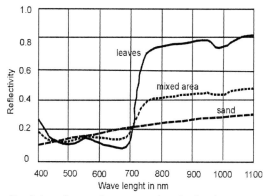

Figure 2: Spectral reflectivity of green leaves, sand and mixed area.

The normalised difference vegetation index NDVI is calculated from the difference of the intensities of the reflected radiation in the near infrared I_{nir} and the red bands I_{red} divided by their sum:

$$NDVI = \frac{I_{nir} - I_{red}}{I_{nir} + I_{red}}. \tag{1}$$

More information about reflectivity of plants and vegetation indices can be found e.g. in Hildebrandt (1996) or Lyon et al. (1998).

The mapping of the storm damage and their combination with a digital elevation model and land use data is achieved in a Geographic Information System (GIS).

Figure 3 shows the total overflown region and its topography in grey. Additionally the mapped windthrow areas are marked with black spots. As an example, the valley around Baden-Baden is shown as an enlarged section, for a closer look on the damaged areas. The damage pattern obviously depends on orographic characteristics. During the storm with westerly wind directions, especially the western and north-western hills of the Black Forest were concerned. Storm damage occurred also on the lee sides of hills, possibly caused by air flow passing the hills, and in valleys potentially due to channelling effects.

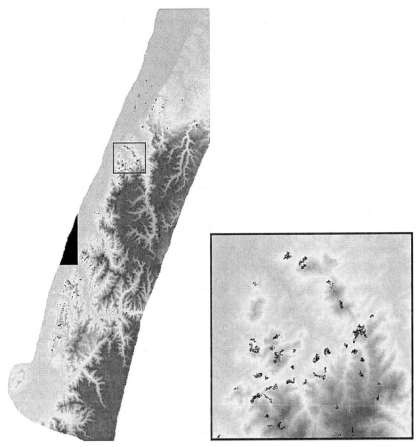

Figure 3: Topography of the overflown region (Schwarzwald, greyscale) and mapped windthrow areas (black spots). On the right side, a smaller section (Oostal, Baden-Baden) of about 15 · 15 km² is shown.

3 Topographic and Soil Parameters

From digital height data at fine spatial resolution (50 m in our case), the slope magnitude and orientation of the terrain are calculated. But these parameters are not decisive whether a damaged area is located on a hilltop, in a valley or on saddles. Therefore the curvature of the terrain is calculated as well. Using the method of Zevenbergen and Thorne (1987), a polynomial of second order is fitted to each grid point of the digital elevation model. With this method the curvature along the gradient of the slope and perpendicularly to it is calculated. From this concave and convex curvature as well as saddle areas and regions with no significant curvature are identified; see Fig. 4 as a sample region.

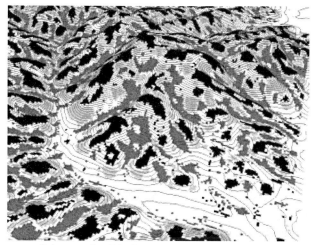

Figure 4: Curvature of the terrain. Black areas are convex, dark grey areas are concave. Saddle regions are marked in light grey; terrain without significant curvature is marked white. Additionally the contour lines of the topography are shown.

Beside the topographic parameters, also soil characteristics are considered within the analysis of the storm damage, since they affect the stability of the trees. These data were made available by the "Forstliche Versuchs- und Forschungsanstalt, Freiburg, FVA". Soil is divided in classes of different sand, clay, tone, rock blocks and organic soil. Additionally a classification for the water budget is performed. The classes range from dry to water logged. These properties may influence the stability of the trees during storm situations and are therefore considered in the analysis.

4 Atmospheric Conditions

The data of the meteorological stations in the area of investigation provide some limited information on the atmospheric conditions during the windstorm. The measured maximum gusts went up to 59 m/s on the hilltops (Feldberg 1493 m MSL) and 42 m/s in the Rhine Valley (Karlsruhe 145 m MSL), but several instruments got damaged or were at their upper limit, set as 50 m/s at the 200 m mast of the Research Center Karlsruhe. The wind directions varied between south-westerly and north-westerly. Generally, the orographic effects on the flow, such as prevailing around or over mountains, lee wave and flow separation may vary significantly with the atmospheric stratification and the wind in the free atmosphere. Thus, vertical profiles of temperature, humidity and wind velocity would be useful for a more detailed analysis. However, because of the windstorm itself, no data could be measured with radiosondes. Hence, only data from ground stations and reanalysed data of the „Lokalmodell" (LM) of the German Weather Service are used to assess the temporal changes of the three dimensional wind field.

5 Results

The damage pattern is analysed in relation to topographic and soil parameters. It is obvious, that the forest was affected for all slope classes, but forests on slopes with inclinations of 15° to 25° were more affected than on flat or steeper terrain (Fig. 5). The distribution of the damaged areas in relation to the slope orientation (azimuth) documents, that north-westerly slopes were affected most, while also forests on south-easterly oriented hills received a lot of damage. The least damage occurred on south-westerly orientated hills. Normalized damage frequencies vary by a factor of 2,5 between most and least affected slopes. Wind vector measurements, being normally performed in rather smooth terrain, show that highest wind speeds tended to occur with rapid wind veering as well as with south-westerly wind directions. But there is no clear wind speed /direction relationship, which would help to decide on the local wind direction during the periods of damage. Orographic steering and accelerated flow around the mountains as well as flow reduction both on the presumed luv and lee sides provides the most consistent idea on flow conditions.

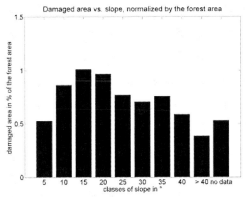

Figure 5: Distribution of the damaged area (normalized by the forest area) with the slope of the terrain.

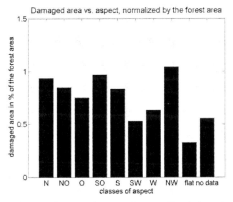

Figure 6: Distribution of the damaged area (normalized by the forest area) with the orientation of the slope.

The curvature as a single factor is not significantly related to the damage pattern as shown in Fig. 7. Only weak dependencies can be seen, as saddles are affected more and areas with little curvature less than others. But when soil parameters are considered additionally, clearer connections can be found. Trees on clay soil located on saddles (Fig. 8) and concave terrain such as valleys were affected most. Organic soil seems to be a less stable underground for forest in convex areas. The soil class of rock blocks provides comparatively stable conditions for trees of all curvature classes. Trees on wet soil were damaged significantly more than trees on dry or moist soils (Fig. 9), which may also be attributed to the soil stability itself. The combination of wet soils and saddle or convex areas results to be less stable for trees in respect of storm situations.

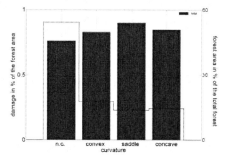

Figure 7: Distribution of the damaged area (normalized by the forest area) with the curvature of the terrain (left axis). The line shows the distribution of the forest with the classes of curvature (right axis).

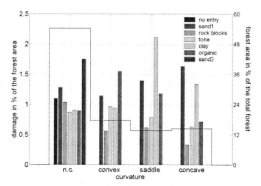

Figure 8: Distribution of the damaged area (normalized by the forest area) with the curvature of the terrain and the soil characteristics (left axis). The line shows the distribution of the forest with the classes of curvature (right axis).

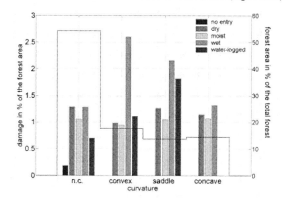

Figure 9: Distribution of the damaged area (normalized by the forest area) with the curvature of the terrain and the water budget of the soil (left axis). The line shows the distribution of the forest with the classes of curvature (right axis).

6 Conclusions and Outlook

The final aim of the combined mapping and modelling of the storm damage is to prove and assess orographic flow effects on the damage pattern in extended forests over low mountain ranges at spatially high resolution. At the actual state of the research work, mapping of damage caused by the severe winter storm LOTHAR and for windthrow areas larger than 10^4 m² is completed. We were able to relate windthrow areas and orographic characteristics empirically, with additional regard to soil characteristics. The complete windthrow pattern can be only explained by multiple and interrelated factors, additionally involving tree heights and types, which are not yet available in digital form for the region. Higher order correlation and cluster techniques will be applied to assess to which extent factors characterising the forest himself and his site stability explain the observed damage.

Based on this analysis further investigation will focus on the orographic effects which additionally affect and modify the wind field in such a way that these large damage can happen. The wind field will be reconstructed using meteorological station data, combined with wind flow over complex terrain using a linearized model. Lateral boundary conditions will be obtained from the weather forecast model LM of the German Weather Service. We then plan to test the applicability of 3-D non-linear flow models (Adrian and Fiedler, 1991) to situations of hazardous winds.

Acknowledgements

This research project is being performed at the Institut für Meteorologie und Klimaforschung, Universität Karlsruhe (TH) and Forschungszentrum Karlsruhe, in collaboration with the Forstliche Versuchs- und Forschungsanstalt (FVA), Freiburg. It is supported by the DFG in the context of the Graduiertenkolleg "Naturkatastrophen", Universität Karlsruhe (TH). The survey flights were funded by the Forschungszentrum Karlsruhe.

References

Adrian, G. and Fiedler, F.: Simulation of Unstationary Wind and Temperature Fields over Complex Terrain and Comparison with Observations. *Beiträge zur Physik der Atmosphäre*, Vol. 64, No. 1, pp. 27 – 48, 1991.

Bochert, A., Hacker, J. M., and Ohm, K.: Color Line Scanner as imaging NDVI sensor. *Second EARSEL Workshop on Imaging Spectroscopy*, Enschede, 11-13 July 2000

Hildebrandt, G.: *Fernerkundung und Luftbildmessung für Forstwirtschaft, Vegetationskartierung und Landschaftsökologie*. Herbert Wichmann Verlag, Hüthig GmbH, Heidelberg, 1996.

Kalthoff, N., Bischof-Gauß, I. and Fiedler, F.: Regional effects of large-scale extreme wind events over orographically structured terrain. *Theroretical Applied Climatology*, Vol. 74, pp. 53 – 67, 2003.

Kottmeier, C., Schmoeckel, J., Schmitt, C., and Bochert, A.: Sturm Lothar: Schadenanalyse und Risikokartierung aus meteorologischer Sicht. In *G. Tetzlaff, T. Trautmann, and K. S. Radtke (ed.), Zweites Forum Katastrophenvorsorge*, pp. 421 – 427. Institut für Meteorologie, Universität Leipzig, 2002.

Lyon, J. G., D. Yuan, Lunetta, R. S., and Elvidge, C. D.: A Change Detection Experiment Using Vegetation Indices. *Photogrammetric Engineering and Remote Sensing*, Vol. 64, No. 2, pp. 143 – 15, 1998.

Zevenbergen L. W., and Thorne C. R: Quantitative analysis of land surface topography. *Earth Surface Processes and Landforms*, Vol 12, pp. 47 – 56, 1987.

Wind Tunnel Experiments with Porous Model Forests

Wilken Agster, Bodo Ruck
Laboratory of Building- and Environmental Aerodynamics, Institute for Hydromechanics, University of Karlsruhe (TH), D-76128 Karlsruhe, Germany

Abstract

This paper describes wind tunnel investigations on flow around porous canopies with regard to windthrow in forest stands. There is a marked edge effect at the leading edges with distinct zonation. Zones of high turbulent shear stress at characteristic distance from the edge indicate high wind loading and provide potential areas for inital failure of trees. This edge effect is more pronounced in less porous models, leading to the assumption that permeability is a main factor concerning windthrow in plant canopies.

Keywords: canopy, edge effect, forest, permeability, windthrow, wind tunnel.

1 Introduction

Windthrow, together with fire and snowbreak is an important and common part of the natural succession in forest ecosystems. Considering managed forests in central Europe, storm damage is by far the main risk factor for spruce stands older than 50 years (Kouba, 2002). Such damage seems to be associated with the structure of exposed forest edges. Often, windthrow is initiated right behind the edge. Typically, the edges themselves are the only undamaged parts of the former stand. (e.g. Fritzsche, 1933). Damage patterns very much alike are found in other plant canopies, mainly cereals. Therefore, very fundamental mechanisms of wind-induced failure in plant canopies can be assumed. Already Woelfle (1941, 1950) mentioned the important role of flow structures above exposed forest edges as a key to the understanding of windthrow.

Since the 19th century, the problem of windthrow is approached by measures of stabilisation of trees at the edge (e.g. Reuß, 1881). Amongst the different measures of technical protection, Hütte 1983 found only top pruning (c.f. Stach, 1925) to be adequately effective. From the aerodynamic point of view, this is remarkable in two aspects. Top pruning results in chamfered edges. In practice, this is combined with removal of stronger branches within the crown (c.f. Yelin, 1886), leading to greater porosity of the edge. These two effects are already part of the advice from Woelfle, 1950, who recommends to design forest edges two stand heights in width with increasing tree heights towards the stand. These edges should be strongly thinned starting early in life, thus achieving stable trees with suffcent porosity.

Figure 1: Typical forest edge in even-aged spruce stands with only one single row of long
crowned trees (left) and other example of such a edge having survived a severe
storm (right).

A marked influence of permeability of forest edges and stands on the location of characteristic zones (e.g. maximum of turbulent shear stress in canopy height, c.f. Ruck and
Adams 1991, Agster and Ruck 2002) or high bending moments on single trees (c.f. Stacey
et al. 1994, Gardiner et al. 1997) seems plausible. For pragmatic reasons, however, in all
known model experiments with regard to permeability of forest stands, only tree densities
were varied. It seems feasible, that the influence on permeability is different to variation of
the permeability of every single crown. In addition the known model forests, show degrees
of canopy closure well below 1.0 and give a quite porous overall impression. There are
uncertainties in modelling and controlled scaling of permeability in vegetative canopies, as
field data of e.g. pressure loss coefficients is difficult to obtain. Therefore we compare two
single-layered models (a 'static' foam model and a 'dynamic' model of flexible foam rubber
rods) with homogeneous but different permeability.

Throughout this work meteorological notation in coordinate systems is followed (x-, y-,
and z-directions in space with corresponding velocities u, v, and w). In the illustrations
airflow is always supposed to come from the left with positive x-direction and wind speed u.

2 Experimental Setup

2.1 Wind tunnel and Simulated Atmospheric Boundary Layer

The experiments were conducted in an atmospheric boundary layer wind tunnel of the
Laboratory of Building- and Environmental Aerodynamics at the Institute for Hydromechanics, University of Karlsruhe. With an overall length of 29 m this wind tunnel has a
test section 8 m in length and 1.5 m in width. Air speed is adjustable within a range from
0...50 m/s.

Modelling is done at a scale 1:200. For simulation of an atmospheric boundary layer
with respect to distribution of wind speed and turbulence, the test section is equipped with
triangular vortex generators ('spires', c.f. Counihan, 1969) and roughness elements regularly
distributed over the floor. In strong wind situations (wind speed u > 10 m/s at 10 m above
the ground) thorough mixing of the atmosphere can be assumed, leading to neutral,
adiabatic conditions. Thus, the power law and the logarithmic law, respectively, provide
suitable descriptions in the simulation of an atmospheric boundary layer during a storm

(Schroers and Lösslein, 1983). The vertical distribution of the mean horizontal velocity $u(z)$ in the test-section, normalized by a reference velocity $u_{ref} = 3.2$ m/s in a height $z_{ref} = 0.05$ m, can be fitted well with the power law

$$u(z) / u_{ref} = (z / z_{ref})^{\alpha} \tag{1}$$

using a profile exponent $\alpha = 0.26$. The profile exponent is typical for suburban terrain and forested areas, respectively (see e.g. Höffer and Koss 1995). The velocity profile also satisfied the logarithmic law-of-the-wall

$$u(z) = (u_* / k) \cdot \ln((z-d) / z_0) \tag{2}$$

where k is the von Kármán constant and has a value of 0.4. The zero displacement d and the roughness length z_0 are 0 mm and 1.55 mm, respectively. z_0 amounts to 0.31 m in nature. The friction velocity u_* amounts to 0.367 m/s and thus the wall shear stress $\tau_0 = u_*^2 \cdot \rho$ is 0.165 N/m^2. Turbulence intensities decrease gradually as the height z increases and agree reasonably with the theoretical distributions described in Höffer and Koss 1995. They amount to $T_u = 0.43$ and $T_w = 0.19$ near the ground ($z/z_{ref} = 0.3$). Thereby, Taylor´s hypothesis is applicable for $T_u(z) < 0.5$, see Stull 1988. The vertical distribution of integral length scales agrees sufficently with Eurocode (1994). A comparison of the dimensionless spectra in the wind tunnel boundary layer with that of the natural boundary layer (von Kármán spectrum) shows, that the spectral density functions agree very well with respect to the position and the value of their maximum. The decay of the spectra is proportional to $f^{-2/3}$ in the higher frequency range. Due to the finite frequency distribution of each measuring device, the measured spectra in the wind tunnel exceed the theoretical values in the high frequency range.

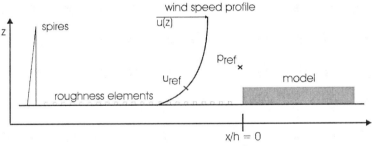

Figure 2: Test section of the wind tunnel, schematic sketch.

2.2 Models

2.2.1 Foam Model

The foam model consists of a block of open-celled polyurethane foam of 200 cm x 137 cm x 15 cm in x-, y- and z-direction. In the scale 1:200 this may correspond to a dense forest stand of 30 m in height. Its density is given with 10 pores per inch (ppi) by the manufacturer (product type: 10-222, vendor Modulor, Berlin). The pressure-loss coefficient of this material has been determined as 303 /m in previous tests. The foam shows a homogeneous and isotropic porosity distribution and is fully penetrable by fluids.

Figure 3: Foam model (left) and dynamic model (right) mounted in the wind tunnel. The white spots on top of each rod support visualisation of dynamic movement.

2.2.2 Dynamic Model

The dynamic model consists of an array of flexible rubber rods. Approximately 14,000 of these rods were glued to a wooden plate (100 cm x 137 cm) using a rectangular pattern with 1 cm distance between the center points. A single rod is 10 cm long and 0.5 cm in diameter. The plate also contains 100 pressure taps along the center line (x-axis) of the model for the measurement of static pressure at the floor ($z/h = 0$). The flexible rods are capable of dynamic interaction with the flow and show collective aeroelastic behaviour. This is visualisable by the movements of the white spots on top of each. With its homogeneous distribution of porosity with height, this model most closely resembles a canopy of cereals. The pressure-loss coefficient is 102 /m.

2.3 Laser Doppler Anemometry

The wind tunnel's test section is equipped with a two-component Laser Doppler Anemometer system (LDA) working with forward scattering principle. It includes an argon-ion-laser (max. 4 Watt) and two double bragg cells for frequency shifting. Particle seeding is done with a fog generator and 1,2-Propandiol as tracer, producing droplets with a diameter of 1-2 μm. These can be assumed to adequately follow the flow. After optical signal detection, the velocity data are processed in TSI signal processing units and the software package FIND 4.5. Typical data rates fall within the range of 2,000-10,000 Hz. The two-component system can measure in the model's x-z-plane, thus, the 2-dimensionality of the flow must be ensured by a great lateral extent of the model when compared to height.

2.4 Pressure Measurements

Static pressure along the floor of the dynamic model was measured against a reference pressure at the test section's wall (Figure 2). Signal processing was performed using a pressure transducer and a digital multimeter, with an integration time of 20 sec. The pressure taps were evenly spaced along the x-direction, each 1 cm apart, with the first being located right in front of the first row of rubber rods at $x/h = -0.05$.

3 Results

The measurements with the two models were conducted with different wind speeds u_{ref} (see Table 1) and deliver similar results as shown in Figure 5. At the sudden (step) change in aerodynamic surface roughness at the wind exposed edge a speed-up in u and w is registered associated with a minimum in turbulent shear stress in canopy height within 0.5-1 x/h (foam model) and 1-2 x/h (dynamic model), respectively. This zone is followed by a zone of deceleration in u and w at relatively low turbulence level. With the foam model, recirculation occurred at 0.63 m/s. This zone also shows a mixing layer, characterised by increased degree of turbulence and especially high horizontal skewness Sk_u. Turbulent shear stress τ in canopy height increases from the exposed edge until it reaches a zone with maximum (absolute) values located at about 3.4 x/h (foam model) and 4.5 x/h (dynamic), respectively. These locations are shown in more detail in Figure 6. This zone also shows high degrees of turbulence and high values for horizontal Kurtosis Ku_u. Farther leewards, the flow slowly adapts to equilibrium conditions.

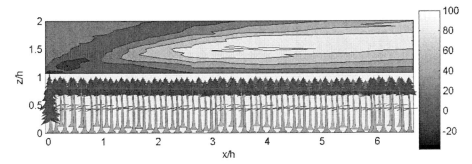

Figure 4: Example of distribution of turbulent shear stress (shown as) $-\tau/u_{ref}$ above the foam model at $u_{ref} = 2.95$ m/s. Maximum in terms of absolute values at 3.62 x/h.

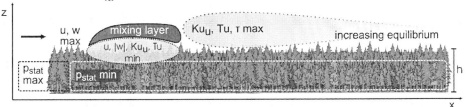

Figure 5: Sketch of fundamental effects at exposed forest edges. Characteristic zones derived from different LDA and pressure measurements for velocities u and w, Skewness Sk (mixing layer) und Kurtosis Ku of velocity distribution, degree of turbulence Tu, turbulent shear stress τ and static pressure at the floor p_{stat} are shown.

This zone of high turbulent shear stress is of special interest as potential zone for initial failure of forest stands. In the foam model (see Figure 6 and Table 1) this region extends from about 2 to 4 x/h with maximum values at 3.4-3.6 x/h (except for the experimental run at 0.63 m/s, where it extends from ca. 1.8 to 2.8 x/h, however, the Reynolds number Re = ul/v (with wind speed u, the stand height h as characteristic length l and kinematic viscosity v of the air) seems to be inadequate in this case (Re = 6,432). The dynamic model exhibits the corresponding zone around 4.5 x/h. The heights of the centres of these zones also differ with about 1.5 z/h in the foam model and 1.25 z/h in the dynamic model, being closer to the canopy's surface in the latter.

Within porous bluff bodies there is a steep gradient in static pressure through the exposed edge. Figure 7 shows measurements along the the floor in the dynamic model. As can be seen, forest edges lead first of all to an increase in pressure, which, then, is reduced with increasing distance behind the edge. Approximately at a distance of 0.3 x/h, all curves reach the value of the static ambient pressure. Characteristically, the pressure drops further until a minimum is reached within a wider region of 3...7 x/h. With increasing wind speed, the gradients get steeper, leading to more pronounced minima inside the model.

Model	u_{ref}	x/h	z/h	τ max	h [m]	Re
Foam	0.63	2.47	1.47	17	0.15	6,342
Foam	2.95	3.62	1.5	98	0.15	29,698
Foam	3.27	3.4	1.49	337	0.15	32,919
dynamic	2.66	4.5	1.25	92	0.10	17,852
dynamic	4.56	4.25	1.2	109	0.10	30,604

Table 1: Location of the zones with maximum turbulent shear stress for different experimental runs. u_{ref} is taken from the approaching flow upstream the model in reference height $h_{ref} = h$.

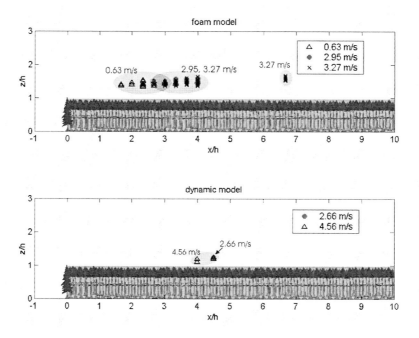

Figure 6: Location and extend of maximum turbulent shear stress above the model canopies. The symbols depict points (x/h and z/h coordinates) measured with LDA with the 10% highest values for τ/u_{ref} taken as threshold. Note the differences in position (x/h and z/h) between the foam and the dynamic model.

4 Discussion

By means of systematic flow measurements above the two models characteristic zones of similar properties could be identified. Between the models, there are visible differences considering the location of such zones, e.g. the maximum in turbulent shear stress. Apparently, variations in wind speed do not change these locations significantly in neither of the models. The edge effect described in Section 3 with three distinct zones can be described as follows:

I) exposed edge with increased velocities u and w within 1-2 x/h; wind loading on the exposed structures (e.g. trees) is high

II) zone of deceleration and low turbulence, showing characteristics of a developing mixing layer; wind loading seems to be low (almost no movements of rods)

III) zone with high degree of turbulence and high turbulent shear stress, leading to high wind loading again, seen as flexible response in the dynamic model.

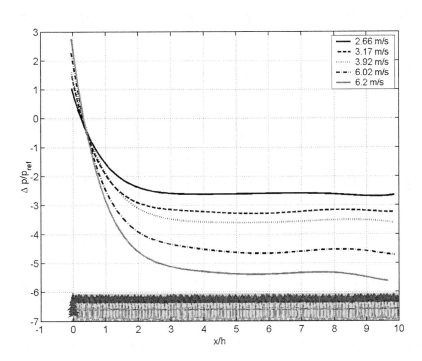

Figure 7: Static pressure along the floor in the dynamic model for different wind speeds u_{ref}

These zones are narrower in the quite dense foam model than they are in the dynamic model. Morse et al. 2002 performed wind tunnel studies on the flow around simulated stands of sitka spruce. Generally, their data show the same overall pattern of distribution of turbulent shear stress, however the maximum is located much farther leewards at about 7 x/h. The model forest used in these experiments is designed to meet dynamic similarity. However, it looks far more porous from visual estimation than the two models presented here. Gardiner et al. 1997 - having worked with the same standard forest - tested the influence of stand density on bending moments measured at single trees along a gradient x/h. While bending moments measured at less exposed trees inside the canopy approach those of the exposed edge trees with increasing porosity, the bending moments are still highest at the edge and steadily decreasing with leeward direction up to the maximum of the data set at 6 x/h, thus not showing the edge effect seen in Figure 3 within 6 x/h. Looking at the data from Morse et al. 2002, one might assume, however, that bending moments will increase again where turbulent shear stress becomes higher, from 7 x/h on.

The results indicate that a more porous plant canopy shows a less distinct edge effect than a dense one, with wider stripes/edges not being affected by high wind loading and potential failure. As a consequence, windthrow should be much more edge-bound in dense stands. This would fit sylvicultural experience and support the recommendations from Woelfle 1950, mentioned in Section 1.

References

Agster, W. and Ruck, B.: Modellierung der Umströmung von Waldkanten in Windkanaluntersuchungen. In *Proceedings "Lasermethoden in der Strömungsmesstechnik"*, pp. 37.1-37.7, Rostock, 2002.

Counihan, J. An improved method of simulating an atmospheric boundary layer in a wind tunnel. *Atmospheric Environment*, 3: 197-214, 1969.

Eurocode 1. Basis of Design and Actions on Structures, Part 2.4: Wind Actions. In *Windlastnormen nach 1992* (Ed. Niemann, H.-J.). Windtechnologische Gesellschaft, Aachen, pp. 15-286, 1994.

Fritzsche, K. Sturmgefahr und Anpassung: Physiologische und technische Fragen des Sturmschutzes. *Tharandter Forstliches Jahrbuch*, 84 (1): 1-94, 1933.

Gardiner, B., Stacey, G., Belcher, R., and Wood, C. Field and wind tunnel assessment of the implications of respacing and thinning on tree stability. *Forestry*, 70 (3): 233-252, 1997.

Höffer, R. and Koss, H. Windkanalversuche in der Gebäudeaerodynamik. In Windprobleme in dichtbesiedelten Gebieten (Ed. Plate, E.). Windtechnologische Gesellschaft, Aachen, pp. 241-285.

Hütte, P. Die Absicherung angebrochener Fichtenbestandesränder gegen Sturmschäden in Abhängigkeit von Durchforstungsstärke und Standort. *Forstwissenschaftliches Centralblatt*, 102: 343-349, 1983.

Kouba, J. Das Leben des Waldes und seine Lebensunsicherheit. *Forstwissenschaftliches Centralblatt*, 121 (4): 211-228, 2002.

Leder, A.: *Abgelöste Strömungen: physikalische Grundlagen*. Vieweg, Braunschweig, 1992.

Reuß, H. Ueber die Bewahrung von Windrissen in werthvolleren Nadelholz-Mittelbeständen. *Centralblatt für das gesammte* Forstwesen, 7: 445-453, 1881.

Ruck, B. and Adams, E. Fluid mechanical aspects of the pollutant transport to coniferous trees. *Boundary-Layer Meteorology*, 56: 163-195, 1991.

Schroers, H. and Lösslein, H. Extremwertextrapolation und Windprofile bei Starkwind und Sturm. *Meteorologische Rundschau*, 36: 205-213, 1983.

Stacey, G., Belcher, R., Wood, C., and Gardiner, B. Wind flows and forces in a model spruce forest. *Boundary-Layer Meteorology*, 69: 311-334, 1994.

Stach, W. Technische Sturmsicherung. *Forstliche Wochenschrift Silva*, 13 (47): 369-373, 1925.

Stull, R.B. *An Introduction to Boundary Layer Meteorology*. Kluwer, Dordrecht, The Netherlands, 1988.

Woelfle, M.: *Waldbau und Forstmeteorologie*. 2. Auflage. Bayerischer Landwirtschaftsverlag, München, 1950.

Thom, A. Momentum absorption by vegetation. Quarterly Journal of the Royal Meteorological Society, 97: 414-428, 1971.

Yelin. Ueber nützliche Astungen. *Forstwissenschaftliches Centralblatt*: 517-522, 1886.

Considerations Concerning Geodetical Long-Term Monitoring Tasks

Thomas A. Wunderlich

Chair of Geodesy, Technical University Munich, D- 80290 Munich, Germany

Abstract

When Geodetic Monitoring has to be realized over a period of almost 20 years, a couple of difficulties have to be mastered which shall be demonstrated on the basis of an exemplary case: a potential landslide area at the alpine mountain Hornbergl in the Tyrolean Lech Valley. Since the observation of the first epoch in 1987 the monitoring task's constraints have changed twice in a fundamental way, caused by technological progress. At first the transition from terrestrial to satellite-supported measurement had to be conquered; now we ought to benefit from current opportunities of continuous observation methods and of providing the necessary independent power supply to gain detailed insight into the motion's kinematics. Severe difficulties also arise with the need to adapt evaluation and analysis processes to overcome loss of reference points, new datum transformations and time scale modifications. Hybrid analyses create additional obstacles, if useful non-geodetic information, e.g. from photogrammetry and remote sensing, or relative movements from geotechnical observations should be taken into account for a general judgement of the hazard situation.

Keywords: landslide monitoring, engineering geodesy, deformation analysis.

1 Introduction

Monitoring of land or rock mass movements by geodetic observation techniques is selected, when absolute displacements shall be derived. In contrast to other, e.g. geotechnical methods, which give relative evidence, the geodetical results always refer to an agreed, common reference surface and coordinate system. Within such a frame – which surveyors call datum – the behaviour of each single point or mutual motions of several points can be investigated from epoch to epoch. Depending on the degree of threat, the funds available, the particular surveying method and the seasonal accessibility of the landslide area, the intervals between consecutive epochs can be irregular or periodic; with interval lengths decreasing to hours or minutes continuous observation becomes possible. Current permanent systems operate automatically and enable a real-time alert capacity.

As the successful employment of the different techniques and instruments demands expert knowledge and experience, the mission, in general, is put in then hands of engineering surveyors. These are used to get out the utmost accuracy down to millimeters.

Nevertheless, even experts have to try hard, if the monitoring mission's design should suit for many years of observation. This is due to the fact that the monitoring constraints as well as the technological potentials will change over the years. It is not a simple attempt to switch over from one system to another. Particular surveying methods have particular claims for visibility and might refer to different datums. Hence, prudent station selection and capable transformation solutions are a must. Equally important is to set a sufficient number of well-monumented control points and creating an adaptable evaluation concept to ensure that results of campaigns by different observation methods will stay comparable.

2 Observation Methods and Corresponding Configurations

During the last three decades geodesy has encountered a tremendous technological (r)evolution. Tacheometres became digital total stations, capable of aiming automatically at reflecting targets, with GPS high precision receivers for satellite-supported baseline determination turned up and recently prismless electronic distance measurement became available in total stations and in specific new instruments: terrestrial laser scanners.

Each observation method is associated with a certain favourite geometric configuration. The classical approach makes use of total stations to measure horizontal directions, vertical angles and spatial ranges between distinct points. As a rule the entire observations form an over-determined network to be evaluated by free adjustment. The resulting coordinates and the variance-covariance matrix of two epochs serve as basis for a rigorous deformation analysis. It usually proves advantageous to carefully select stable control points along the outer edges of the network, surrounding the central object points, which should indicate distinct displacements in a most characteristic manner for the whole area monitored.

Basically, the same network could be surveyed by GPS-receivers, which deliver three-dimensional vectors between simultaneously occupied network stations. The fundamental distinction between the terrestrial and the satellite-supported approach is caused by the totally different visibility claims. The terrestrial measurements need free sight between points to be connected, whereas the undisturbed reception of the satellite signals calls for an unobstructed view to the sky for all points participating in a certain observation session.

Thus, switching from terrestrial to satellite-supported observations can only be managed, if both conditions have been considered during initial station selection. The same rule holds for a combination of the methods or additional incorporation of aerial images. As a matter of fact a lot of monitoring networks have been established before GPS became an every day tool of the surveyor and therefore lack good upward viewing conditions. Sometimes a better situation can be found in the vicinity of existing points so that eccentric measurements become possible, but the original accuracy level can only be maintained when the eccentricities are observed and processed with particular care.

Since a couple of years continuous monitoring is possible by means of both, terrestrial and satellite methods (Wunderlich, 1995). On one hand we now dispose of robotic servo total stations, pointing automatically at prisms by digital image processing techniques. Such automatic polar systems (APS) use an infrared beam whose reflected radiation can be detected by a CCD-array with high precision. They can be controlled by on-board programmes or by an external computer. As indicated by the designation APS, a single robot tacheometre observes a number of polar target points fitted with prisms. On the other hand, the same configuration can be used for permanent observation by GPS receivers with real-time kinematic option. One reference station on a stable location receives phase data from several rover receivers at the moving object points via a robust telemetric link.

Finally, current research projects try to apply prismless electronic distance measurement with capable total stations or terrestrial laser scanners. Instead of specified points the object is covered with a dense array of (arbitrary) points with good reflecting property. It is evident, that this method can only be applied at slopes with almost no vegetation. Moreover, solutions have to be found how to compare the results of subsequent scans without identical points. The great advantage of scanning is that we do not have to enter the threatened area and that no instruments or equipment is risked to get lost in course of a landslide event.

For GPS pillars have to be erected on all points, for APS a pillar for the robot will be sufficient. For economical reasons the target prisms are usually mounted on metal poles. On creeping slopes the poles might tilt with time; this inclination has to be monitored too.

In polar configurations we lose the geometric conditions that granted control and adjustment procedures in networks. That is the reason why we now need to establish suitable algorithms to constantly test the quality of our continuous measurements. In APS we have to model refraction influences and mitigate observation interruptions during periods of insufficient visibility, in GPS we must exclude signal multipath and diffraction errors as well as results deteriorated by temporary effects of bad satellite geometry. The latter counteractions are very pretentious, particularly in real-time monitoring scenarios (Wieser and Brunner, 2002).

3 Datum Transformations and Parameter Determination

Apart from the problems due to configuration changes we have to face the fact that terrestrial and satellite methods refer to completely different datums. Terrestrial observables are related to gravity – practically, because we level our instruments and theoretically, because we want to examine the motion separately as horizontal and vertical displacements.

If we dispose of observed or computed deflections from the vertical, we can apply corrections and link the observed quantities rigorously to the national reference spheroid and can further on derive plane coordinates by means of a projection process. Heights are commonly referred to a certain physical level surface in mean sea level, the geoid and called orthometric. Spheroidal (ellipsoidal) height differences may be converted to orthometric ones by adding the respective undulation differences, i.e. the change in height differences between geoid and spheroid along the vertical profile between two points. To do this we need a detailed knowledge of the geoid.

Baselines derived from GPS phase measurements have no relation to gravity. They refer purely geometrically to an international mean earth spheroid, agreed upon within the WGS84 datum frame (World Geodetic System 1984). The relation to this datum comes implicitly with the satellite's ephemeris which is computed in this system.

Obviously, the transition from WGS 84 to a national datum is a demanding procedure. In practice, the conversion is managed easily by a spatial transformation with 7 parameters. These 7 parameters are determined by a least squares algorithm based on the coordinate values of in minimum 3 points known in both systems. It is of utmost importance that a sufficient number of such points is included in the control points. Otherwise point losses could make it impossible to maintain consistent transformation parameter computation, which would prevent to proceed rigorous deformation analysis (Niemeier, 1992).

Although it is possible to use fixed parameters or perform the processing directly in the WGS84, the evaluation will either lack rigour or leave only approximate horizontal and vertical components. The worst case happens, when just that control point is lost that has been used to connect ellipsoidal and orthometric heights. Because of the limited accuracy of existing geoid models, any substitute will result in a small, but unwelcome constant shift.

4 Power Supply Requirements

When a system change aims at continuous monitoring the most drastic readjustment concerns power supply for instruments and data transmission. The power demands for conventional observation campaigns is met by accumulators loaded overnight or for long sessions and short walking distances by car batteries. Permanent monitoring, in contrast, has long-term energy supply requirements that can only be fulfilled by advanced independent power supply units like e.g. combined solar-wind generators. Compared with APS, GPS installations have the disadvantage that every station has to be equipped with power supply.

5 The Hornbergl Landslide Area: A Long-Term Monitoring Mission in Transition

The Hornbergl monitoring network has been established in 1987 for classical triangulation and trilateration (Fig. 1). From a frame of stable control points a subnetwork of object points in between stations 1 and 2 was surveyed to indicate the movements there. After two terrestrial epochs the change to GPS application was intended in 1995. Because of signal obstructions (topography, tree canopy) on most of the points (Fig.2), exacting forecasts (Fig. 3) became necessary to determine suitable observation periods for couples or groups of points (Jobst, 1994). These had to be carefully arranged to allow for station changes during the intervals between the session windows – a demanding problem of logistics. We preferred this strategy to eccentric observation or expensive 24 hour sessions with receivers on all points

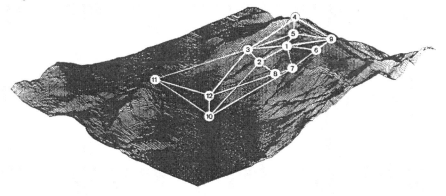

Figure 1:   Main Monitoring Network (Stations 3 to 12 considered stable) with Subnetwork between Stations 1 and 2.

The evaluation concept pursued the classical Hanover approach by comparing the results of free adjustments by a congruency test based on statistics (Wunderlich, 1990). For the first two epochs standard methods could be applied, whereas the hybrid analysis between epoch 2 (terrestrial) and epoch 3 (satellite) became a challenge. At first we had to ensure similar accuracy levels in both epochs which could be achieved by variance component estimation and adaptation of weights. Besides, all terrestrial values were corrected for deflections from the vertical. Then we had to care for datum parameter compatibility, i.e. to prevent network distortions through unnoticed datum constraints. A recommended way to handle this problem (Niemeier, 1992) is to include the effective 4 of the 7 transformation parameters (3 small rotations, 1 scale factor) as additional unknowns in the free adjustment of the GPS epoch. Moreover, preprocessing and input of the baselines must follow a particular scheme.

Having conquered the difficulties caused by the transition from terrestrial to satellite-supported network observation, the next 4 epochs till 2003 could be surveyed and processed easily as all were executed with GPS. Even the loss of a fundamental control point in course of construction works and its substitution could be met, because we had allowed for a sufficient number of control points and had invested in a splendid local geoid model. Fig. 4 gives an impression of the horizontal and vertical displacements revealed by our geodetical monitoring campaigns from 1987 to 2003. The alarming amounts of up to 3 meters in total and the rather long, irregular intervals between campaigns, which prevented detailed study of the motions now led to a joint effort to install a permanent GPS monitoring facility.

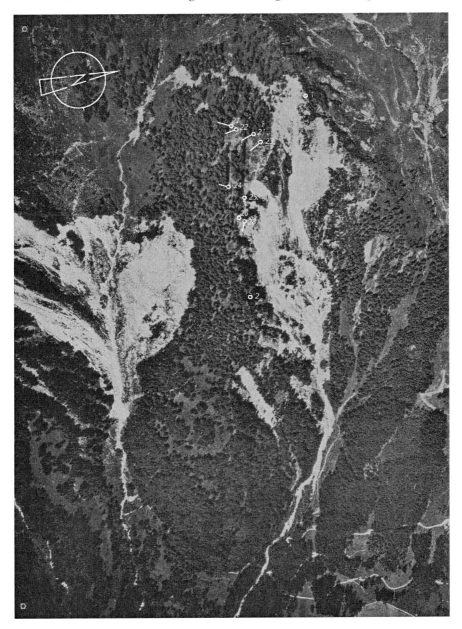

Figure 2:   Aerial Image of the Hornbergl Landslide Region with marked Subnetwork Points and Directions of the Movements.

Partially funded by the "Lawinen- und Wildbachverbauung Außerfern" (Dr. Dragositz), geodesists of the "Technische Universität München" (Prof. Wunderlich) and the "Universität der Bundeswehr München" (Prof. Becker, Prof. Heunecke) will set up and operate a continuous monitoring system with 3 single frequency GPS receivers in 2004/05.

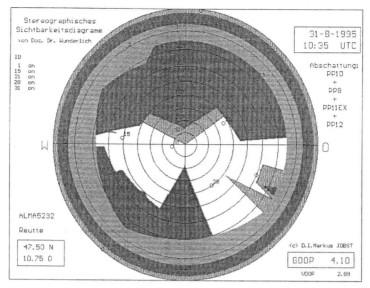

Figure 3: Stereographic Diagram of the Sky to Show a Superposition of Different Obstructions of the Horizon for a GPS Session with 4 Receivers. At the Selected Time 4 Satellites Can be Tracked Simultaneously.

The system will collect and preprocess phase data every 15 minutes and transmit the values via WLAN and IT to both universities for evaluation and analysis. Two rover stations on selected object points will be monitored by a single reference station on stable ground. All the three have to be reliably supplied with energy for receiver and transmitter operation. It is planned to employ capable solar-wind generators for this purpose. The rover stations cannot be accessed for months as the alpine region there keeps its snow cover very long.

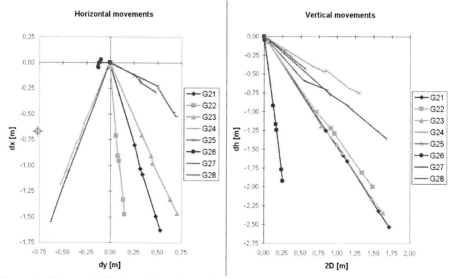

Figure 4:   Detected Horizontal and Vertical Movements from 1987 to 2003. Note the Changing Behaviour of Point 28 before Sliding and Loss.

Concerning the future way of processing and analysing, we have to make use of robust error detection procedures (Brunner et al., 2003) and to change from comparing two distinct epochs to a Kalman-Filter concept (e.g., Foppe and Matthesius, 1998) permitting steady displacement evidence and prediction. Transformation from WGS84 to the used horizontal and vertical frame can only be done by means of fixed parameters with the drawbacks mentioned above. From time to time extra control procedures must ensure that the reference station itself does not move at all. We are now preparing this second fundamental system change in our long-term monitoring project.

6 Conclusion

Geodetical monitoring of landslide areas over many years must face system changes to benefit from technological progress. The changes will most certainly imply consequences concerning survey configuration, visibility claims, datum transition, control procedures, analysis concepts and power supply requirements. The author's intent was to disclose these problems and recommend solutions; the theoretical discussion is accompanied by a practical example from a long-term landslide monitoring project in the Tyrol to prove the findings.

References

Brunner, Fritz K., Zobl, Fritz, and Gassner, Georg.: Monitoring eines Rutschhanges mit GPS-Messungen. *Felsbau*, 21 (2): 51-54, 2003.

Foppe, Karl and Matthesius, Heinz-J.: DFG-Projekt Geotechnisches Informationssystem, Final Report, Wissenschaftliche Arbeiten der Fachrichtung Vermessungswesen der Universität Hannover, No. 228, Hanover, Germany, 1998.

Jobst, Markus: Erfassung und Beurteilung von Abschattungen bei GPS-Messungen. Diploma Thesis, Department of Engineering Geodesy, Technical University, Vienna, Austria, 1994.

Niemeier, Wolfgang: Zur Nutzung von GPS-Meßergebnissen in Netzen der Landes- und Ingenieurvermessung. *Zeitschrift für Vermessungswesen*, 117 (8/9): 542-556, 1992.

Wieser, Andreas and Brunner, Fritz K.: SIGMA-F: Variances of GPS Observations Determined by a Fuzzy System. In *International Association of Geodesy Symposia*, Vol. 125, pp. 365-370, Adam and Schwarz (eds.), 2002.

Wunderlich, Thomas A.: The Hanover Quasistatic Deformation Analysis Approach Applied to a Tyrolean Mountain Threatened with Landslide. In *Proceedings of the Workshop on Precise Vertical Positioning*, pp. 16, Pelzer (ed.), Hanover, Germany, 1990.

Wunderlich, Thomas A.: Die geodätische Überwachung von Massenbewegungen. *Felsbau*, 13 (6): 414-419, 1995.

On the Formation of Wind Speed-ups Caused by Typical Terrain Structures

Ralph Lux [1,2]**, Franz Fiedler** [2]

[1] *Postgraduate Collegue "Natural Disasters", University of Karlsruhe (TH), D-76128 Karlsruhe, Germany;*
[2] *Institute for Meteorology and Climate Research, University of Karlsruhe (TH), D-76128 Karlsruhe, Germany*

Abstract

For agricultural areas as well as for urban zones, heavy storms can provoke considerable damages. To reduce the amount of damages, a sufficient knowledge of the influence of orographic structures on the wind field must be given. While forecasts in the synoptic scale are fairly reliable, predictions in the regional scale or even in the micro scale cannot be performed in operational model calculations yet. Thus, the aim of this project is providing information about the impact of mesoscale orographic structures on the wind field. The quantifying of the changes due to the orographic effects is given by speed-up factors, which represent the relation of the disturbed to the undisturbed states. To obtain the wind field, calculations with a numerical model are done. From these results, regions with an elevated wind risk are deduced.

1 Introduction

For the insurance industry, storm damages are the most expensive damages caused by natural disasters in Germany. A part of these damages may be prevented if a more comprehensive knowledge about the areas with the risk of higher wind speeds would be available. Approaches to asses such areas are existing i.e. in the DIN 1055-4, which is the German standard to calculate the wind loads on buildings. Though various factors are considered in this norm like the roughness length, the gustiness and even local elevations above the surrounding terrain, important parameters do not appear:

- non-linear behaviour in certain areas (i.e. recirculation zones)
- influence of the thermal stability
- influence of steep topography
- special topographic structures like valleys, high elevations, wind exposed slopes

However, these parameters can have an enormous effect on the regional prevailing wind speeds and wind directions. To quantify the wind field, there exist several approaches:

a) Statistical models originate by analyzing and extrapolating former measurements. An assessment of future wind risk is possible. Detailed information about certain wind speeds at certain situations cannot be made at small scales.

b) Linear models are obtained by the linearization of the flow equations and do not require much computer resources for the calculation. The disadvantage of this method is that no information about non-linear processes and the flow behaviour in steep terrain can be served.

c)   With wind tunnel experiments, a real flow can be examined. But with this
      method, problems of scaling and measurement arise.

d)   Numerical models claim to solve the flow equations as exactly as possible by
      discretizing them. The solution is including non-linear phenomena and can
      handle complex terrain. The limiting factor is the power of current processors.

In this project, a numerical model is used to perform calculations. Typical terrain
structures are identified and model runs are made for different meteorological situations to
spot the regions with an elevated wind risk. Calculating the ratio of the speed of the
undisturbed situation to the wind speed of the area affected by the topography in the same
height above ground, a speed-up value can be given.

2 Previous Work

Various experiments (numerical or linear model runs as well as wind tunnel experiments)
have been made to investigate the change of the wind field caused by topography. Most of
these configurations were tested with a hill in two and in three dimensions. Though the
results of different methods do not match exactly, approximate values can be taken as
confirmed in cases without stalling flow and under neutral stratification conditions.
Technical papers give simple approximations against geometrical factors (Lemelin et al.,
1988).

    For valleys, the knowledge is less. A complementary behaviour to elevations is only
valid for linear cases, where the slope and the deepness are slight. Furthermore, wind
directions parallel to the valley can be observed (Dorwarth, 1986). Therefore, this work is
emphasising on topographic structures which has not been investigated sufficiently yet.

3 Topographic Structures

In naturally formed terrain, all kinds of geometries, length scales and orientations can be
found. An example of the classification of the most common topographic structures can be
seen in figure 1. The area shows the surrounding of Baden-Baden in the northern Black
Forest.

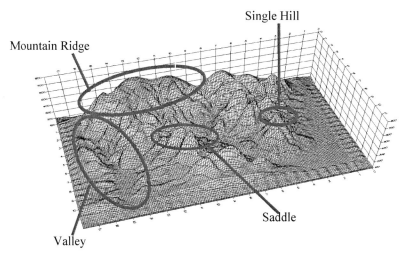

Figure 1: Identifying topographic structures

4 The Numerical Model KAMM2

The numerical model KAMM2 is a mesoscale model used for regional prediction. It was developed at the Institute for Meteorology and Climate Research (IMK) of the University of Karlsruhe. The model is non-hydrostatic, fully compressible and works with a time-splitting scheme (Klemp and Wilhelmson, 1978). Since it is a regional model with open lateral boundaries, the numerical treatment there is of particular importance. It could be noticed, that especially for idealized cases, a simple zero-gradient condition tends to fail. For that reason, a radiation scheme with an additional damping layer has been applied (Davies, 1976). Furthermore, an analytically obtained wind profile is fixed at the inflow, to which the values in the calculation area are nudged. So, a fully developed wind profile is simulated at the inflow, while the values can be calculated freely in the inner region.

5 Model Runs

As input variables, the model needs, among other things, the land use data, the geostrophic wind and the thermal stratification of the atmosphere. The latter two parameters are varied in each model run, so different scenarios are possible by varying the following variables:

- wind speed
- wind direction for non-symmetric obstacles
- thermal stratification
- length scales of the geometry
- steepness

The land use was assumed to be homogeneous. From this parameter, the roughness length is calculated, which is for all cases 0.02 m, which is approximately according to grassy terrain.

In a stable atmosphere, the influence of gravity waves can reach up the several thousands of meters. As these waves can take effect on the windward conditions, the modelling domain should comply with this fact and has to be chosen high enough. In some cases, the results did not seem to be resolved correctly, so multiple model runs were necessary for one single configuration. The horizontal resolution is adapted from case to case, depending on the length scale of the geometry.

Before testing the model with not well investigated structures like valleys, it was tested with a single hill at moderate wind speeds. The results match well with the results calculated by other numerical models. Another way of testing the model output is, to compare it with the linear theory, keeping in mind the restrictions demanded by this method (Jackson and Hunt, 1975).

6 First Results

Because of the lack of information of the impact of valleys on the wind field, the effect of this topography should be described more detailed. The aim is to execute model runs varying the geostrophic wind speeds, the angles of the approaching flow the classes of thermal stratification and the geometry. It has to be noted, that a comprehensive analysis cannot be given here, because the model is still in development and some configurations could not be calculated yet.

The first case considered here is a valley with dimensions as sketched in figure 2. The valley has no diminution in the direction of its axis, so the calculation can be considered as two dimensional. Thus, processor time will be saved. These dimensions are fitting approximately with the northern Rhine Valley. The flow direction is from the left to the right.

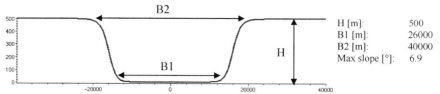

Figure 2: Modelled geometry of a valley

For this setup, it can be noted that there is a little speed-up near ground level at the windward edge of the valley at perpendicular approaching flow. But significant values only occur for the case of stable thermal stratification, where the values are up to 30% over the reference flow. If the angle of incidence is varied towards the valleys axis, the canalising effect of the valley increases and a second speed up can be observed at the downwind edge of the valley. At the same time, the speed down in the valley increases because the shading become less.

By changing the height H of the valley to 1000 meters and with it increasing the maximum slope to 12.7°, the dimensions are fitting approximately to the southern Rhine Valley. For this case, the flow conditions change remarkable. As shown in figure 3 on the left picture, a vortex formed out in the middle of the valley. Furthermore, the velocity has a component in the northern direction, so that the air is transported in this direction. In the right figure, additional streamlines can be seen (black coloured), which are placed above the others. The direction of the in- and outflow of these streamlines is indicated by the thick arrows. Obviously, from a certain height, the air overflows the valley without getting in the vortex.

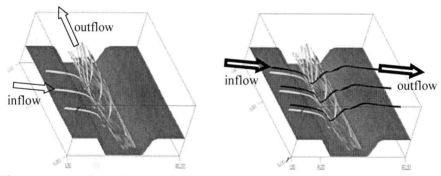

Figure 3: Streamlines of the flow in the valley.

This effect only occurs for stable stratified cases (at $0.0065 Km^{-1}$). The case was calculated with a geostrophic wind until the ground, which has the direction of 300 degrees. Thus, the pressure gradient has a component to the north with a value of $0.0042 N/m^3$ near the ground. Because of the dimensions of the valley, coriolis effects can be neglected, so that the pressure gradient can drive the flow in its direction (Fiedler, 1983). This hypothetical statement still has to be proofed by other simulations. In this case, a relative high speed-up was calculated at the left edge of the valley.

7 Conclusion

The method of calculating the speed-ups for certain topography seems to serve a concise overview about areas of higher wind risk for city planners, constructors and the forestry. Because of the multitude of possible setups and flow conditions, it must be shown if there is a possibility for maintaining the idea of quick and easy manageable information about the high-risk zones. But also a general gain of knowledge about the behaviour of the wind field under the influence of mesoscale obstacles would be desirable. Further model work is planned to investigate the influence of certain parameters. So, it may be that for high Froude Number flows, the results can be extrapolated to higher wind speeds. Other topographic structures than those presented here will be analyzed too.

References

Davies, H. C., 1976: A lateral boundary formulation for multi-level prediction models. *Quart. J. Roy. Meteor. Soc.* , 102, pp 405-418, 1976.

Dorwarth, Gerard: Numerische Berechnung des Druckwiderstandes typischer Geländeformen. *Wissenschaftliche Berichte des Instituts für Meteorologie und Klimaforschung Nr. 6*, Karlsruhe, 1986.

Fiedler, F: Einige Charakteristika der Strömung im Oberrheingraben. *Wissenschaftliche Berichte des Instituts für Meteorologie und Klimaforschung Nr. 4*, Karlsruhe, 1983.

Jackson, P.S., Hunt, J.C.R.: Turbulent wind flow over a low hill. *Quart.J. R. Met. Soc.* 101, pp 929-955, 1975.

Klemp, J. B., and R. B. Wilhelmson, 1978: The simulation of three-dimensional convective storm dynamics. In *J. Atmos. Sci.*, 35, pp 1070-1096, 1988.

Lemelin, D.R. ,Surry, D., Davenport, A. G.: Simple Approximations for Wind Speed-up over Hills. In *Journal of Wind Engineering and Industrial Aerodynamics International*, 28, pp. 117- 127, 1988.

Experiments on Overtopped Homogeneous Embankments: Soil/Water Interactions and Breach Development

Gerd Pickert[1], Gerhard H. Jirka[1], Andreas Bieberstein[2], Josef Brauns[2]
[1] *Institute for Hydromechanics, University of Karlsruhe (TH), D-76128 Karlsruhe, Germany*
[2] *Division of Embankment Dams and Landfill Technology, Institute of Soil Mechanics and Rock Mechanics, University of Karlsruhe (TH), D-76128 Karlsruhe, Germany*

Abstract

When a flood exceeds the design level of an embankment dam, the dam will be overtopped and breaching can occur. Consequences are major flooding of the area beyond the dam, which can at least cause high economical damage. Residents, local governmental administrations and environmental agencies are highly interested in knowing the precise description of possible breach development and outflow hydrograph Q(t). Many publications and models are available on this topic and model outputs can vary up to $\pm$ 50% from the original data. This is due to a limited understanding of the processes involved in the breaching of embankments and the lack of fundamental experimental data.

Keywords: embankment, breaching, overtopping, physical experiments

1 Introduction

When a flood exceeds the design level of an embankment dam, the dam will be overtopped and breaching will occur. Consequences are major flooding of the area beyond the dam, which can at least cause high economical damage. Residents, local governmental administrations and environmental agencies are highly interested in knowing the precise description of the breach development and outflow hydrograph Q(t). Appropriate emergency plans and scenarios could be developed based on this information and the loss of life and/or properties could be minimized.

Existing models, for example the NWS-BREACH-model developed by Fread (1991), describe overtopping events by hydrodynamic equations, erosion equations and simple geotechnical considerations. These simplifications of the breaching process, and especially the processes for the breach widening, have negative influence on the modelling results. Breach widening strongly depends on the neglected soil/water interactions in the soil matrix. These interactions, for example the tension in partly saturated soils have a strong influence on the erodibility of the soil and on the stability of the breach slope.

The present research is focussing an overtopped earthen embankment for flood protection and has the following aims: First, to generate a reliable data base by carrying out physical experiments and measure the discharge hydrograph as well as the "erosion curve". Second, to improve the knowledge of the soil/water interactions in partly saturated soils, represented by the water tension on the breach slopes. Third, to provide a 3D elevation picture of a breaching geometry in homogeneous embankments. And fourth, improving the geotechnical approach for the breach slope failure as well as the breach growth.

2 Background

2.1 State of the Art

A comparison of different dam breach models by Morris (2000) done within the EU workshop of Concerned Action on Dambreak Modelling (CADAM) showed the wide variability of model results. Figure 1 shows the computed breach discharges over time (Q(t)) of six models. Apart from the results of the Broich2D model which was calibrated against their own measured data, the output variation is rather high. Broich (1998) reports that the accuracy of predicting the peak discharge is perhaps ± 50%, and considered the accuracy in predicting the onset of breach formation as even worse. Morris (2000) published a list of existing dam breach models specifying the implemented breach morphology, flow equations, sediment transport equation and the geomechanics of the breach-side slope. Other summaries and analyses of dam breach models were done by Wahl (1998), Broich (1997), Lecoint (1998) and Mohamed et al. (2000).

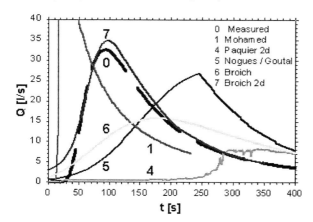

Figure 1: Breach discharge of six models compared to experimental data (Morris, 2000)

Graham (1998) stated that the modelling of breach formation through embankments is done by using process based models. Most breach models are based on steady state sediment equations related to homogeneous banks and adopted breach growth mechanisms. The modellers therefore need a significant number of assumptions and simplifications in order to simulate the breach, all of which can greatly affect the modelling results. The breaching process is dominated by the interaction of following three factors:

- hydraulics of the flow over the embankment and through the breach
- erosion process
- soil properties and geotechnical considerations

Simplification in one of these processes will certainly affect the overall modelling results. Especially the geotechnical simplification concerning the breach widening processes, affects the modelling results. Breach widening depends mainly on two effects. First the erosion of embankment material due to the transport capacity of the breach flow, and second on the failure of the breach side slopes.

2.2 Erosion Process

Mohamed et al. (2002) stated that the implemented sediment transport equations are derived for subcritical steady state flow conditions, for a specific type of sediment, and for a certain range of grains sizes. These conditions are likely to be violated during the breaching process since conditions change from subcritical flow to unsteady supercritical flow and back. Soil types do also vary widely in embankment construction. But in the absence of any other method to predict the sediment transport, a careful selection from the existing sediment transport formulas should be done based on dam breach experimental data, by considering their applicability to flow on steep slopes and for supercritical flow. As stated above, the choice of a sediment transport equation requires professional judgment and knowledge about the breaching mechanisms. In addition to the above stated considerations following verifications of the transport equation can be done:

- Revise the developed model on very good documented historical dam failures, for example the Teton dam failure 1975 (USA), and compare the model results with the actual breach width and depth.
- Verify the developed model on experimental data of the erosion curves for a falling water level in the reservoir (Broich, 1997) or on the explicit breach width and depth for different breaching times (Coleman, 2000).
- Carry out experimental models.

2.3 Soil Water Interactions

The analysis of the breaches in dams and embankments showed a wide variation of the breach slope, some are very steep others are flat (Bücker, 1998). This is mainly due to the dam material parameters and the effluent water volume. As stated before the water content of the soil has a great influence on the slope stability. First, as a destabilising factor if the water content is too high (excess pore pressure), and second, as a stabilising factor in unsaturated soils with a moisture content w < 1 [-] as tensile strength. This stabilising force introduced on the soil matrix by the soil/water interactions is based on the soil-moisture tension or suction power and increases the effective stresses and so the apparent cohesion of the soil.

All water movements in the soil depend directly on the soil-moisture tension, since water will tend to flow from areas of high potential to those with lower potential. The soil-moisture tension reflects the sum of the water holding forces of the soil. With tensiometers, this potenital can be determined directly:

$$\psi_m = m \times g \times h \qquad (1)$$

where ψ_m = matrix potential [Nm]; m = mass [kg]; g = gravity [m/s^2]; and h = height [m] above level of saturation (Schachtschabel, 1998).

3 Experimental Setup and Procedure

In the research laboratory of the Institute for Hydromechanics at the University of Karlsruhe a 1 : 10 scaled model of a homogeneous embankment has been realized in a flume of 15 m length, 1 m in width and 0.7 m in depth. The model has a height of 0.3 m, has a base width of 1.9 m and is built over the whole flume width of 1 m (Figure 2).

With respect to common embankments for flood protection the upstream as well as the downstream slope is 1 : 3, which ensures the geometrical similarity of the embankment model. In fact the slopes angles are smaller than the angle of friction of the used soil. Thus the slopes are stable.

The embankment is constructed of several layers of sand. In all experiments the mean moisture content of the embankment material is w = 5 % and the mean degree of densification of D = 0.5 [-]. Within the embankment five tensiometer probes are installed during the construction. The tensiometer probes are located in different levels and in different distances from the breach. A drainage toe ensures that the embankment withstands the hydraulic load and that seepage is discharged safely at the downstream toe.

The homogenous embankment model is alternativly built up with three different sands on a fixed bed (see Table 1):

Label	Type	d_{50} [mm]	grain size distribution
Sand 1	coarse sand	2	uniform
Sand 2	medium sand	0.22	uniform
Sand 3	coarse silt	0.035	uniform

Table 1. Dam materials used in the experimental models

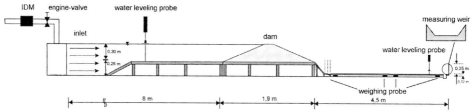

Figure 2: Sideview of the experimental setup

Before filling the reservoir, the initial breach plug is located at one of the side walls and is used to initialise the overtopping at this location. This plug is a wooden block with a length of 0.1 m (crest width) and a width of 0.02 m. With this plug a rectangular initial breach channel of 0.02 m width and 0.02 m depth over the total width of the embankment crest of 0.1 m is initiated. The reservoir is filled up slowly, so that a constant seepage line will arise in the embankment after a while. The experiment is started by withdrawing the plug, when the reservoir is filled up to water level of 0.29 m (0.01 m below the embankment crest) and a constant seepage is reached.

During the breaching process the outflow hydrograph, the erosion rate and the suction pressure are measured. Furthermore, the breaching process is recorded for a detailed analysis.

4 Measurement Technique

4.1 Erosion and Flow Measurement

The experimental setup described in chapter 3 and in Figure 2 is designed in such a way that it is possible to measure the erosion of the dam material with constant water level in the reservoir and no backwater influences on the breaching process. Right behind the dam toe a sill leads to a channel within the flume where the flow velocity is reduced by three grids in such a way that the eroded dam materials will deposit. This channel is bedded on weighing sensors so that the instantaneous weight of the eroded material and the outflow of the reservior is measured as W_{total} [kg]. The outflow from the breach is measured at the end of this channel by a measuring weir.

The "erosion curve" is obtained by subtracting the weight of the water W_{wat} [kg] from the total weight W_{total} [kg] for each time step, see equation 2.

$$W_{sed}(t) = W_{total}(t) - W_{wat}(t) \ [kg]$$ (2)

During the calibration phase, a deposition rate of 95% of the eroded material was achieved as well as a measurement error of ± 1% of the total weight.

Water levelling probes measure the reservoir water level and the water level in the sedimentation channel. In addition to the reservoir level, the inflow to the reservoir is measured with a inductive flow meter (IDM).

4.2 Fringe Projection

The breaching of an embankment is a highly transient process. Due to that, it is difficult to receive a profile of the breach itself. Coleman (2000) has achieved very good results by stopping the breaching process and mapping the dried breach at certain time steps. But to evaluate or even to model the different stages of collapse of the breach slopes another nonintrusive mapping technique has to be used for continuous measurements. The technique of Moiré fringe projection technique was adopted to solve this problem and obtain a a 3D elevation picture of a breach.

The simplest Fringe Projection for contouring an object is to project interference fringes or a grating onto an object and then view from another direction. Figure 3 shows a definition diagram for Fringe Projection and in Figure 4 the present experimental setup.

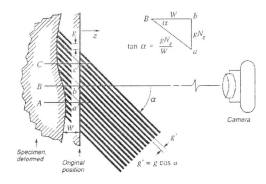

From the triangle B-a-b in Figure 3, we find the out-of-plane displacement equation (3) at point B with:

$$W = \frac{g \times N_z}{\tan \alpha}$$ (3)

where W = displacement; N_z = the fringe order at point B, g = grading pitch in figure 3; and α = angle.

Figure 3: Optical setup for Fringe Projection (Post, 1994)

The easiest way to produce a stripe pattern on the surface is the direct projection using a light source and an alternating black/transparent stripe onto the surface. Most convenient a slide projector is used with a framed stripe image of black and transparent stripes of a distinct stripe width. The camera must have a distinct angle relative to the projection axis, most convenient is a angle between 10° and 80°.

As the stripe width is very large compared to the wave length of the light, even very rough surfaces can be inspected. This is the case in the conducted experiments, preliminary experiments showed that the breach slope is rough with overhang. The analysis system consists of a slide projector which produces the pattern of white and black stripes on the dam profile.

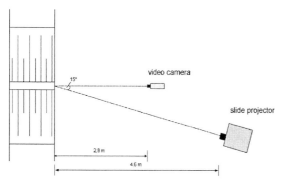

An IEEE 1394 CCD video camera with a resolution of 1280*980 pixels records the breaching process through the glass wall of the flume. This system has the advantage that video films can be recorded directly on hard disc up to the maximum hard drive space of 80 GBytes, which is very useful for cohesive materials where the breaching process is very slow.

Figure 4: Fringe Projection setup

5 Experimental Results

The test embankment was built with sand 2 from table 1 and has the dimension as descriped in chapter 3 and in figure 2. The embankment was built with a degree of densification of D = 0.5 [-] and a moisture content of w = 15 %, the total weight of the embankment was approximately 650 kg. Unfortunately the fringe projection was not used during this experiment.

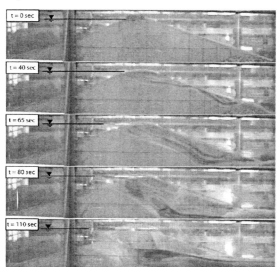

From removal of the initial breach plug to the end of measurements with an unaffected flow 110 sec elapsed. Figure 5 shows a time series of pictures from this test. At t = 0 sec the embankment is intact, the reservoir is filled up to the crest and the initial breach plug is still in place. The second picture (t = 40 sec) shows the initial breach phase. This phase is dominated by set of vertical steps which are merging together. This type of erosion is introduced by Hanson (1997) as headcut erosion in cohesive materials. The breaching phase begins in picture 3 (t = 65 sec). All steps are merged and an erosion channel parallel to the downstream

Figure 5: Breaching sequence

embankment slope is formed. The breach channel has been cut upstream, across the total width of the embankment crest. At t = 80 sec breach crest is migrating upstream on the embankment slope, whereas the discharge funnel migrates with the breach crest. The breach inclination increases referred to the downstream embankment slope. At t = 110 sec the breach crest migrates further upstream and the inclination of the reach increases.

Figure 6 shows the breached embankment viewed from upstream looking downstream after the depletion of the reservoir. The hourglass shape with the bottle neck at the embankment crest of the breach well defined by Coleman (2000) is recognisable.

Figure 6: Breached test embankment (view from upstream)

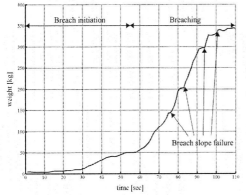

Figure 7: Measured erosion curve of breached embankment

Additionally the streamlines of the breach flow are cleary visible in the embankment material.

The geotechnical aspect of this research is also evident. Looking at the bottle neck the overhang of the breach slope is stable due to the negative soil-moisture tension. A dry medium sand ($d_{50} = 0.22$ mm) is cohesionless and has an angle of friction of $\varphi \approx 30°$ with a standard slope stability, which would make an overhang impossible. But taking the negative soil-moisture tension into account the so-called apparent cohesion stabilises this overhang. Thus soil mechanical and geotechnical considerations must be included in the understanding of breaching processes.

Figure 7 shows the graph of the correspondent erosion curve. A quantitative comparison of the photograph and the erosion curve indicates that half of the embankment was eroded. The mass of the eroded material was measured in the sedimentation channel. This graph (Figure 7) with the increasing mean sediment weight per time is in good accordance to the video taped breaching process (Figure 5). The initial breaching phase and the main breaching phase can be identified. Additionally to this, major collapses of the breaching channel due to failure of the breach slope can be correlated in the graph and in the video.

6 Conclusion

This paper shows first results of ongoing work on the understanding of breach development and the aim of developing a fundamental data base for modellers. The preliminary experimental results show good support for the concept of a breaching process occurring in three phases:

- breach initiation
- breaching
- depletion of the reservoir

But not only the breaching phases correlate with literature results (Coleman, 2000). The breach development and the breach propagation also show good agreement with results of other experiments (Hanson, 1997; Coleman, 2000). With respect to the erosion curve it can be stated that the breaching process is measured quite well, but further improvement on the analysis is necessary. Based on this experiments, the failure mode of the breach side slope can be identified as a fall failure. The Fringe Projection technique in order to receive a 3D model of a breach and the measurements of negative soil-moisture tension are introduced in this paper.

A drawback of this experimental setup is, that the breaching process herein can only be observed in the initial phase and in the beginning of the breaching phase, which is due to the width of the experimental flume. The further breach development with the possible maximum breach outflow cannot be observed.

Acknowledgements

This project is sponsored by the German Research Council (DFG), with in the Postgraduate College "Natural Disasters"

References

Broich, K.: Computergestützte Analyse des Dammerosionsbruchs; Universität der Bundeswehr München, Institut für Wasserwesen, München, 1997.

Bücker, M.: Breschenbildung in Erdämmen – Auswertung historischer Dammbrüche und Zusammenstellung wesentlicher Einflussparameter – Literaturstudie, Institut für Bodenmechanik und Felsmechanik, Abteilung Erddammbau und Deponiebau (unpublished), Karlsruhe, Germany, 1998.

Coleman, S. E., Andrews, D. P.: Overtopping Breaching Of Noncohesive Homogeneous Embankments, Department of Civil and Resource Engineering, Auckland, New Zealand, 2000.

Fread, D. L.: BREACH: An erosion model for earthen dam failures, Silver Spring, USA: National Weather Service, 1991.

Graham, W.: Dam failure inundation maps – are they accurate? Proceedings of the 2nd CADAM workshop, Munich, Germany, 1998.

Hanson, G. J., Robinson, K. M., Cook, K. R.: Headcut migration analysis of a compacted soil, Transaction of the American Society of Agricultural Engineers, 1997.

Kast, K., Bieberstein, A.: Detection and assessment of dambreak-scenarios, Dams and Safety Management at Downstream Valleys, Balkema, Rotterdam, Netherland, 1997.

Lecoint, G.: Breaching mechanisms of embankments – an overview of previous studies and the models produced. Proceedings of the 2nd CADAM workshop, Munich, Germany, 1998.

Mohamed, M. A. A., Samuels, P. G., Ghataora, G. S., Morris, M. W.: A new methodology to model the breaching of non-cohesive homogeneous embankments, Proceedings of the 4th CADAM Concerted Action Meeting, University of Zaragoza, Spain, 1999.

Mohamed, M. A. A., Samuels, P. G., Morris, M. W.: Improving the Accuracy of Prediction of Breach Formation through Embankment Dams and Flood Embankments HR Wallingford, Wallingford, UK, 2002.

Morris, M. W.: Concerted Action on Dambreak Modelling Final Report – CADAM, http://blakes.hrwallingford.co.uk/projects/CADAM/CADAM/index.html, 2000.

Post, D., Han, B.; Ifju, P.: High Sensitivity Moiré, Springer Verlag, New York, USA, 1994.

Schachtschabel, P., Scheffer, F.: Lehrbuch der Bodenkunde, Enke, Stuttgart, Germany, 1998.

Wahl, T. L.: Predicting Embankment Dam Breach Parameters - A Needs Assessment, XXVIIth IAHR Congress, San Francisco, California, USA, 1998.

The Role of Detailed Hydrological Investigation for the Identification of Dominating Structures and Processes which Lead to Mass Movement in Mountainous Regions

Falk Lindenmaier[1], Erwin Zehe[2], Jürgen Ihringer[1]
[1] *Institute for Water Resources Planning, Hydraulics and Rural Engineering*
Universität Karlsruhe (TH), Germany, Kaiserstrasse 12, 76128 Karlsruhe
Email: lindenmaier@iwk.uka.de, ihringer@iwk.uka.de
[2] *Institute for Geoecology*
Universität Potsdam, Germany, Karl-Liebknecht-Strasse 24-25, 14476 Golm
Email: ezehe@rz.uni-potsdam.de

Abstract

While hydrological processes of small slope failures is widely described and modelled, detailed hydrological behaviour studies of large mass movements are often omitted or neglected. This is usually due to insufficient data availability or simply due to the fact that failure already occurred. Besides more classical approaches towards landslides behaviour like mapping of landslide prone areas or landslide-triggering threshold evaluation with the help of antecedent rainfall sums, a detailed and highly resolved hydrological study is of vast importance for understanding landslide failure mechanisms. For giving advances into detailed modelling of landslide hydrology a joint project at the Universität Karlsruhe (TH) was initiated in 1998 to investigate a slowly creeping slope as an example system prior to failure with geological, geodetical and hydrological aspects.

The study slope is located in the Vorarlberg Alps near Bregenz, Austria. The slope extends over 1800 x 500 m and has an elevation of about 400 m; movement rates are up to 10 cm a year. Buildings of a vacation village on top of the slope are twisted and show cracks or even had to be torn down.

On the slope "classical" information for mass movements like geology, geotechnology and movement characteristics were gathered. The hydrology group also studied soil-type distribution combined with a geobotanical analysis and gathered hydrological and climatological data for the past six years. In our presentation we want to show how a multidisciplinary approach lead to the identification of dominating processes and structures of mass movements and how physically based hydrological modelling can help in understanding these processes.

Keywords: identification of dominating structures and processes, physically based hydrological modeling, hydrological induced landslides

1 Introduction

Slope instabilities both in artificial slopes and in mountainous regions are a major concern for mankind. Not only because of possible human losses but also through a continuously rising financial threat evolving from a more and more intensive use of our environment. Major triggers for slope failures can be for example earthquakes or rising water tables which in turn may be influenced through precipitation and infiltration into the unsaturated zone. Hydrology plays an important role in triggering mass movements, especially when its impact is seen in changing climatic frameworks (Delmonaco & Margottini, 2004). More intensive rainstorms or higher amplitudes of dry and wet climatic periods can possibly lead to more slope failures as well. Both landslides and slope failures are complex phenomena which usually show a long enduring time span for the evolution towards failure and a short event duration after the initiating trigger process. Hence, it is necessary to investigate slopes which are still prior to failure to better understand the interactions between hydrological processes and the development of shear zones.

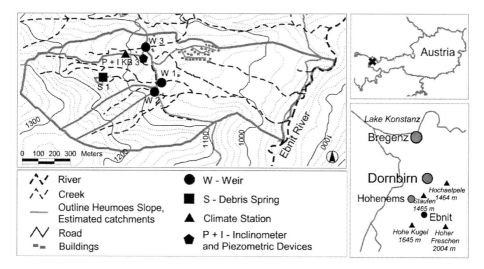

Figure 1: Location of research slope and overview of measurement set up

As we set our goal to improve the understanding of landslide hydrology, we chose a creeping but not yet failed slope near Bregenz, Austria (Figure 1). We will differ between a slope comprising the total catchment size and the moving part of the slope itself. The size of the hydrological important slope extends over 1800 x 500 m and has an elevation of about 400 m. The moving part of the slope body is less wide, showing sizes of 1800 x 200 m and has an estimated depth of up to 60 m. A precipitation scheme with high yearly precipitation sums and a high climatic variability seems to be the trigger for movement rates of the slopes of up to 10 cm a year. The research stands in close cooperation with the local authorities whose goal is to reduce creeping in builded areas. Data acquisition started in 1998 with geological and geotechnical equipment and observations and geodetical observations of surface points with GPS. In addition a hydrological measurement network has been installed with a meteorological station and several precipitation and runoff gauges. The system has been enlarged several times with additional measurement equipment but also suffered data due to natural restrictions.

2 Dominating Structures and Processes

We first want to present dominating structures, moving from large structures towards small scale structures. The same we will do with dominating fast and slower processes concerning hydrology and mass movement. Detailed soil mechanical studies, as indicated in Figure 2, are part of our partners work and so we do not go into detail here.

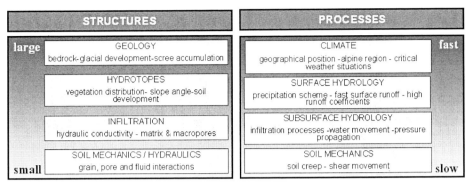

Figure 2: Scales of encountered structures and processes in landslide hydrology.

2.1 Geological Structures

Through geological and geotechnical mapping (Schneider, 1999) we achieved knowledge about the major geological features on the study slope (Figure 3). The slope is surrounded and underlain with marls and limey marls of upper cretaceous formations. On top of these relatively hydraulically inactive bedrocks there is subglacial till from the last glacial episode. The subglacial till is finely grained and highly compacted but is considered to be in saturated conditions and in weathered areas shows high plasticity. After meltdown of the Rhine glacier in this part of the Alps, there was accumulation of scree material on top of the slope. With variable accumulation rates and weak soil development we can find about 8 meters of a highly heterogeneous scree material on the moving slope. This material shows a high variety of hydraulic conductivities due to coarser scree from rock fall and debris flows and even embedded wood and on the other end highly weathered marl and scree. A borehole (KB 3, Figure 3) and geophysical investigations could not shed light towards the real depth of the subglacial till – bedrock transition, but it is considered to be in up to 60 m depth.

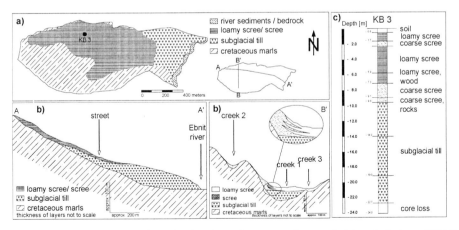

Figure 3: a) Geological overview, b) sketches of cross sections and c) borehole log of KB 3; subglacial till and scree belong to the moving slope; the marls can be added for the hydrological catchment area; changed, from Schneider (1999); Lindenmaier et al. (2004).

2.2 Soil and Vegetation Structures

For additional information about local surface and near subsurface structures we decided to use a combination of soil and vegetation analysis to shed light onto hydrological active areas of the slope. The soil analysis showed that soil development is not advanced and due to marls which weather easily the grain size distribution would suggest low conductivities on the moving slope. Conductivity tests on hydrotope 2 (Figure 4) showed a high range of hydraulic conductivity (from 10^{-4} to 10^{-8} m/s) of similar soil probes which confirms that influence of cracks and macropores play an important role (Lindenmaier et al. 2004). On top of the marls we found residual soils with low depths. It was not possible to use the point information of the soil and conductivity analysis to interpolate towards a spatial distribution of hydrological features. But plants, with similar biological needs, do form vegetation societies which can be used to gather information about soil properties in general and to define areas with similar soil moisture distribution (Waldenmeyer, 2003). This we could use to define local hydrotopes with similar hydrological behaviour. Through an upscaling towards larger areas by adding information about topography, slope angle and geological structures we achieved information about dominating runoff and infiltration processes, which we call composed hydrotopes (Figure 4).

Figure 4: Local and composed hydrotopes: shaded areas are derived from soil and vegetation analysis. The composed hydrotopes have white numbers; changed, from Lindenmaier et al. (2004).

2.3 Hydrological Surface and Subsurface Dynamics

Meteorological, precipitation and soil moisture data is provided by a station on top of the slope (Figure 1), additional data is provided by a precipitation station at the nearby village Ebnit and a meteorological station on top of a surrounding mountain (Hochälpele). Three discharge gauges and a debris spring gauge on the slope itself and one gauge of the adjacent Ebnit river provide runoff data. In mountainous regions, data uncertainty rises due to bypass of water at the gauges and through continuously changing rating curves due to high sediment loads in the creeks. Nevertheless data is considered to be reliable and good for interpretation.

In summer, precipitation is dominated by events with short duration and high intensities. Most of the 100-year average precipitation of about 2100 mm falls from May through October (~60%) and also there is a high variability in the precipitation scheme. Possible hydrological trigger for movement are expected then, despite that soil moisture measurements with TDR-probes (Time-Domaine-Reflectrometry) show only low variety, indicating low infiltration activity on the moving slope. In winter snowfall and rainfall with low intensities dominate precipitation. Soil moisture is high, without freezing grounds under snow cover. Base flow is higher than in summer but still low. Runoff curves which present a fast rise towards the maximum discharge and also a fast decline afterwards with almost no base flow are in correspondence with steep angles and low infiltration capacities on most of the slope (Figure 5a).

The debris spring, which is located at the bottom of hydrotope 1 shows a quick rise of discharge similar to the creek, but in contrast to the creeks a prolonged tailing afterwards (Figure 5a). This is in agreement with pressure head data measured with piezometric devices at borehole KB 3 in the middle of the moving slope (Figure 1, 5b). Similar behaviour of the spring and the piezometric device in combination with the composed hydrotopes lead to the hypothesis that pressure changes in the moving body are caused through preferential infiltration on hydrotope 1 and that a pressure propagation from the debris material towards the less conductive moving slope is responsible for the piezometric reaction.

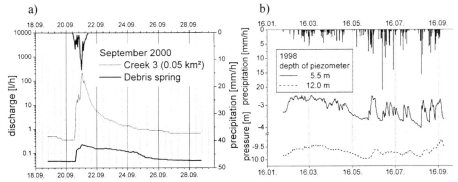

Figure 5: a)Discharge of creek 3 (hydrotope 2) and debris spring (bottom of hydrotope 1); b) pressure reaction on hydrotope 2; changed, from Lindenmaier et al. (2004).

2.4 Movement Characteristics

Inclinometer measurements in borehole KB 3 show shear movement in 7.5 to 8.5 m depth of about 12 mm in 1.5 years (1997-1998; Schneider, 1999), which is the scree-subglacial till transition. GPS and terrestrial survey (Depenthal, 2003) gave surface movement rates of more than 10 cm per year in the upper and lower part of the slope and up to 5 cm in the middle part of the slope. Surface movement is also marked through small scale slope failures and an uneven morphology. The surface movement did not show any correlation to slowly varying hydrological signals such as variation of base flow in the creeks or adjacent river, which is partly due to the long measurement intervals of three to six months each.

3 Verification of Dominating Structures and Processes with Physically based Modelling

The detailed investigation of structures and processes lead to the identification of the dominating features which cause pressure rise and possible movement of the slope. We chose a physically based hydrological model which was created and used in several studies at our institute (Zehe et al., 2001; Casper, 2002), because it is capable of simulating in a 2,5 dimensional way surface and (unsaturated) subsurface water dynamics.

3.1 Hydrological model CATFLOW

The essential model philosophy of CATFLOW (Zehe *et al.*, 2001) is to subdivide a catchment into a number of hillslopes that are interconnected via a drainage network. Each hillslope is discretized along the main slope line into a 2-dimensional vertical grid using curvilinear orthogonal coordinates. Each model element, as defined by the grid, extends over the width of the hillslope. The widths of the elements can vary from the top to the foot of the hillslope. Surface runoff is routed on the hillslopes, fed into the channel network and routed to the catchment outlet based on the convection diffusion approximation of the 1-dimensional Saint-Venant-Equation. The hillslope module can simulate infiltration excess runoff, saturation excess runoff and return flow. Evapotranspiration is represented using an advanced SVAT model based on the Penman-Monteith approach that accounts for plant growth, soil albedo as a function of soil moisture and the impact of local topography on wind speed and radiation. Soil water dynamics at each individual hillslopes is assumed to be independent to adjacent slopes and is modelled using Richards Equation in the pressure based form. The soil hydraulic functions are parameterized after van Genuchten (1980) and Mualem (1976). Preferential flow may be represented using a simplified approach to switch to a higher hydraulic conductivity when soil saturation exceeds field capacity.

3.2 Rainfall Runoff Response on Moving Slope

We used the 0.05 km^2 large sub catchment of creek 3 which completely lies on the moving body. We subdivided it into 12 hillslopes with related discharge channels. Each slope was discretized in a 2-dimensional grid consisting of 11 by 14 nodes. The vertical extend of the hillslope elements was given a thickness of 1 m. We used a simple, two layered build up of the soil-profile with a silty loam for the upper and a silty clay for the lower horizon of a stagnic gley soil encountered on most of the subcatchment. Soil hydraulic parameters were estimated after Carsel and Parrish (1988). The whole model catchment was covered with grass, the corresponding parameters such as leaf area index, plant height and root depth, Manning's roughness coefficient as well as the stomata resistance were also taken from the literature (Zehe *et al.*, 2001). For modelling we used a 85 day period over the largest ever observed discharge event in May 1999. Figure 6 shows an underestimation of the calculated runoff but we consider this as good enough because of the error prone discharge measurements (sediment transport). We basically can show that hydrological behaviour is runoff dominated due to the fact that we encounter steep angles and soils with small infiltration capacities – therefore pressure build up underneath cannot come from infiltration on hydrotope 2.

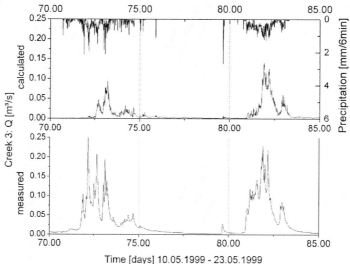

Figure 6: Measured and calculated runoff time series for the period from 10.05.1999 to 23.05.1999 for creek 3; changed, from Lindenmaier et al. (2004).

3.3 Subsurface Water Dynamics on Hydrotope 1

Secondly we used the slope module of CATFLOW to simulate a single slope of hydrological active hydrotope 1 with parts of hydrotope 2 to show how pressure propagation could function on the slope. We used a sandy clay loam to represent the flat soil layer on the steep slope and also the coarse scree layer, which is overlain by a silty clay loam representing the loamy scree of the moving slope (Figure 7). Underneath the sandy clay loam we added a silty clay to represent impermeable marl. The dimension of the modelled slope was 279 m in length and 8 m in depth, including 150 m of elevation difference. Using the observed precipitation and meteorological data as input, the water cycle at the idealised hillslope was simulated for the period of January 1998 to December 2000. Figure 7b gives a snapshot of the distribution of pressure head. As expected, a temporary water body developed in the less conductive silty clay loam layer in the lower part of the hill exhibiting more than 2 m of pressure head which is of the same magnitude as observed at KB 3. However, as soil hydraulic functions were only estimated on pedotransfer functions, a quantitative comparison between observed and simulated subsurface dynamics is not feasible.

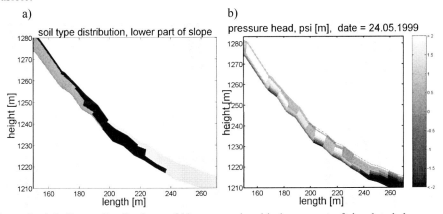

Figure 7: a) Soil type distribution and b) pressure head in lower part of simulated slope.

4 Discussion and Conclusions

The presented field evidence and simulation suggests that subsurface water dynamics at the slope is dominated by a 3 dimensional pressure system, which is triggered by fast infiltration of surface water in hydrotope 1, subsequent lateral water flow and related pressure responses. This signal seems to propagate further downhill to the piezometer on hydrotope 2. The results hint that the fast pressure reactions there belong to a temporary continuous water body, which could further extend downhill. Specific climatic conditions and rainfall patterns, especially in summer time, are thought to be able to build up pressure signals that could likely lead to further development of the shear zone. The relaxation time of the pressure signals is approximately 4 days, which is in our opinion short compared to the time scales where elastic and plastic or the porous medium develop (Lindenmaier et a.., 2004). So, a definite link between pressure signals and soil mechanical deformations has still to be developed. One step in moving towards this is to concentrate on better and more continuous movement observations on the slope, e.g. with TDR-Technology. The second step certainly has to deal with the improvement of the interfaces between hydrological process models, which deal with the atmosphere-plant-unsaturated soil zone interactions, with models based on unsaturated/saturated multi phase fluid dynamics in the underground such as MUFTE UG (Helmig et al. 1998), and on coupling these models with models for mechanics in soil continua such as PANDAS (Ehlers and Ellsiepen, 1998). With advanced hydrological understanding of mass movements and better modeling solutions we can move closer towards the prediction of trigger situations for large scale slope failures.

Acknowledgements

We would like to thank the Deutsche Forschungsgemeinschaft and the local Austrian authorities and energy providers for support of this project.

References

Casper M: Die Identifikation hydrologischer Prozesse im Einzugsgebiet des Dürreychbaches (Nordschwarzwald). *Diss. Univ. Karlsruhe, zugl. Univ. Karlsruhe, Inst. f. Wasserwirtschaft u. Kulturtechnik*, Mitteilungen H. 210, 2002.

Carsel RF, Parrish RS: Development of joint probability distributions of soil water retention characteristics. *Water Resources Research* **24**: 755 – 769, 1988.

Delmonaco G, Margottini C: Meteorological Factors Influencing Slope Stability. *Natural disasters and sustainable development. Editors: Casale, R., Margottini, C., Springer Verlag Berlin Heidelberg New York*, pages: 397, 2004.

Depenthal C, Schmitt G: Monitoring of a landslide in Vorarlberg/Austria. In *Proceedings of the 11th FIG Symposium on Deformation Measurements, Santorini, Greece, 25-28 May 2003*. http://www.fig.net/figtree/commission6/santorini/ (January 2004), 2003.

Ehlers W, Ellsiepen P: PANDAS: Ein FE-System zur Simulation von Sonderproblemen der Bodenmechanik. *In P. Wriggers, U. Meißner, E. Stein, W. Wunderlich (eds.): Finite Elemente in der Baupraxis - FEM '98*, Ernst & Sohn, Berlin, 391-400, 1998.

Helmig R, Bastian P, Class H, Ewing J, Hinkelmann R, Huber RU, Jakobs H, Sheta H: Architecture of the Modular Program System MUFTE-UG for Simulating Multiphase Flow and Transport Processes. In *Heterogeneous Porous Media. Mathematische Geologie*, Berlin, Vol.: 2, p. 123-131,1998.

Mualem Y.: A new model for predicting the hydraulic conductivity of unsaturated porous media. *Water Resources Research* **12**: 513 –522, 1976.

Lindenmaier F, Zehe E, Dittfurth A, Ihringer J: Process identification at a slow moving landslide in the Vorarlberg Alps. *Hydrological Processes*, accepted in January 2004.

Schneider U: Untersuchungen zur Kinematik von Massenbewegungen im Modellgebiet Ebnit (Vorarlberger Helvetikum). Ph.D. thesis, In *Schriftenreihe Angewandte Geologie Karlsruhe, University of Karlsruhe (TH)*. ISSN 0933-2510. **57**: 149, 1999.

Van den Ham G, Czurda K: Numerical modelling of a slowly deforming slope in the Vorarlbergian Alps, Austria. In *Geophysical Research Abstracts. European Geophysical Society 27th General Assembly*. **4**, 2002.

Van Genuchten MT: A closed-form equation for predicting the hydraulic conductivity of unsaturated soils. *Soil Science Society America Journal* **44**: 892 – 898, 1980.

Waldenmeyer G: Abflussbildung in einem forstlich genutzten Einzugsgebiet (Dürreychtal, Nordschwarzwald). Ph.D. Thesis. In *Karlsruhe Schriften zur Geographie und Geoökologie, University of Karlsruhe*. ISSN 3-934987-09-5. **20**: 277, 2003.

Zehe E, Maurer Th, Ihringer J, Plate E: Modelling water flow and mass transport in a Loess catchment. *Physics & Chemistry of the Earth, Part B* **26**: 487 – 507, 2001.

3D Object Classification in Laserscanner Data

Dániel Tóvári, Thomas Vögtle
Institute of Photogrammetry and Remote Sensing, University of Karlsruhe, 76128 Karlsruhe, Englerstr. 7

Abstract

Digital Terrain Models (DTM) play a very important role in hydrological numerical models. A very accurate DTM is needed for a run-off determination. Besides topography, terrain objects also have influence on the flow. Various kinds of objects like vegetation or buildings have different impact on this.

Nowadays, airborne laserscanning becomes a common tool of data acquisition for DTM generation purposes. This technology provides an extremely high density of measurement points with a planimetric accuracy of approx. 30 cm. Additionally, data acquisition can be realized during night or by poor weather conditions as well. This feature is not typical for traditional methods (e.g. photogrammetry). In disaster management, the data acquisition is a time critical task and should not depend on suitable lighting conditions, therefore, integration of different kinds of data (e.g. multi-spectral data) has been avoided, exclusively laser data has been used.

By utilization of point coordinates and intensities, a 3D object classification method has been developed that detects and classifies objects into 3 main categories. After classification of buildings, vegetation and terrain objects, vegetation may be sub-segmented into single trees or bushes (i.e. connecting the points which belong to the same plant) by the well-known watershed algorithm or by a method based on local maxima and minima respectively.

In this phase the method is based on raster data. Using laser scanning data, the pixel-wise classification is limited in terms of reliability of its results. Therefore, first of all a segmentation of 3D objects has been realized. For each segment, object-specific features are extracted and used for the above mentioned classification process. Different kinds of object oriented features are calculated for each segment, like height texture, border gradients, first/last pulse height differences, shape parameters or laser intensities. The fuzzy logic method is used to obtain a reliable classification of building, vegetation and terrain based on these features. The vegetation is detected from first and last pulse height difference data and segmented by the watershed algorithm.

This methodology and its results for test area 'Salem' will be presented in this paper.

Keywords: laserscanning, classification, detection, fuzzy logic, disaster.

1 Introduction

Nowadays, airborne laserscanning (ALS) becomes a common tool of data acquisition for DTM generation purposes. Starting from the extraction of digital surface and terrain models (DSM, DTM) a great variety of applications has been developed, like creation of 3D city models, determination of tree parameters in forestry or control of power lines, to name only a few (Lohr 1999). ALS is a very efficient tool of fast topographic mapping. Since it is not strongly dependent on weather conditions and can be used at night as well, it is very appropriate for flood plain mapping or disaster response and damage assessment. At our institute we use laserscanning data in two different projects. On one hand detection and modeling of buildings is based on these data to recognize and classify rough damages after strong earthquakes. On the other hand a high resolution terrain models including the determination of vegetation areas (position, size, density and height of trees etc.) has to be extracted from airborne laserscanning data to support the modeling of hydrologic processes, e.g. runoff models for simulations of floods.

For these purposes it is necessary to classify all 3D objects on the surface of the earth, i.e. mainly buildings and trees/bushes, in some cases also terrain objects like rough rocks which may be additionally included in the detected objects. Such a classification is a precondition for a class-specific modeling of buildings as well as vegetation objects. On the other hand the knowledge about the object type can be used for a significant improvement of the extraction of terrain models by a class dependent filtering of the original laser point cloud.

The first step of this approach is a segmentation of laserscanning data for detecting 3D objects on the terrain. Inside these segments object-specific features will be extracted which are used in the subsequent classification process. Fuzzy logic is used as classification method, since former investigations show that it is easy to implement, fits very well to the characteristics of the data and produces at least the same or higher correctness than maximum-likelihood statistical method (Tóvári, Vögtle, 2004).

The classification process can classify only objects that can be segmented from last-pulse DSM (e.g. buildings, terrain objects, dense vegetation). The rest of the vegetation can be detected by means first/last pulse differences and may be segmented by the watershed algorithm.

2 Data

At this state of our approach all features are derived exclusively from laserscanning data itself without additional information like spectral images or GIS data. This is caused by specific restrictions in context of disaster management - as mentioned above - where data acquisition has to be carried out also during night time and poor weather conditions. On the other hand the potential as well as the limitations of analyzing airborne laserscanning data should be investigated.

For this approach data of TopoSys II sensor in raster format (grid size=1.0m) for two different test areas are used, e.g. Salem test area near Lake Constance (rural environment, hilly terrain, size: approx. 2km x 1km). The test site was captured in first and last pulse mode and additionally laser intensity was registered. Figure 1 shows a subset of this test site. The data set was used by kindly permission of TopoSys (Germany).

3 Classification of 3D Objects

3.1 Definition of Object Classes

On the basis of the former experiences in classification, 3 main object categories are defined: buildings, dense vegetation and terrain. These objects can be classified with a high rate of correctness, however some objects are mixed and more categories exist in reality. These categories can be subdivided in further steps.

3.2 Segmentation of 3D Objects

Although this approach analyses raster data not the commonly used pixel based classification was preferred but an object oriented method based on the segmentation of 3D objects. Some other works in this direction can be found (e.g. Hofmann, Maas, Streilein, 2002; Schiewe, 2001; Lohmann, 2002). In most cases the image processing system eCognition (Definiens, 2001) is used. In opposite to these our approach is not based on general standard features but on a-priori knowledge about the characteristics of the relevant 3D objects, i.e. about their specific appearance in laserscanning data (Voegtle, Steinle, 2003).

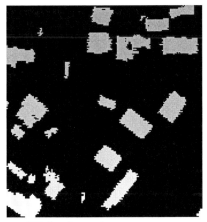

Figure 1: Segmented objects of 'Salem' test site

In a first step of this approach a so-called *normalised digital surface model* (nDSM) is created to exclude the influence of topography (e.g. Schiewe 2001). For this purpose a rough filtering of the original laserscanning data (DSM) is performed to extract exclusively points on the ground (DTM). This filtering is based on our *convex concave hull* approach (von Hansen, Voegtle 1999) which results – by an accordant choose of the filter parameters - in a rough trend surface of the terrain (rough DTM) without vegetation or building points. Now the resulting nDSM is calculated by subtracting this DTM from the DSM. In this data set all 3D objects on the surface of the terrain remain, in some cases also a few terrain objects are included caused by rough rocks or sharp terrain edges. It is evident that this result hasn't to be perfect because non-relevant objects – in this case the terrain objects – can be excluded after subsequent classification process.

Favourably, the segmentation of relevant 3D objects is carried out in such a normalised surface model (nDSM) by a specific region growing algorithm which extracts and separates 3D objects. The procedure results in separated areas of 3D objects while very small and low objects are excluded. Figure 1 shows the segmented objects of our test site.

3.3 Feature Extraction

Inside each segmented object area specific features for distinction of the relevant classes *buildings*, *vegetation* and *terrain* are extracted:
- Gradients on segment borders
- Height texture

- First/last pulse differences
- Shape and size
- Laser pulse intensities

Significant gradients along the border of segmented 3D objects contribute mainly to a discrimination of buildings/vegetation on one hand and terrain objects on the other hand. While buildings and trees generally show a high amount of border gradients in laserscanning data (70% - 100%) most segmented terrain objects – even if sharp relief edges are included – have at least at some parts of the segment borders smooth transitions to the surrounding terrain model. Therefore, the amount of significant border gradients decreases below 50% in most cases.

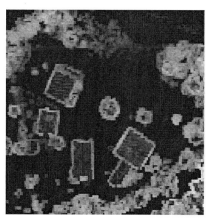 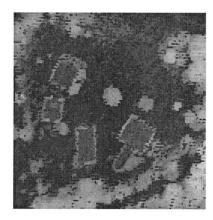

Figure 2: Grey coded height texture (root mean square) and grey coded first/last pulse differences

In contrast height texture and first/last pulse differences allow the distinction of vegetation and buildings. Taking the shape of building roofs into account exclusively those height texture parameters seem to be useful that model the deviations from oblique planes which fits very well to the characteristics of buildings in laserscanning data. Suitable results can be obtained by the *Laplace operator* (Maas, 1999), the *root mean square* of the heights or by *local curvature* (Steinle, Voegtle, 2001), i.e. the difference of subsequent gradients in the four directions across a raster point. Inside the roof planes of buildings small height texture values will be obtained while vegetation objects causes significant higher values (Figure 2 left.). The differences of first and last pulse measurements show a similar characteristic (Figure 2 right). Building roofs normally consist of solid material, so - dependent on the slope of the roof plane - no or only smaller differences between first and last pulse measurements can be observed. In contrast at vegetation objects with its canopy partly penetrable for laser beams larger differences will occur. On principle high texture values as well as high first/last pulse differences can be observed at the border of both, buildings and vegetation. Therefore, only the interior part of the segment areas can be used for determination of these parameters to avoid disturbances by this effect.

The shape of segmented object areas may contribute to the discrimination of artificial (man-made) objects (e.g. buildings, bridges etc.) and natural ones (e.g. trees, groups of trees, rough terrain or combination of both). For determination of shape parameters the contour lines of each segment has to be extracted. Former investigations have shown that commonly used standard parameters like *roundness*, *compactness* etc. don't fulfil the requirements, which are necessary to distinguish between the object shapes in this application. Therefore, alternative parameters had been developed like *geometry of the n longest lines*, where at first the *n* longest lines of a contour polygon are selected (e.g. n=4). These lines are analysed in terms of parallelism and orthogonality. A measure is calculated which is 100 for perfect parallel or orthogonal lines and decreases proportional to increasing deviations from that. This shape parameter has proved to be suitable to distinguish artificial and natural objects in most cases, i.e. if their area is large enough. Small object sizes lead to ambiguities. Figure 3 shows examples of filtered contour polygons of typical building and vegetation objects respectively.

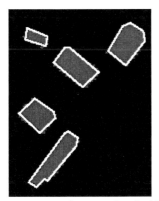 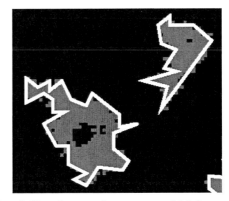

Figure 3: Contour polygons of typical building (left) and vegetation segments (right)

Laser intensities were also available which are recorded by the new TopoSys II sensor. This additional information was also included in the test program. The registered intensity of laser pulses depends highly on the characteristic of the reflecting material. In most cases buildings with commonly used rooftiles cause very high or -in the other case- nearly the same intensity values as vegetation. An example of typical intensity characteristics of buildings and vegetation can be seen in Figure 4.

Some statistical values like minimum, maximum, average and RMS was determined for all features mentioned above. In every case the average value was selected for classification purposes as it has proved to be the most suitable one.

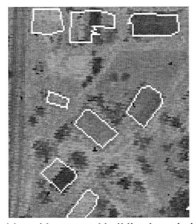

Figure 4: Laser pulse intensities with extracted building boundaries

4 Fuzzy Classification

The subsequent classification and its results depend on the preceding segmentation process because only segmented objects can be classified. The *fuzzy logic* classification is based on the extracted features, which have been described above. Fuzzy logic presents an opportunity to get answers to questions with a truth value in a range of 0 and 1. Fuzzy logic has been used in a wide range of applications, mainly in system controlling, and supports classification processes as well. The uncertain and often contradictory information can be handled and quite accurate results may be obtained. The fuzzy theory tries to blur the boundary between membership and non-membership. Therefore, the elements can be members, non-members and partially members as well. The basic idea is to model this uncertainty of classification parameters (features) by so called *membership functions*. A user has to define such a membership function for every parameter and every class (fuzzification). They may be built up by straight line sections in order to make computation easier, but also functions of higher degree can be defined dependent on the respective application. But in practice it has been proved that different approaches don't influence the results too much. Normally, membership functions are defined in an empirical way by means of training samples visually selected and interpreted by an operator. In this case about 25 segments have been chosen for each class.

A concrete value of feature i leads – by means of the corresponding membership function – to the related degree of membership $\mu_{i,j}$ for every class j, in this project j=3 (buildings/vegetation/terrain). All membership values for the same class j have to be combined for a final decision (inference process). Former investigations show, the most appropriate operator is the product operator which can be defined for a class as:

$$\mu_{(ABC)}(x) = \mu_A(x) * \mu_B(x) * \mu_C(x),$$

where A, B, C = extracted features
μ_A, μ_B, μ_C = degree of membership of the features

The inference procedure results in a crisp value for each segment and class. In every case the final decision is based on the maximum method, i.e. the class of highest probability will be assigned to the corresponding segment. As an example for the obtained classification results the confusion matrix for the product operator in shown in Table 1.

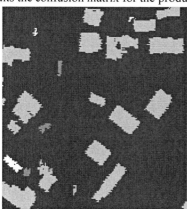 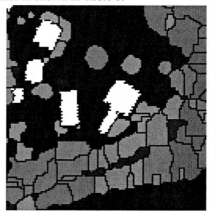

Figure 5: Classified segments on the left (red- light grey, blue- dark grey, green- white) and segmented vegetation (grey) on the right

Product operator	Buildings	Vegetation	Terrain
Buildings	95	5	0
Vegetation	4	96	0
Terrain	0	7	93

Table 1: Confusion matrix of classification rates [%] for the product operator

5 Vegetation Segmentation

As it is mentioned above, not all vegetation objects can be segmented and classified using last pulse data. In most cases the remaining vegetation is not dense enough; therefore the last pulse is reflected by the ground. In this case, the trees don't appear in the nDSM, but the first/last pulse difference is much higher than in the case of buildings or terrain objects. However building edges can cause a similar effect, but they appear as lines while vegetation cause larger areas. Since the buildings are already classified, it is possible to mask out the building edges from the height difference data. After this step the height differences show locations of the poorly dense vegetation and possibly overhead power lines, which can be classified and filtered out in subsequent steps.

Vegetation can be segmented e.g. by the well-known watershed algorithm (Figure 5. right). Since the segments are usually parts of a whole canopy, these need to be merged. Taking segment shape, relative location and height of local maximums into consideration, the segments that possibly belong to the same tree can be joined.

6 Conclusions

Airborne laserscanning itself provides appropriate data for object classification, so a fast weather independent data acquisition can be performed. Using a priori knowledge about the characteristics of 3D objects in laserscanning data for definition and extraction of object-relevant features suitable results can be achieved using fuzzy logic classification. A classification may significantly improve the quality of DTM, by remaining the terrain objects but exclude buildings and vegetation objects. The further steps of development can be a more accurate vegetation segmentation and an estimation of tree trunks.

References

Definiens, 2001. www.definiens.de

Douglas, D., Peucker, T., 1973. Algorithms for the reduction of the number of points required for represent a digitized line or its caricature. *Canadian Cartographer*, 10(2), pp. 112-122.

von Hansen, W. & Voegtle, T., 1999. Extraktion der Geländeoberfläche aus flugzeuggetragenen Laserscanner-Aufnahmen. *PFG*, Nr. 4/1999, pp. 229-236.

Hofmann, A. D., Maas, H.-G., Streilein, A., 2002 Knowledge-Based Building Detection Based on Laser Scanner Data and Topographic Map Information. *ISPRS Comission III, Vol.34, Part 3A "Photogrammetric Computer Vision"*, Graz, Austria, A169-174

Lohmann, P.: Segmentation and Filtering of Laser Scanner Digital Surface Models, *Proc. of ISPRS Commission II Symposium on Integrated Systems for Spatial Data Production, Custodian and Decision Support, IAPRS*, Volume XXXIV, Part 2, pp. 311-315, Xi'an, Aug. 22-23, 2002

Lohr, U., 1999. High resolution laser scanning, not only for 3D-city models. *Fritsch, D. and Spiller, R.: Photogrammetric Week '99*, Wichmann, Karlsruhe, Germany

Maas, H.-G., 1999. The potential of height texture measures for the segmentation of airborne laserscanner data. In: *Fourth International Airborne Remote Sensing Conference and Exhibition / 21st Canadian Symposium on Remote Sensing*, Ottawa, Ontario, Canada.

Schiewe, J., 2001. Ein regionen-basiertes Verfahren zur Extraktion der Geländeoberfläche aus Digitalen Oberflächen-Modellen. *PFG*, Nr. 2/2001, pp. 81-90.

Steinle, E. & Voegtle, T., 2001. Automated extraction and reconstruction of buildings in laserscanning data for disaster management. In: *Automatic Extraction of Man-Made Objects from Aerial and Space Images (III)*, *E. Baltsavias et al. (eds.)*, Swets & Zeitlinger, Lisse, The Netherlands, pp. 309-318.

Tilli, T. 1993. *Mustererkennung mit Fuzzy-Logik.* Franzis-Verlag GmbH, München

Tóvári, D., Vögtle, T., 2004. Classification methods for 3D objects in laserscanning data. *ISPRS Conference*, Istanbul Turkey, 12-23 July 2004.

Voegtle, T., Steinle, E., 2003. On the quality of object classification and automated building modeling based on laserscanning data. *The International Archives of Photogrammetry, Remote Sensing and Spatial Information Sciences*, Dresden, Germany Vol. XXXIV, Part 3/W13, 8-10 October 2003, ISSN 1682-1750

Hazard and Risk Assessment

Vulnerability Assessment:
The First Step Towards Sustainable Risk Reduction

Janos J. Bogardi[1], Jörn Birkmann[2]

[1] *United Nations University, Institute for Environment and Human Security, Bonn, Germany;*
[2] *University of Dortmund, Faculty of Spatial Planning, from September onwards United Nations University, Institute for Environment and Human Security Bonn, Germany*

Abstract

The increasing frequency and magnitude of water-related extreme events worldwide brought floods and droughts into the focus of research interest. In this context both the driving natural phenomena and the corresponding social response and vulnerability are to be investigated. The importance of reducing disaster risks as a challenge for the sustainable development of affected countries is duly reflected in the recent global report of UNDP (UNDP, 2004). The present paper proposes a comprehensive view on the "flood hazard – flood risk – flood vulnerability chain" and its possible occurrence as the "flood event – flood damage – flood disaster" sequence. Within this context, social vulnerability and coping capacity are analyzed in their interconnection. Social vulnerability, disaster awareness and coping capacity need to be measured and monitored. They may serve as indicators for prioritization of investments in flood preparedness but also to gauge risk perception. The paper outlines the inherent theoretical and practical difficulties. It underlines the importance of vulnerability assessment for sustainable development. The authors introduce potential options to be explored towards the development of policy-relevant vulnerability indicator(s).

Key words: vulnerability, coping capacity, assessment, floods

1 Introduction

Recently published statistics (UNDP, 2004, WMO and UN/ISDR, 2004) show an increase in the frequency and magnitude of extreme events, especially those of hydro-climatic origin. Climate change and variability, induced either by unsustainable human practices or/and by natural forces, are frequently blamed for this trend. Deforestation and land degradation contribute to worsen the situation. Population growth, from the present 6 billion people to 9 – 11 billion in two generations´ time (Vlek, 2004), is a clear indicator that the pressure on the Earth will further increase.

Unsustainable and declining livelihoods in rural communities drive millions every year to migrate into cities. People move to unsafe, marginal land: unstable slopes and inundation-prone areas. As most of the major urban agglomerations are situated along rivers, in deltas or in coastal zones, the further concentration of people living in these environments augments the number of people susceptible to being exposed to the consequences of extreme events like storms, typhoons, hurricanes, subsequent landslides, inundations, storm surges, tsunamis and similar events. Most of the newly arrived migrants lack the indigenous experience to live with these hazards. Thus the individual and social vulnerability of those displaced will increase considerably. The term "vulnerability" triggers reactions of human solidarity and therefore it is used frequently. However, it is far from being clear just what "vulnerability" stands for as a potential scientific definition, not to mention the subsequent task to develop adequate techniques and indicators to measure it in order to facilitate political decision-making.

2 Vulnerability: the "Key" to Human Security

The occurrence of extreme events, their superposition with the creeping environmental deteriorations, is usually a local or regional phenomenon. Thus they may be better defined within the context of human (in)security considering those actually impacted.

The concept of human security focuses on threats that endanger the lives and livelyhoods of individuals and communities. Safeguarding these requires a new approach, a better understanding of many interrelated variables – social, political, economic, technological and environmental factors that determine the impact of extreme events when they occur.

It is important to understand the logical sequence and the "stochastic" nature of the "hazards – risks – vulnerability" chain. These terms and their connections do not seem to be well defined. While there is a fair consensus regarding the terms "hazard" (the occurrence of an extreme event with an estimated (low) frequency) and "risk" (associating the occurrence probability of the extreme event (hazard) with the (economic and financial) losses it would imply (Plate, 2002, Kron, 2003)), there is much uncertainty about what the term "vulnerability" covers. One of the simplest and, at the same time, quite comprehensive definitions of vulnerability identifies it as the

"likelihood of injury, death, loss, disruption of livelihood or other harm in an extreme event, and/or unusual difficulties in recovering from such effects" (Wisner, 2002).

In its global review of disaster reduction initiatives, ISDR defines vulnerability as

"a set of conditions and processes resulting from physical, social, economical, and environmental factors, which increase the susceptibility of a community to the impact of hazards" (ISDR, 2002).

Similarly, UNDP's report on Reducing Disaster Risk underlines the societal connotation of vulnerability by defining it as

"a human condition or process resulting from physical, social, economic and environmental factors, which determine the likelihood and scale of damage from the impact of a given hazard" (UNDP, 2004).

These definitions focus on human society. Implicitly they reflect the "stochastic" nature of vulnerability. Terms like "likelihood" and "susceptibility" describe vulnerability as an unwanted opportunity, which might – or might not – become manifest. Vulnerability can be seen as a hidden weakness which may remain undetected for a long time, but could demonstrate itself viciously once a vulnerable community is exposed to a hazard, above a specific – albeit ill-defined - threshold. This linkage of susceptibility with its possible exposure suggests that vulnerability should always be analyzed in its dynamic relation with other elements of the hazard-risk-vulnerability chain. The ISDR definition precisely juxtaposes vulnerability with its complementary component "capacity", which is defined as

"a combination of all strengths and resources available within a community or organization that can reduce the level of risk, or the effects of a disaster" (ISDR, 2002).

There is thus a fundamental question to be clarified as to whether (social) vulnerability can adequately be characterized without considering simultaneously the response (coping) capacity of the same social entity. While adequate coping capacity can mitigate vulnerability, there are reasons suggesting that they should not be seen simply as the positive and negative sides of the same phenomenon. Technical or social coping capacities, awareness and resourcefulness can reduce vulnerability for economic losses or social disruptions. Thus, coping capacity and vulnerability may have different dimensions and, as such, be incommensurable. Likewise there is some fuzziness in distinguishing risks and vulnerabilities. Furthermore, vulnerability itself tends to be analyzed and disintegrated into its possible components (Alexander, 2000). This illustrates the comprehensiveness of the term "vulnerability" but does not facilitate its quantification.

The multifaceted nature of vulnerability is captured by Bohle, who introduced an external (environmental) and internal (human) side of vulnerability (Bohle, 2002), thus clearly identifying vulnerability as a potentially detrimental social response to environmental events and changes. Vulnerability can cover susceptibilities to a broad range of possible harms and consequences; it implies a relatively long time period, certainly exceeding that of the extreme event itself, which might have triggered its exposure. This interpretation of vulnerability is unavoidably related to resilience, the ability to return to a state similar to the one prevailing prior to the disaster. Thus, vulnerability is not only ill-defined, but its manifestation and magnitude depend on many partially unknown factors and their coincidence.

Since vulnerability is defined within the context of human (in)security, there are additional – but very much policy relevant questions - to be clarified. What is the proper scale to capture and to quantify vulnerability? While household level vulnerability assessments are not new (Currey, 1984), aggregation of vulnerability measures or their assessment at lower spatial and social resolution (thus at the policy-making level) are still not satisfactory. As people may be affected by various hazards, their respective vulnerabilities may have to be assessed separately and then aggregated to vulnerability indices. To what extent vulnerability should be seen as the "susceptibility" alone, rather then as the product of hazard exposure and that very susceptibility, is also still being debated.

This undeniable fuzziness associated with vulnerability is also a sign of the need for focused yet interdisciplinary research to explore features and to develop assessment methods. With respect to the ongoing climatic, environmental, but also socioeconomic and political changes, an additional important research question is to clarify how (social) vulnerability is affected by these trends and fluctuations.

3 Vulnerability and Sustainable Development

Before presenting ideas for measuring vulnerability, it is important to underline the close connection between vulnerability and sustainable development. To show the importance of vulnerability assessment (especially social vulnerability, disaster awareness and coping capacity) for sustainable development, it is essential to focus on the theoretical bases of the concept of sustainability.

In this regard the debate about sustainable development is characterized by two different "pre-analytic visions": the "triangle of sustainable development" and the "egg of sustainability". While the triangle places the three dimensions of sustainable development on different corners and does not answer the question as to how the economic, the social and the environmental systems are related to each other, the "egg of sustainability" defines a clear hierarchy between the dimensions (see Fig.1). The triangle model implies an isolated goal definition for each of the three dimensions of sustainable development, neglecting the linkages that exist between them. When it comes to implementation, traditional conflicts between the social, the economic and the environmental spheres become apparent (Birkmann, 2004). In contrast, the egg of sustainable development underlines the hierarchy and interdependency of the different dimensions and helps to define goals that respect these linkages and to put them into the right balance (Presscott-Allen, 1995; Busch-Lüty, 1995). Goals for sustainable economic development need to take into account goals of the social sphere as well as goals of the surrounding environmental sphere. The vulnerability and the

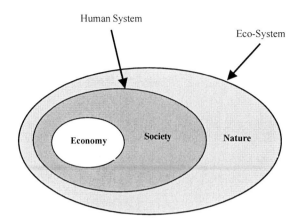

sustainability of the human system depends on the conditions of the surrounding environmental sphere as well as on the inner conditions of the socio-economic system. If disaster vulnerability, as Wisner at al. argue, can also be seen as a function of the way in which humans interact with nature (Wisner et al., 2004), the "egg of sustainability" is a good theoretical basis to start from.

Figure 1: The "Egg of Sustainable Development" (Source: own diagram following Busch-Lüty, 1995)

In relation to natural risk like floods, this pre-analytic vision outlines the necessity also to consider the socio-economic system and its influence on the carrying capacity of the surrounding environmental sphere when it comes to the definition of goals regarding risk and vulnerability reduction. The degradation of the environmental sphere, especially through creeping processes like the ongoing climate change and land degradation caused by unsustainable land use, production- and consumption patterns, increases the risk of disasters for the inner human sphere. Unsustainable development due to higher risk and natural disasters can be interpreted in this regard as the loss of the ability of a (sub-)system (economic, social or environmental layer) to return to a state similar to the one prevailing prior to the disaster, or more precisely to a status where the basic functions of the system are not irreparably damaged.

4 How to Measure Vulnerability?

How far would this vaguely defined concept of vulnerability exceed the coping capacity with a flood calamity and subsequently the resilience of the population to bounce back to its normal daily routines once a catastrophe is over? Preferably a single number is derived, making it possible to estimate the vulnerability of a certain group. Comparison of vulnerability indices of different groups should make a ranking possible. While this single indicator is desirable, vulnerability is very much an aggregate measure. It depends on the intricate interactions of economic dynamism and status, robustness of ecosystems and land-based production systems, but also on solidarity, response capacity of the people and authorities, social memory and psyche, trust, aspirations and dedication to succeed. Fig.2 shows a concept of how vulnerability could be interpreted and assessed. Fig.2 displays the potential event, or "opportunity axis" of the "hazard – risk – vulnerability" sequence for floods.

The "reality axis" demonstrates how the potential elements of the "hazard – risk – vulnerability" chain occur as "flood event", "flood damage" and "flood disaster". The two axes penetrate three cycles, representing the natural event sphere, the impacts of any disturbance in this sphere upon the economic sphere, and finally on the social sphere, implying negative consequences beyond the economic ones. This social sphere is the centerpiece of Fig.2. Whether a flood event becomes a "disaster" (and the vulnerability of the society is exposed) depends almost as much on the preparedness and response (coping) capacity of the affected society as on the magnitude and nature of the flood event itself. The three capacity circles indicated in Fig.2 correspond to different levels and versatility of capacities. Adequate coping capacity (C1) means that neither does the social vulnerability become apparent, nor would the flood trigger a disaster.

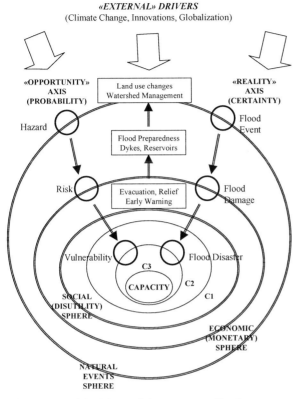

Figure2: Model of the social response to floods

C2 corresponds with a reduced social capacity. Vulnerability, and consequently disaster features, become obvious. Finally, C3 depicts the case where the available social capacities are entirely inadequate to deal with the flood event. A fully-fledged disaster would occur. Fig.2 also depicts some aspects of the social preparedness and response capacities influencing the magnitude of a (potential) event or/and the mitigation of potential damage once the flood has occurred. A disaster is defined as a serious disruption of the functioning of society, causing widespread losses which exceed the ability of the affected society to cope using its own resources. A disaster coincides with the propensity of vulnerability. The distinction (and relation) between risk and vulnerability on the one hand and between damage and disaster on the other implies the possible consideration of two spheres (and corresponding measurement scales) of economic and social consequences. According to this model, the more comprehensive social vulnerability should be measured on a scale incorporating its monetary dimension (risk), but also "intangibles' like confidence, trust, fear, apathy and other features of the social evaluation of an event and its consequences. The model of Fig.2 does not account for environmental vulnerability. Disutility functions have already been tested how far they could be used to capture all these features (Bogardi, 1981, Huang, 1989). But even if accepted as a consensus, they do not mean that vulnerabilities can be easily measured. Their abstract "dimension" renders them cumbersome for political deliberations. Therefore surrogate indices which are "relatively" simple to derive and which represent politically conceivable targets (like health) may be explored for their feasibility.

Before going deeper into the question of how to develop and structure indicators for vulnerability and coping capacity, it is important to consider that all approaches in this area are dealing with a paradox in that they aim to measure vulnerability, but cannot precisely define it. Moreover, it has to be recognized that the number of necessary and desirable indicators to measure vulnerability and coping capacity is not only limited by "intangible" aspects like trust and confidence, but also by limited available data. To illustrate this, two examples will be given.

The UNDP report on "reducing disaster risk" includes a Disaster Risk Index (DRI). The Index provides decision-makers with an overview of levels of risk and vulnerability in different countries in terms of the risk of death in disasters. It is primarily based on mortality data. This is because other features of disaster risk are not available in global level disaster databases (UNDP, 2004). Mortality can be a key indicator for flood disasters, especially for developing countries. Nevertheless, it is hard to judge different prevention strategies or instruments with it, particularly in those countries where the vulnerability is much more evident in other features.

Plate, for example, recommended the development of a critical index of vulnerability measured as the distance (value of human security) between the part of the GNP per person needed for maintaining minimum social standards and the available GNP per person (Plate, 2002). This index would focus on the financial resources available within a society or a community, or even an individual household, that can reduce the effect of a disaster. The index has a strong connection to the question of coping capacity and should be integrated into an approach for vulnerability assessment. Nevertheless, this vulnerability measure would also cover only part of the problem; other aspects, like the environmental dimension, can not adequately be expressed in monetary terms.

More direct indicators of national and regional scale are needed which could be linked to strategic goals and instruments of vulnerability assessment.

A lesson which can be learned from the development process of sustainability indicators since 1992, as well as from the debate on social indicators in the 1960s is the fact that general data on the macro level do not give sufficient information regarding regional and local features of vulnerability, coping capacity, or, for instance, of the success of the implementation of certain policies.

In this regard there is a need to examine indicators which do reflect more strategic developments and trends and those which already provide information on specific policies and instruments. Vulnerability assessment will therefore require research on general goals of vulnerability reduction and coping capacity as well as an analysis of specific and operative goals and instruments available in the specific region. The distinction between strategic and operative goals and indicators in terms of vulnerability assessment could be an interesting option to achieve this. It will not be easy to classify available indicators, and especially to measure the effectiveness of political strategies and instruments. Vulnerability assessment in this regard is confronted with the "stochastic" nature of vulnerability and the problem that many unknown factors and their coincidence can have an important impact on vulnerability. Therefore it will be an important task to achieve a consensus on key indicators of vulnerability for the strategic level, as well as on the formulation of recommendations regarding how the impact of policies and instruments on the operative level could be taken into account. Besides this distinction, it would be important to develop a framework that supports a process perspective. Vulnerability should be captured in its dynamic relation with other elements of the "hazard-risk-vulnerability" chain and could potentially be separated into driving forces, state and response indicators, without the intention to measure linear causalities. The distinction between driving force, state and pressure indicators could also outline options to influence vulnerability, as well as make them more applicable for the integration into public-awareness-raising strategies.

An interdisciplinary approach will be essential to take into account economic, social and environmental consequences as well as different objects of protection (individual, community features). While the potential economic losses caused by floods can often be quantified and estimated, methods and data to measure social, cultural, institutional and environmental features of vulnerability and coping capacity are still not sufficiently developed.

Furthermore, research is needed for example into the measurement of group and individual vulnerabilities. First of all because:

- There are no agreed indicators to measure the individual "components" of vulnerability
- There is no agreed methodology as to how to collect the (subjective) information on vulnerability and how to aggregate them into a single composite index.

Research is also needed on the representativity of this composite index.

Concerning floods, there is empirical evidence that even the affected people themselves may not be able to assess their own vulnerability/coping capacity adequately (Rozgonyi et al, 2000). Vulnerability can substantially, however unnoticed, deteriorate due to "hidden" creeping processes (climate change, land and environmental degradation, economic tendencies, political situation and social tensions). Vulnerability assessment and mapping therefore require interdisciplinary research involving the social and natural sciences, engineering and management. The development of adequate methodology and the practical proof of its applicability are needed to predict the potential consequences of flood events. Thus vulnerability assessment is part of early warning. It helps to identify target areas and groups for the most urgent pre-disaster interventions and capacity-building.

5 Conclusions

- The last three decades revealed our increasing vulnerabilities. This paper presented some reasons for concentrated, interdisciplinary scientific efforts to support knowledge-based policy analysis and decisions mitigating human insecurity.
- The United Nations University Institute for Environment and Human Security (UNU-EHS) has been established to investigate how environmental events and changes influence and are influenced by humanity. Its scope includes degradation processes, hazards, risks, vulnerabilities and coping capacities related to and influencing human security.
- The Institute will work to anticipate the impact on risk and vulnerability of 'creeping' environmental and climate change, land degradation, population pressure and migration, as well as changing resource availability and quality. An important task will be the development of indicators for vulnerability assessment.
- Ensuring human security requires a paradigm shift in the concept of disaster prevention and preparedness. Instead of starting with a focus on natural hazards and their quantification, the assessment and ranking of the vulnerability of affected groups and objects of protection should serve as the starting point in defining priorities and means of remedial interventions. It is with this perspective that UNU-EHS will undertake its work.

References

Alexander, D.: *Confronting Catastrophe*, Terra Publishing, 282 p., Harpenden, 2000

Birkmann, J.: *Monitoring und Controlling einer nachhaltigen Raumentwicklung* (Monitoring and Controlling of Sustainable Spatial Development). Dortmunder Vertrieb für Bau- und Planungsliteratur, 371 p., Dortmund, 2004

Bogardi, J. J.: *Efficiency Measurement of Flood Control Development Schemes by Using Disutility Functions*. Proceedings, XIXth IAHR Congress, Vol. IV. Pp.319-326, New Delhi, 1981

Bohle, H.-G.: Land Degradation and Human Security. In *Environment and Human Security Contributions to a workshop in Bonn*, Plate, E. editor, Bonn, 2002

Busch-Lüty, Ch.: Nachhaltige Entwicklung als Leitmodell einer ökologischen Ökonomie. In *Nachhaltigkeit: in naturwissenschaftlicher und sozialwissenschaftlicher Perspektive*. Fritz, P., Huber, J., Levi, H. editors, Stuttgart, 1995

Currey, B.: Bangladesh Vulnerable Rural Households, maps, Shomabesh Institute, 1984

Huang, W. Ch.: Multiobjective Decision Making in the On-line Operation of a Multipurpose Reservoir. PhD dissertation, Asian Institute of Technology, Bangkok, 1989

International Strategy for Disaster Reduction (ISDR): *Living with Risk* (A global review of disaster reduction initiatives), Preliminary version, Geneva, 2002

ISDR/UN/WMO: *Water and Disasters - Be Informed and Be Prepared*. WMO Publication No. 971, Geneva, 2004

Kron, W.: High water and floods: resist them or accept them? pp 26-34 in *Schadenspiegel* (Losses and loss prevention) 46th year No 3. Munich Re Group, Munich, 2003

Plate, E.: Environment and Human Security, Results of a workshop held in Bonn on 23-26 2002. In *Environment and Human Security, Contributions to a workshop in Bonn*. Plate, E. editor, Bonn, 2002

Plate, E.: Towards Development of a Human Security Index. In *Environment and Human Security, Contributions to a workshop in Bonn*. Plate, E. editor, Bonn, 2002

Presscott-Allen, R.: *Barometer of Sustainability*: A method of assessing progress toward sustainable societies, Contribution to the IUCN/DRC Project on Monitoring and Assessing Progress Toward Sustainability, British Columbia, 1995

Rozgonyi, T., Tamás, P., Tamási, P., Vári, A: A TISZAI ÁRVÍZ Vélemények, kockázatok, stratégiák, 175 p. (in Hungarian) Magyar Tudományos Akadémia Szociológiai Kutatóintézet, Budapest, 2000

UNDP: *Reducing Disaster Risk a Challenge for Development*, a Global report, UNDP Bureau for Crisis Prevention and Recovery, New York, 2004

Vlek, P.: Nothing Begets Nothing: the Threat of Land Degradation, Keynote lecture at the official opening of UNU-EHS, Bonn, 2004

Wisner, B.: Who? What? Where? When? In an Emergency: Notes on Possible Indicators of Vulnerability and Resilience By Phase of the Disaster Management Cycle and Social Actor. In *Environment and Human Security, Contributions to a workshop in Bonn*, Plate, E. editor, Bonn, 2002

Wisner, B., Blaikie, P., Cannon, T., Davis, I.: *At risk, Natural hazards, people's vulnerability and disasters*, second edition, London, New York, 2004

Seismic Risk Analysis for Germany: Methodology and Preliminary Results

Rutger Wahlström[1,2], Sergey Tyagunov[1], Gottfried Grünthal[2], Lothar Stempniewski[1], Jochen Zschau[2] and Matthias Müller[1]
[1] CEDIM, University of Karlsruhe (TH), D-76128 Karlsruhe, Germany;
[2] GeoForschungsZentrum Potsdam (GFZ), Telegrafenberg F454, D-14473 Potsdam, Germany

Abstract

The main objective of the earthquake risk sub-project of CEDIM is assessment and mapping of seismic risk for Germany. There are several earthquake prone areas in the country, producing ground shaking intensity up to grade VIII (EMS-98). The seismicity is highest in parts of the Federal States of Baden-Württemberg, Rhineland-Palatinate, North Rhine-Westphalia, Saxony and Thuringia, which all are densely populated, industrialized and have a high concentration of developed infrastructure. This implies a challenge for future disaster preparedness and risk mitigation activities. The seismic risk in Germany represents typical features with a low earthquake occurrence probability, yet potentially high consequences. Therefore, the results of seismic risk analysis are indispensable for planners and decision-makers for preventing possible future seismic disasters. The paper describes a methodology of seismic risk analysis, including hazard, vulnerability and assets, and presents preliminary results.

1 Introduction

Germany has several seismic prone zones, where earthquakes can produce shaking intensity up to grade VIII. The seismic prone zones are in part densely populated, industrialized and have a high concentration of developed infrastructure, which implies a challenge for future disaster preparedness and risk mitigation activity.

Seismic risk consists of the components seismic hazard, seismic vulnerability and value of elements at risk (both in human and economic terms). The proper approach to the problem of risk assessment and risk management should include consideration of all the contributing components. Countries of low and moderate seismicity can still have high risk values. The seismic risk assessment in Germany represents a typical problem with a low earthquake occurrence probability but potentially high damaging consequences.

The Center for Disaster Management and Risk Reduction Technology (CEDIM), established as a joint initiative of the GeoForschungsZentrum Potsdam and the University of Karlsruhe (TH) conducts an interdisciplinary study aimed at assessment and mapping of different kinds of risks for the territory of Germany, including the earthquake risk. The Potsdam team is concentrated on the hazard aspects and the Karlsruhe team on the vulnerability aspects The GIS technique is utilized in combining different layers of information. The paper presents current results of the earthquake risk subproject of CEDIM.

The used technique is preliminary. Other techniques will be considered in the continued work.

2 Seismic Risk Analysis at Various Scales

There is a general agreement that the term "earthquake risk" refers to the expected losses of a given element at risk over a specified time period. The seismic risk can be specified for different types of elements at risk: economic loss, loss of human lives, physical damage to property, etc., where appropriate measures of damage are available. The risk may be expressed as the average expected loss or damage or in a probabilistic manner and should include proper consideration of hazard, vulnerability and exposed values.

The approach to risk analysis depends on the geographical scale. For individual existing buildings or construction sites, the analysis can be conducted in detailed manner, taking into account geotechnical information about the site, location of probable hazard sources and estimated seismic influence, using advanced numerical or simplified methods of structural analysis and considering all relevant elements at risk. Obviously, this is a finance- and time-consuming procedure and it is applicable only for individual sites, in particular for critical buildings and facilities. At the next level for local, often urban areas, microzonation maps and building stock inventory are used. The inventory is often implemented using visual screening procedures and selecting representative buildings. In the same manner, the distribution of the exposure at risk can be estimated. Yet one level up, at a regional or national scale, another set of input data and more generalized methods of analysis are used. This is the final level aimed at in the CEDIM project and the developed and applied GIS-based procedure is described more in detail below.

3 Seismic Hazard

At the first stage of the study we use the German part of the so-called D-A-CH map (Grünthal et al., 1998), a seismic hazard map for Germany, Austria and Switzerland expressed in intensity for a non-exceedence probability of 90% in 50 years. The seismic intensities of the D-A-CH map are interpolated from a grid of points over the territory of Germany. The intensity is now assigned for the centre of each community. The corresponding GIS layer is shown in Figure 1. Some of the most densely populated areas, in particular the Lower Rhine Embayment and parts in the south-west of the country (Baden-Württemberg), coincide with earthquake prone zones.

In the framework of CEDIM, a new national hazard map will be calculated based on a new earthquake catalogue (Grünthal and Wahlström, 2003), a seismotectonically better founded source zonation model, a revised technique for calculation of the maximum magnitude, and consideration of the aleatory and epistemic uncertainties in the input models and parameters. This will give uncertainties in the hazard output and subsequently in the risk values.

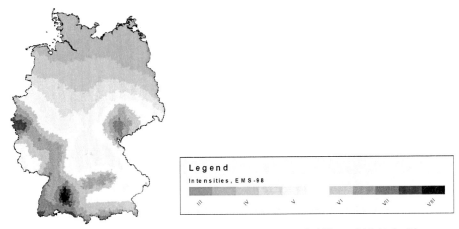

Figure 1: Seismic hazard distribution for non-exceedence probability of 90 % in 50 years; after Grünthal et al. (1998) with intensity contours modified according to community borders.

4 Seismic Vulnerability of the Building Stock

The seismic vulnerability implies the expected degree of damage to a given element at risk resulting from a given level of seismic hazard. There are two principal approaches to vulnerability assessment - observed and predicted vulnerability. Observed vulnerability refers to assessment based on statistics of past earthquake damage. Predicted vulnerability refers to assessment of expected performance of buildings based on engineering computation/judgement and design specifications. Obviously the second way is more suitable for areas of low and moderate seismicity, where, as a rule, there are few or no data of observed earthquake damage. In Germany, despite the damaging Albstadt (1978) and Roermond (1992) earthquakes, there are not sufficient data about the seismic performance of the existing building stock. On the other hand, there is growing interest in the national engineering community to address the problem of vulnerability assessment of the building stock (e.g., Sadegh-Azar, 2002, Schwarz et al., 2002a and Meskouris and Hinzen, 2003).

Analysis of seismic vulnerability in our study is conducted using the classification of buildings in terms of the European Macroseimic Scale (EMS-98; Grünthal, 1998), where six vulnerability classes were introduced and for different types of structures a most probable class and a probable range of classes are given. For Germany, the classes A-D apply.

Taking into account the national scale of the task, the communities of Germany are taken as units at risk. The approximately 14,000 communities are classified in five population size classes: P1 (less than 2,000 inhabitants), P2 (2,000 – 20,000), P3 (20,000 – 200,000), P4 (200,000 – 800,000) and P5 (more than 800,000). In our simplified study, the building stock is considered similar in each community belonging to a certain population class. As prototypes we took into consideration the communities listed in Table 1. All these communities are located within seismic prone zones and we assume that they are representative for the five population classes given above.

For some of the prototype communities, existing information was used; in particular for Cologne and Schmölln, where study cases of the recent DFNK (Deutsches Forschungsnetz Naturkatastrophen - German Research Network Natural Disasters) project with detailed vulnerability analyses were conducted for the building stock (Schwarz et al., 2002a,b, 2004). Also information from Stricker (2003) was used for Cologne. For the other communities, information about the building stock was collected using simplified visual screening procedures and other available data. Generally, the most probable vulnerability class of the EMS-98 schedule was assigned.

Community	Population / class
Cologne	1 020 000 (P5)
Schmölln	13 000 (P2)
Albstadt	48 000 (P3)
Lörrach	47 000 (P3)
Karlsruhe Stupferich (Karlsruhe)	283 000 (P4) 3 000 (P2)
Ettlingen Schluttenbach (Ettlingen) Schöllbronn (Ettlingen) Spessart (Ettlingen)	39 000 (P3) (P1) (P2) (P2)

Table 1:   Considered prototype communities

Based on available information and using engineering judgement, vulnerability composition models for the building stock of German communities corresponding to different population classes were constructed (Table 2). Table 2 gives the vulnerability composition models as percentage of buildings of different vulnerability classes. Probable ranges are given, not only to depict the uncertainty but also to emphasize that the composition of the building stock of individual communities in the same population class is different. The contents of Table 2 are illustrated in Figure 2.

Population classes (number of inhabitants)	Percentage of buildings of different vulnerability classes (EMS-98)			
	A	B	C	D
P1 (< 2,000 inhabitants)	Few	Most	Few	Very few
P2 (2,000 – 20,000)	Few	Most	Many	Very few
P3 (20,000 – 200,000)	Very few	Many	Very many	Very few
P4 (200,000 – 800,000)	Very few	Many	Most	Few
P5 (> 800,000)	Very few	Many	Most	Few
Very few - (0-5%); Few - (5-20%); Many - (20-40%); Very many - (40-65%); Most - (65-100%)				

Table 2:   Vulnerability composition models of the building stock of communities

Damage probability matrices (DPM) were constructed following the ideas of the European Macroseismic Scale (EMS-98), where the description of damage distribution in terms of "few", "many", "most" is given in the definition of the highest damage grades. Supplementing with results of other studies, e.g., ATC-13 (1987) and Nazarov and Shebalin (1975), the descriptions are here extended also to lower damage grades.

Vulnerability functions were constructed for each of the considered vulnerability classes (from A to D) in terms of the mean damage ratio (MDR) versus intensity of ground shaking (Table 3). For computation of the MDR, which is considered as the cost of repair over the cost of replacement, the damage ratio range was assigned to the damage grades classified in the EMS-98 based on many earlier studies. The resulting vulnerability functions for the vulnerability classes A-D and for the considered interval of seismic intensities from V to IX (EMS-98) are shown in Figure 3.

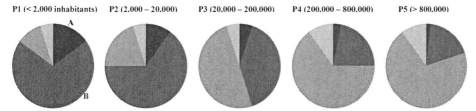

Figure 2: Building stock vulnerability models for different communities; composition of the vulnerability classes A, B, C and D

Classification of damage; after Grünthal (1998)	MDR %	Mean value %
Grade 0: No damage	0	0
Grade 1: Negligible to slight damage (no structural damage, slight non-structural damage)	0-1	0.5
Grade 2: Moderate damage (slight structural damage, moderate non-structural damage)	1-20	10
Grade 3: Substantial to heavy damage (moderate structural damage, heavy non-structural damage)	20-60	40
Grade 4: Very heavy damage (heavy structural damage, very heavy non-structural damage)	60-100	80 (100)
Grade 5: Destruction (very heavy structural damage)	100	100

Table 3: Classification of damage and damage ratio

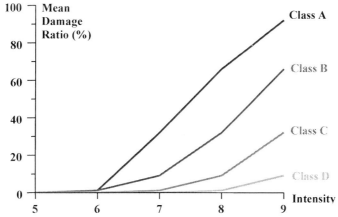

Figure 3: Vulnerability functions for the different vulnerability classes according to EMS-98

Combining the vulnerability curves (Figure 3) with the building stock vulnerability models (Figure 2), the expected damage can be plotted versus seismic intensity for the different population classes (Figure 4). From the curves in Figure 4 and the hazard map of Figure 1, we can make rough judgements of the earthquake damage potential of different German communities.

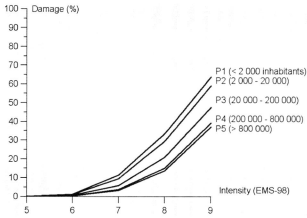

Figure 4: Vulnerability functions for the building stock of different population classes

5 Damage and Risk Estimations

Combining the seismic hazard (Figure 1), the distribution of communities of different population classes and the vulnerability models for these classes (Figure 4), the distribution over
German communities of the specific damage of buildings is obtained as a GIS layer and
shown in Figure 5. The resulting range of estimated damage values is from 0 to about 37%
for the assumed probability of non-exceedence of 90% in 50 years. The map is constructed
without consideration to the exposed values.

At the present, time the CEDIM team is engaged in collecting data about the distribution
of values at risk (assets) in the country, which are necessary for seismic risk analysis and
will be used for assessment of other kinds of risks as well. Awaiting this information for the
final risk assessments, a preliminary concept, "seismic risk potential", is defined and
calculated as the product the specific damage (Figure 5) and the number of inhabitants in the
communities. The outcome is shown in Figure 6. This map provides a first indication of
seismic risk distribution over Germany, although its provisional character must be pointed
out.

6 Conclusions and Future Tasks

The principal emphasis of the first stage of the CEDIM study from the earthquake group was
to work through the methodology of seismic risk assessment at the national scale. A new
hazard assessment based on modern data and techniques (see Chapter 3), more detailed
vulnerability data and future access of asset data will improve the yet in several respects
simplified approach and its preliminary results.

It is interesting to compare the maps of seismic hazard (Figure 1), specific damage
(Figure 5) and risk potential (Figure 6). Although the distributions of specific damage and
risk potential generally follow that of the hazard, there are also clear distinctions. The
estimated specific damage to the building stock, which is a combination of hazard and
vulnerability and shows the percentage of damaged buildings, does not consider the number
of buildings and other values at risk in the community. Therefore, the picture is rather
smooth and provides no idea about potential losses in the area. On the other hand does the
map of the risk potential, where the distribution of exposed values is taken into
consideration in a rough manner, give at least a hint of the main features in a future risk
map.

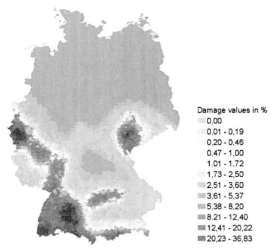

Figure 5: Distribution of the estimated specific damage (percentage of damaged buildings) based on a non-exceedence probability of 90% in 50 years

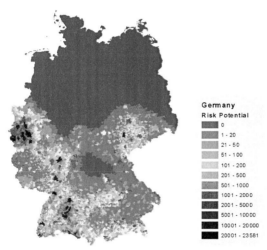

Figure 6: Distribution of the estimated "seismic risk potential" (relative scale; see text for explanations) based on a non-exceedence probability of 90% in 50 years

The obtained results show that, on the one hand, the smaller communities are characterized by more vulnerable composition of their building stock and, therefore, a higher percentage of damaged buildings can be expected there than in more populated communities in the case of a damaging earthquake. On the other hand, the larger communities located in earthquake prone areas, even with more favourable building stock composition and smaller estimated damage percentage, can have a higher level of risk due to the higher concentration of exposed values.

The future steps of the seismic risk program of CEDIM include:

- Improvement of the seismic hazard input data as outlined in Chapter 3.
- Improvement of the vulnerability input data, meaning improvement of the used generalized vulnerability models and development and application of

vulnerability analysis on the basis of available GIS data of the actual building stock distribution in the communities.

- Collection and analysis of data on values at risk. This will be done in conjunction with the other working groups of CEDIM.
- Testing other techniques for risk calculation for possible use.

All these activities are directed towards the main goal of the project – assessment and mapping of seismic risk for Germany.

References

ATC-13: Applied Technology Council: Earthquake Damage Evaluation Data for California. Redwood City, California, 1987.

Grünthal G. (Ed.): European Macroseismic Scale 1998. Cahiers du Centre Européen de Géodynamique et de Séismologie, 15, Luxembourg, 1998.

Grünthal, G. and Wahlström, R.: An Mw based earthquake catalogue for central, northern and northwestern Europe using a hierarchy of magnitude conversions. Journal of Seismology, 7: 507-531, 2003.

Grünthal G., Mayer-Rosa D and Lenhardt W.A.: Abschätzung der Erdbebengefährdung für die D-A-CH-Staaten – Deutschland, Österreich, Schweiz. Bautechnik, 10, 19-33, 1998.

Meskouris K. and Hinzen K.-G.: Bauwerke und Erdbeben. Vieweg Verlag, 2003.

Nazarov A.G. and Shebalin N.V. (Eds): The seismic scale and methods of measuring seismic intensity. Moscow, 1975 (in Russian).

Sadegh-Azar H.: Schnellbewertung der Erdbebengefährdung von Gebäuden. Dissertation am Lehrstuhl für Baustatik und Baudynamik, Rheinisch-Westfälische Technische Hochschule Aachen, 2002.

Schwarz J., Raschke M. and Maiwald H.: Seismische Risikokartierung auf der Grundlage der EMS-98: Fallstudie Ostthüringen. Zweites Forum Katastrophenvorsorge, DKKV, Bonn und Leipzig, 325-336, 2002a.

Schwarz J., Raschke M. and Maiwald H.: Seismic risk studies for central Germany on the basis of the European Macroseismic Scale EMS-98. Proceedings of the 12th European Conference on Earthquake Engineering, Elsevier Science Ltd., Paper 295, 2002b.

Schwarz, J. and Maiwald, H. and Raschke, M.: Erdbebenszenarien für deutsche Großstadträume und Quantifizierung der Schadenpotentiale. In B. Merz and H. Apel, editors: Deutsches Forschungsnetz Naturkatastrophen (DFNK) Abschlussbericht, 188-200, 2004, in press.

Stricker E.: Schadenprognose für den Großraum Köln bei Erdbeben mit besonderer Berücksichtigung der direkten wirtschaftlichen Kosten. Diplomarbeit, Institut für Massivbau und Baustofftechnologie, Universität Karlsruhe, 2003.

Drought and Desertification Hazards in Israel: Time-Series Analysis 1970-2002

Hemu Kharel, Hendrik J. Bruins
Ben-Gurion University of the Negev, Jacob Blaustein Institute for Desert Research
Department Man in the Desert, Sede Boker Campus, 84990, Israel

Abstract

Drylands comprise more than one third of the Earth's terrestrial surface, providing habitation to some 1.2 billion people. All dryland zones – *hyper-arid, arid, semi-arid* and *dry sub-humid* – are vulnerable to the hazards of drought and desertification. In this article we present time-series analyses of annual changes in the aridity index of six meteorological stations in Israel for the period 1970-2002. Our research is the first to investigate annual variations in the dryland zones of Israel with the aridity index (P/PET), according to the United Nations Convention to Combat Desertification. P is precipitation and PET is potential evapotranspiration calculated with the Thornthwaite formulas and adjusted to Penman with a correction factor. Our results show a slight downward trend towards increased aridity, which is more significant in the southern part of the country. The most severe drought period, as well as the wettest year, generally occurred in the period 1990-2001, indicating climatic conditions becoming more extreme. Drought occurs in Israel once every 2 or 3 years or as more hazardous multi-year droughts. The trend to increased aridity may lead to climate-induced desertification and increased drought frequency and severity.

Keywords: aridity Index, dryland zone classification, drought, desertification, hazard assessment.

1 Introduction

Drylands cover about 40% of the total land area of the world (Hulme, 1996). This evaluation is based on the relationship between annual precipitation (P) and annual potential evapotranspiration (PET): the Aridity Index P/PET (UNESCO, 1979; Middleton and Thomas, 1997; Bruins and Lithwick, 1998; Bruins and Berliner, 1998). Drought and aridity constitute two different aspects of drylands. Increased aridity or drought may trigger or accelerate the phenomenon of desertification (Le Houérou, 1996).

In this paper we present for the first time detailed aridity index (P/PET) results of Israel, covering the period 1970-2002. The Penman formula (Penman, 1948) is generally preferred in terms of accuracy to calculate PET. But the required physical and meteorological parameters needed for its calculation are usually not recorded in most meteorological stations. Therefore, the geographical density of stations allowing for Penman PET calculation will be very low, which is unsuitable for evaluation and mapping purposes on a national or global scale. For this fundamental reason the more simple Thornthwaite approach was selected to calculate potential evapotranspiration by the authors of the World Atlas of Desertification (Middleton and Thomas, 1997) and by the UN Convention to Combat Desertification (UNCCD, 1999).

The Thornthwaite method (Thornthwaite, 1948; Thornthwaite and Mather, 1955, 1957) enables the calculation of PET on the basis of only two parameters: mean monthly temperature data and the average number of daylight hours per month, which is related to geographical latitude. The drawback of the Thornthwaite method is that it systematically underestimates PET in dry climates and overestimates PET for moist and wet environments. Consequently, an empirical adjustment factor is applied to the data to bring the Thornthwaite values more closely in line with those of the Penman method (Hulme *et al.*, 1992; Middleton and Thomas, 1997).

We present the annual fluctuations of the aridity index during the period 1970-2002 for six selected meteorological stations in Israel, ranging from the north to the south of the country. Hazard assessment of drought is evaluated in terms of frequency and severity. The linear trend of the aridity index (P/PET) is calculated for the last 30 years (regression line), which is shown in each of the six figures.

2 Methodology

The climate in Israel is characterized by a completely dry summer without any rainfall usually in June, July and August. The rainfall and agricultural year begins more or less in September-October. Therefore, the annual division we have adopted starts on September 1st and ends on August 31st, which is a common approach in Israel for studies involving meteorological and climatic data. The meteorological stations analyzed in this article are the following, going from the wetter north of the country to the dry south: Har Kenaan, Tel Aviv Sede Dov, Jerusalem Airport, Be'er Sheva, Sede Boqer and Elat.

The Thornthwaite method uses the following formulas (Thornthwaite, 1948; Thornthwaite and Mather, 1955, 1957):

$$PET= 16C(10Tm/I)^{a} \qquad (1)$$

$$I = sum\ (Tm/5)^{1.51} \qquad (2)$$

$$a = (67.5 * 10^{-8}\ I^{3}) - (77.1 * 10^{-6}\ I^{2}) + (0.0179\ I) + (0.492) \qquad (3)$$

PET is potential evapotranspiration (in mm), C is the daylight coefficient, Tm is the average monthly temperature (°C), a is an exponent derived from the heat index (I).

Following the monthly calculations, the annual Thornthwaite PET is then adjusted with an empirical correction factor to obtain a result for PET that approaches Penman PET, using the following formula:

$$PET^{P} = 1.3*(PET^{T}) - 0.428*(P) + 246 \qquad (4)$$

PET^{P} is the annual Penman adjusted PET derived from annual Thornthwaite PET.

PET^{T} is annual Thornthwaite PET calculated with the Thornthwaite formulas (1, 2, 3).

P = Annual value of precipitation.

Subsequently, the annual aridity index (P/PET) is calculated for each of the six stations. Then it is possible to classify each meteorological station, according to the four dryland zones (Table 1). The average dryland classifications, as well as the annual fluctuations through time-series analyses, were established for each station, as shown in the figures.

Dryland Zones	P/PET Range
Hyper-arid	<0.05
Arid	0.05 to <0.20
Semi-arid	0.20 to <0.50
Dry sub-humid	0.50 to <0.65

Table 1: The division of the four dryland zones, based on the P/PET index, as calculated with the Thornthwaite formulas (Middleton and Thomas, 1997; UNCCD, 1999):

3 Research Results

3.1 Har Kenaan in the Upper Galilee Region

The Har Kenaan meteorological station is situated in northern Israel in the Upper Galilee Region at an altitude of 934 m. The average annual value of the aridity index P/PET at Har Kenaan is 0.67, *i.e.* the area is not a dryland according to Table1, but belongs to the sub-humid zone. The wettest year in the area was 1991-92 (P/PET = 1.34), accompanied by a lower PET value and increased precipitation. During this year Har Kenaan was situated in the humid zone.

The most severe drought year in this region occurred in 1972-73 with a P/PET value of 0.33, about half the average value for the aridity index. Har Kenaan was *de facto* situated in the semi-arid zone during this drought year, two zones below its normal climatic position. There were 10 drought years during 1970-2002 in which the aridity index was located in the semi-arid zone (P/PET in the range 0.20 to <0.50) for the Har Kenaan area: 1972-73, 1978-79, 1981-82, 1985-86, 1988-89, 1989-90, 1990-91, 1993-94, 1998-99, 2000-01. Other drought years, not yet mentioned, with a P/PET value of below 15% are 1971-72 and 1999-2000. The boundaries between the sub-humid, dry sub-humid and semi-arid zones shifted significantly southward during these drought years. Altogether 12 drought years can be distinguished at this station in the period 1970-2002, according to our methodology.

The linear trend line for Har Kenaan (Fig. 1) shows a very slight downward tendency towards increased aridity. Though this trend is very small indeed, it confirms the pattern clearly visible for the stations in the drier south of the country.

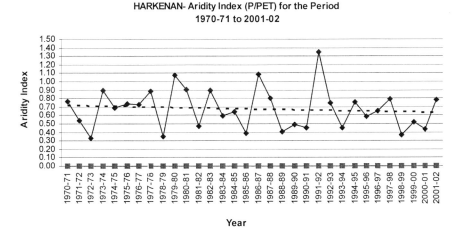

Figure 1:   Annual aridity indices and linear trend line for Har Kenaan.

3.2 Tel Aviv Sede Dov

The Tel Aviv Sede Dov meteorological station is situated along the Mediterranean coast at an altitude of 4 m. The average aridity index value of 0.40 (Fig. 2) shows that the area is normally semi-arid. Wet years in which the area experienced a dry sub-humid climate occurred during 1971-72, 1973-74, 1989-90, 1994-95. The wettest year was 1991-92 (P/PET = 0.99) when the area had a humid climate.

Severe drought years, in which the Tel Aviv Sede Dov area experienced an arid climate, occurred during 1983-84 and 1998-99. There were 11 drought years with a P/PET value of at least 15% below the average are: 1972-73, 1975-76, 1978-79, 1981-82, 1983-84, 1984-85, 1985-86, 1990-91, 1993-94, 1998-99, 2000-01. Like in Har Kenaan, the driest year (1998-99) and the wettest year (1991-92) happened in the 1990s. The linear trend line is almost flat, showing only a very slight downward tendency towards increased aridity.

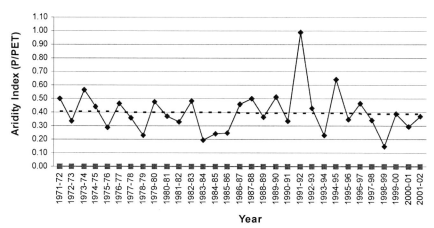

Figure 2:   Annual aridity indices and linear trend line for Tel Aviv Sede Dov.

3.3 Jerusalem Airport

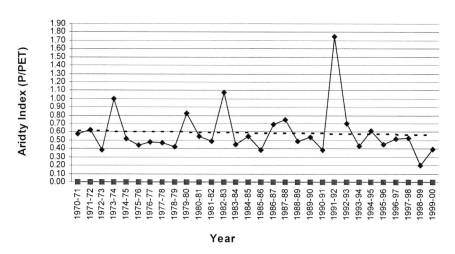

Figure 3:   Annual aridity indices and linear trend line for Jerusalem Airport.

The meteorological station at Jerusalem Airport is situated in the centre of the country in the Judean hills at an altitude of 755 m. The average P/PET value is 0.59 (Fig. 3), placing the area in the dry sub-humid zone. Wet years in which the area experienced a humid climate (P/PET 0.75 or higher) occurred during 1973-74, 1979-80, 1982-83, 1987-88, 1991-92, while a sub-humid climate (P/PET 0.65 to < 0.75) prevailed during 1986-87 and 1992-93. The wettest year was 1991-92 (P/PET = 1.74).

The most severe drought year, in which the Jerusalem Airport area experienced an arid climate, occurred during 1998-99. Other drought years, in which the area had a semi-arid climate, while the P/PET value was at least 15% below the average, are: 1972-73, 1975-76, 1976-77, 1977-78, 1978-79, 1981-82, 1983-84, 1985-86, 1988-99, 1990-91, 1993-94, 1995-96, 1999-2000. Thus the total number of drought years for the above period is 14, according to our criteria. The driest year (1998-99) and the wettest year (1991-92) occurred in the 1990s. The linear trend line shows a very slight downward tendency towards increased aridity.

3.4 Be'er Sheva

The Be'er Sheva meteorological station is situated in the northern Negev desert at an altitude of 280 m. The average aridity index value is 0.14 (Fig. 4), which places the area in the arid zone. Wet years with a semi-arid climate occurred during 1971-72, 1973-74, 1979-80, 1982-83, 1991-92, 1994-95. The wettest year was 1971-72 (P/PET = 0.25), which is different from the previous stations in the centre and north of the country.

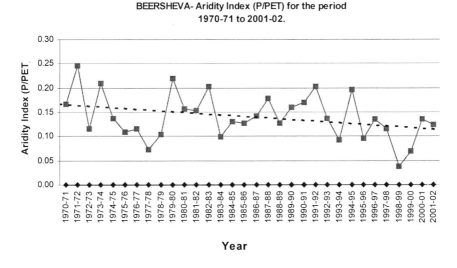

Figure 4:   Annual aridity indices and linear trend line for Be'er Sheva.

The most severe drought year was 1998-99, in which the Be'er Sheva area experienced a hyper-arid climate. Other drought years with a P/PET value of at least 15% below the average are: 1975-76, 1976-77, 1977-78, 1978-79, 1983-84, 1993-94, 1995-96, 1999-2000. The total number of drought years is 9, according to our criteria. The linear trend line for the period 1970-2002 (Fig. 1) shows a most significant downward trend towards increased aridity.

3.5 Sede Boqer

The meteorological station at Kibbutz Sede Boqer is situated in the northern part of the hilly central Negev desert at an altitude of 470 m. The average aridity index value is 0.07 (Fig. 5), which places the area in the dry part of the arid zone, rather close to the boundary (P/PET = 0.05) with the hyper-arid zone. The climate did not cross the boundary into the semi-arid zone during rainy years. The wettest year was 1991-92 (P/PET = 0.14), in which the Sede Boqer climate became similar to the average Be'er Sheva climate within the arid zone.

The most severe drought with an aridity index of 0.02 (hyper-arid) occurred in two successive years: 1998-99, 1999-2000. All drought years, in which the Sede Boqer area experienced a hyper-arid climate: 1972-73, 1977-78, 1983-84, 1987-88, 1995-96, 1998-99, 1999-2000. Other drought years within an arid climate and with P/PET values of 15% below the average: 1978-79, 1986-87, 1992-93, 1993-94. Thus the total number of drought years is 11. The linear trend line for the period 1970-2002 (Fig. 5) shows a significant downward trend towards increased aridity.

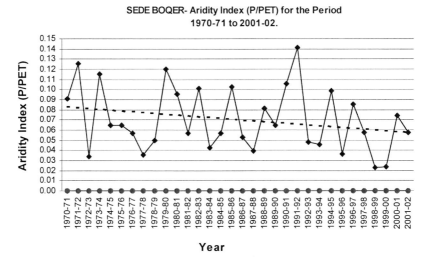

Figure 5:   Annual aridity indices and linear trend line for Sede Boqer.

3.6 Elat

The Elat meteorological station is situated near the Red Sea in the southernmost part of the Negev desert at an altitude of 11 m. This is the driest region in Israel, having an average aridity index value of 0.01 (Fig. 6), which shows that the region is extremely hyper-arid. Even during years with the highest rainfall the climate in the area remained in the hyper-arid zone. The wettest year occurred during 1974-75 (P/PET = 0.03). The most severe drought years were 1995-96 (P/PET = 0.00029) and 1999-2000 (P/PET = 0.00088), both being extremely hyper-arid.

There were altogether 14 drought years in which the P/PET value was at least 15% below the average (P/PET < 0.0085): 1975-76, 1976-77, 1979-80, 1982-83, 1983-84, 1984-85, 1986-87, 1988-89, 1991-92, 1994-95, 1995-96, 1996-97, 1999-2000, 2001-02. The linear trend line for the period 1970-2002 (Fig. 6) clearly shows a significant downward trend towards increased aridity.

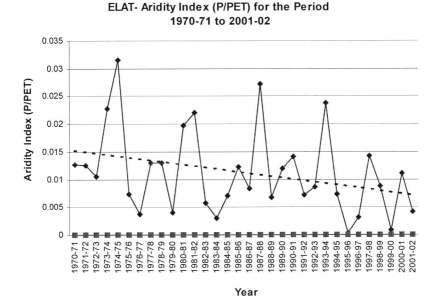

Figure 6:   Annual aridity indices and linear trend line for Elat.

4 Conclusions

One of the most important results is that the three southern stations (Be'er Sheva, Sede Boqer and Elat) show a marked downward trend towards increased aridity. The two stations in the centre of the country (Jerusalem Airport and Tel Aviv Sede Dov) only show a very slight downward trend, which is a bit more pronounced in the northern station of Har Kenaan. Therefore, the tendency towards a drier climate seems apparent in the entire country, but particularly in the southern half.

Both the wettest year and the most severe drought year occurred in the period 1990-2001 for most stations. This may indicate that the climate is becoming more variable and extreme, albeit with an overall trend towards a drier climate. Increased aridity may lead to climate-induced desertification.

Our calculations show that a decrease in the value of the aridity index for all the six stations results from an increased PET value, due to a rise in temperature, and a decreased precipitation value as compared with the average values. The reverse holds true for increased P/PET values.

The year 1991-92 was an exceptionally wet year with the highest P/PET values in all stations, except Be'er Sheva and Elat, showing again a somewhat different climatic pattern for the south of the country.

We noticed that the P/PET aridity index functions also an excellent indicator of drought severity in time-series analysis of each meteorological station. The most severe drought year with the lowest P/PET values in most stations was 1998-99, except for Elat and Har Kenaan, although the 1972-73 drought for the latter station was only slightly more severe than the 1998-99 drought. The number of drought years, according to our criteria, ranged from 9 to

14 during the period 1970-2002. Thus there is a clear drought hazard in Israel once every 2 or 3 years, but multi-year drought may also occur, which is most hazardous.

Acknowledgements

We thank the Israel Meteorological Service (IMS), in particular Dr. Avner Furshpan and Mr. Amos Porat, for providing us with the available temperature and rainfall data needed for this research. We also thank the Ben-Gurion University for providing travel grants that enabled us to participate in the International Conference "Disasters and Society: From Hazard Assessment to Risk Reduction" (University of Karlsruhe).

References

Bruins, H.J. and Berliner, P.: Bioclimatic aridity, climatic variability, drought and desertification: Definitions and management options. In H.J. Bruins and H. Lithwick (eds.) *The Arid Frontier: Interactive Management of Environment and Development*, pp. 97-116, Kluwer Academic Publishers, Dordrecht, 1998.

Bruins, H.J. and Lithwick, H.: Proactive planning and interactive management in arid frontier development. In H.J. Bruins and H. Lithwick (eds.) *The Arid Frontier: Interactive Management of Environment and Development*, pp. 3-29, Kluwer Academic Publishers, Dordrecht, 1998.

Hulme, M.: Recent climatic change in the world's drylands, *Geophysical Research Letters*, 23:61-64, 1996.

Hulme, M., Marsh, R. and Jones, P.D.: Global changes in a humidity index between 1931-60 and 1961-90. *Climate Research* 2:1-22, 1992.

Le Houérou, H.N.: Climate change, drought and desertification, *Journal of Arid Environments*, 34:133-185, 1996.

Middleton, N. and Thomas, D.: *World Atlas of Desertification*. UNEP; Arnold, London, 1997.

Penman, H.L.: Natural evaporation from open water, bare soil and grass. *Proceedings of the Royal Society, Section A*, 193:120-145, 1948.

Thornthwaite, C.W.: An approach toward a rational classification of climate, *Geographical Review*, 38(1), 55-94, 1948.

Thornthwaite, C.W. and Mather, J.R.: The water balance, *Publications in Climatology*, 8(1): 14-21, 1955.

Thornthwaite, C.W. and Mather, J.R.: Instructions and tables for computing potential evapotranspiration and the water balance, *Publications in Climatology*, 10(3): 205-241,1957.

UNCCD: United Nations Convention to Combat Desertification in Those Countries Experiencing Serious Drought and/or Desertification, Particularly in Africa. Permanent Secretariat for the UNCCD, Bonn, 1999.

UNESCO: Map of the World Distribution of Arid Regions, UNESCO, Man and the Biosphere (MAB) Technical Notes 7, Paris, 1979.

Developing a Methodology for Flood Risk Mapping: Examples from Pilot Areas in Germany

Bruno Büchele[1], Heidi Kreibich[2], Andreas Kron[1], Jürgen Ihringer[1], Stephan Theobald[1], Annegret Thieken[2], Bruno Merz[2], Franz Nestmann[1]

[1] *Institute for Water Resources Management, Hydraulic and Rural Engineering (IWK), University of Karlsruhe, D-76128 Karlsruhe, Germany*
[2] *GeoForschungsZentrum Potsdam (GFZ), Section Engineering Hydrology, Telegrafenberg, D-14473 Potsdam, Germany*

Abstract

The paper focuses on flood-risk quantification. The work is divided into three parts: hazard assessment, in particular of extreme situations like a 100-year event and above, vulnerability assessment with an improved method of damage estimation, and risk quantification focusing on techniques for risk mapping. The main goal is the improvement of existing methods in order to achieve more reliable risk analyses on different scales, from microscale (e.g., damage at single buildings) to mesoscale approaches (damage to land use units). Results from pilot areas are presented: 1. regionalization of hydrological extremes, 2. hydrodynamic-numerical simulation of extreme floods, 3. flood damage estimation. The results are a basis for more comprehensive risk analyses and flood management strategies on regional and national level.

Keywords: flood, extreme events, risk, mapping, damage mitigation, vulnerability.

1 General Framework

In CEDIM (Center of Disaster Management and Risk Reduction Technology, a co-operation of the University of Karlsruhe and GFZ Potsdam, http://www.cedim.de), researchers from different disciplines are working together on the project 'Risk Map Germany'. The objective is to quantify and map the risk due to different perils. The methods and results have to be oriented on the specific requirements and working standards of the users (e.g., responsible Water Management authorities of the federal states).

The existing methods for flood risk quantification, i.e. for the estimation of monetary loss and for extreme events, are quite uncertain. Thus more reliable methods for risk assessment are needed, since they contribute substantially to flood damage mitigation and an effective flood management.

The objective is to assemble and improve the modules in the risk quantification procedure. This can be best realised in pilot areas where a good information basis is given. At the same time, a very detailed information basis is already available or planned for all German rivers (e.g., flood hazard maps). However, CEDIM aims to map flood risk on large scale. Therefore simplifications and aggregation of detailed models and simulation results have to be made. The uncertainty of the calculated risk for the pilot areas will be analysed. The transferability of the developed methodology to other regions will be tested, so that the processing of an integrated risk map Germany is possible.

2 Examples from Pilot Areas

2.1 Regionalization of Hydrological Extremes

Until 1999, a regionalization model for flood parameters was developed mainly by the author Ihringer (compare LfU, 1999). As results, mean annual peak discharges MHQ and annual peak discharges HQ-T for return periods T for T = 2, 5, 10, 20, 50, 100 years are available for all watercourses (catchment area > 10 km²) in the Federal State of Baden-Württemberg (LfU, 1999 & 2001). The regionalization model is based on the flood statistics at 335 gauges, respectively on the available time series (between ca. 10 and over 100 years, average: 45 years). At gauges with short time series, the analyses were harmonised with neighboured gauges. The gauges are covering catchment areas from < 10 km² (ca. 7 % of all gauges) up to > 1000 km² (ca. 7 %, as well), summing up to a total area of over 35000 km² and divided into ca. 3400 official hydrological sub-areas. That means that flood parameters can be determined for a multitude of locations without gauges within a coherent approach.

The regionalization model consists of a multiple linear regression model, including the following spatial information for each hydrological unit:

- A_E Catchment area [km²]
- S   Coverage of urban area in relation to A_E [%]
- W   Coverage of forest area in relation to A_E [%]
- I_g Weighed slope of the watercourse [%]
- L   Length of flow path from watershed to gauge [km]
- L_C Length of flow path from center of the catchment to the gauge [km]
- hN_G Average annual precipitation [mm]
- LF Landscape factor [-] (empirical, explaining geological characteristics)

As documented in LfU (1999), the model provides very good results in comparison with the flood statistics at the gauges ($R^2 > 0.994$ for all HQ-T). The deviation is < 2.5 % at ca. 40 % of the gauges, < 7.5 % at ca. 75 % and < 12.5 % at ca. 85 %. The deviation is > 20 % only at ca. 5 % of the gauges, which are influenced by human activities or karst conditions.

Since 2002, the regionalization model has been extended to return periods between 200 and 10000 years (IWK, 2003). In particular, peak discharges of 1000- and 10000-year floods are required by the new design concept for dam safety (DIN E 19700). Moreover, parameter of hydrological extremes exceeding the 100-year level are needed according to the new programs of the state authorities which aim to launch hazard maps for all rivers (> 10 km²) by 2008. Thus the improvement of the regionalization model supplements these programs. As the extrapolation for such extreme return periods is problematical, emphasis is placed on the uncertainties, especially in small catchment areas where the data availability is critical (few gauges, short time series). Therefore, the model results are compared with approved statistical methods and precipitation-runoff models.

Figure 1 shows that the results of the regionalization model are very similar to the statistical analysis of discharge data at the gauge Heidelberg/Neckar for the observation period 1850-1998 (annual peak discharges with return periods up to 10000 years).

Figure 2 shows the spatial distribution of the 1000-year runoff yields in the state of Baden-Württemberg, resulting from the model application for each hydrological unit. As can be seen, the highest specific discharges (Hq1000 in [m³/(s*km²)]) are occurring in the mountainous parts of the Black Forest (Upper Rhine Basin) and the upper Neckar Basin.

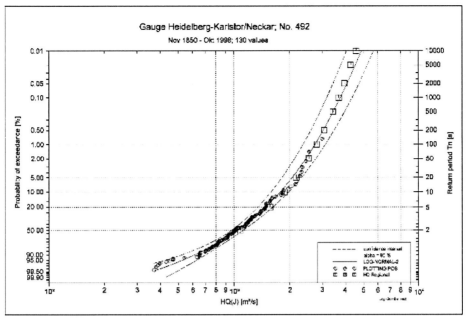

Figure 1: Probability of exceedance [%] respectively return periods Tn [a] of annual peak discharges HQ(J) at the gauge Heidelberg/Neckar (13760 km², compare Fig. 2): comparison of results from the flood statistics for the period 1850-1998 (plotting positions, Log-Normal distribution as solid line, confidence intervals (95%) as dashed lines) with the regionalization model HQ-Regional (squares).

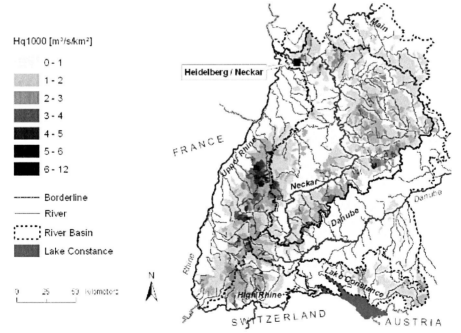

Figure 2: Regionalized 1000-year runoff yields Hq1000 [m³/(s*km²)] in the state of Baden-Württemberg.

2.2 Hydrodynamic-Numerical Simulation of Extreme Floods

To determine the hazard and quantify the flood risk for cities or communities nearby a river, the relevant hydraulic processes of stream flow have to be analysed by hydrodynamic-numerical models (HN-models). For assessment of flood hazard and dimensioning of flood protection measures these models are usually applied for events with a defined recurrence interval, normally up to the discharge with an 100 or 200 years return period. Normally there is no quantification of residual risk in case of failure of protection measures. Floods greater than the design discharge may be rare but can cause severe damages and fatalities in case of overtopping or even destruction of protection structures. But even for floods below the design discharge, there is a residual risk due to dike breaches in consequence of soaked dikes in long term flood events or delay in the operation of mobile protection elements. For a comprehensive flood hazard assessment of extraordinary hydraulic situations such as clogging of bridges due to debris need to be analysed.

The hydraulic calculations of the risk-scenarios are carried out with a HN-Flood-Simulation-Model. This model was developed at the IWK for the river Neckar on behalf of the state authorities of Baden-Württemberg (Oberle et al., 2000).

Beside the statistical flood scenarios, single (realistic) events have to be analysed. For example, when documented historical floods were higher than nowadays flood protection level, they may be reconstructed and serve as reference scenarios. For theses analysis the quality of the data have to be taken into account as well as the hydraulic boundary conditions concerning the historical situation. HN-calculations have shown, that e.g. for the river Neckar, observed historical floods were much higher than todays design discharges. Figure 3 shows the hydraulic situation for the community of Offenau (at Neckar-Km 98.0). The water level of the historical flood of 1824 is app. 2.5 m higher than the dikes that have been built for a 100-year-floodprotection.

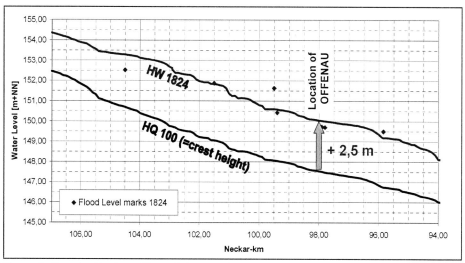

Figure 3: Longitudinal profiles of the maximum water level of the River Neckar near Offenau (km 98.0): calculated 100-year flood and historical event 1824 as reconstructed from water-level marks.

2.3 Flood Damage Estimation

The comprehensive determination of flood damage includes direct and indirect as well as tangible and intangible losses. Although we acknowledge that intangible and indirect damages play an important role, we concentrate on direct damage. On the basis of the accumulated values at risk and the functional relationship between the flood parameters and the resulting damage, risk potentials can be identified, quantified and the expected damage can be estimated. For small areas with detailed information about type and use of single buildings microscale analysis can be undertaken. For greater areas a mesoscale approach with aggregated land use information is advantageous. Within the CEDIM project a microscale and a mesoscale approach are followed, and later on their results and uncertainties compared.

For the microscale damage estimation, the common approach of most damage models is used. The direct monetary damage is estimated from the type or use of the building and the inundation depth (Wind et al., 1999, NRC, 2000). This concept is supported by the observation of Grigg and Helweg (1975) "that houses of one type had similar depth-damage curves regardless of actual value". Such depth-damage functions are seen as the essential building blocks upon which flood damage assessments are based and they are internationally accepted as the standard approach to assess urban flood damage (Smith, 1994). For the damage assessment the following procedure was used:

- Identification and categorization of each building in the regarded area
- Estimation of the Flood-Sill of each structure (lowest damaging water level)
- Estimation of the values of building-structure and contents (fixed/mobile inventory)
- Calculation of the water-level for each building
- Estimation of the damages to buildings and contents for different water-levels based upon the type and use of each building.

Based on the HN Flood-Simulation-Model for the Neckar (Oberle et al., 2000), specific model components for damage estimation have been developed. Thus a comprehensive tool for practical flood-risk analyses can be provided. Fig. 4 shows the results of the calculation of direct damages to residential buildings in Offenau due to the flood 1993.

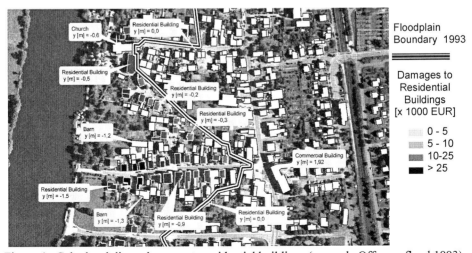

Figure 4:   Calculated direct damages to residential buildings (example Offenau, flood 1993).

Since flood losses are also influenced by other factors besides the water depth, more knowledge about the connections between actual flood losses and damage-determining factors is needed for the development of better tools for mesoscale damage estimation. Therefore, during April and May 2003 approximately 1700 private households along the Elbe, the Danube and their tributaries were interviewed about the flood damages to their buildings and inventory caused by the August 2002 flood as well as about flood characteristics, precautionary measures, warning time, social-economic variables, regional- and use-specific factors. In the affected areas, a building specific random sample of households was generated, and always the person with the best knowledge about the flood damages in a family was interviewed. The standardised questionnaire comprised around 180 questions. An average interview lasted about 30 minutes. The computer aided telephone interviews were undertaken by the SOKO-Institute, Bielefeld, within a project cooperation between the GFZ Potsdam and the Deutsche Rück.

The damage-determining factors can be divided into impact factors like water depth, contamination, flood duration, flow velocity and resistance factors like type of building, preventive measures, preparedness, early warning. The most important damage-determining factor is the water depth, but also other factors have significant influence on the losses.

For example, contaminations with oil, sewage or chemicals lead to significantly higher damage ratios (fraction of the flood damage in relation to the total value) to buildings and contents. During the flood 2002 in Saxony and Saxony-Anhalt contaminations increased the mean damage ratios by 9 to 16% (Fig. 5). During the 1999 flood in Bavaria, contamination with oil led on average to a three times higher damage to buildings, in particular cases even to total loss (Deutsche Rück, 1999).

On the resistance side, precautionary measures significantly reduce the flood loss even during extreme floods like the one in 2002. One-factorial analysis showed, that adapted use and contents reduced the mean damage ratios of contents by 12% and 19%, respectively (Fig. 6). Adapted use and contents means that mainly cellars are not used cost-intensively and no expensive upgrading is undertaken. In cellars and ground floors only waterproofed building material and movable small interior decoration and furniture should be used (MURL, 2000).

Figure 7 shows, that the more factors are specified, the lower is the coefficient of variation within the data. Therefore, to reduce the uncertainty in damage estimation, more factors besides the water depth need to be taken into consideration.

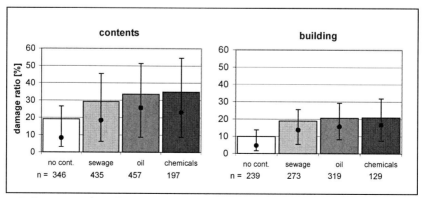

Figure 5: Damage ratios of residential buildings and their contents influenced by different contaminations, whereat double and triple contaminations occurred as well (means, medians and 25-75% percentiles).

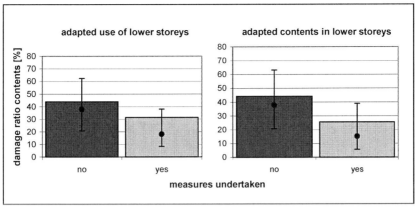

Figure 6: Contents damage mitigation via flood adapted use and contents (all cases with water in cellar and ground floor (n=338), means, medians and 25-75% percentiles).

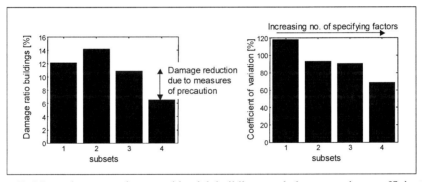

Figure 7: Mean damage ratio to residential buildings and the respective coefficients of variation of different, more and more specified subsets: 1: All private houses (n = 946); 2: Private houses, Water depth 50-150 cm (n = 293); 3: Private houses, Water depth 50-150 cm, No contamination (n = 101); 4: Private houses, Water depth 50-150 cm, No contamination, Heating installation in upper storeys (n = 9).

With this approach a multifactorial damage-estimation model will be developed. Uncertainty in damage estimation will hopefully be reduced and a contribution to the development of more reliable methods for risk assessment will be made. Reliable risk appraisals are much-needed, since they contribute substantially to flood damage mitigation and an effective flood management.

3 Conclusions and Outlook

Microscale analyses and differentiated approaches developed in pilot areas are necessary for the development of more reliable methods on an aggregated spatial level. Thus, the presented results are a basis for the improvement and development of reliable methods for flood risk assessment needed for the processing of an integrated risk map Germany. Future work will include uncertainty analyses of the flood risk assessment.

Acknowledgements

The study is carried out within CEDIM (Center of Disaster Management and Risk Reduction Technology), supported by the GFZ Potsdam and the University of Karlsruhe.

References

Deutsche Rück (Deutsche Rückversicherung AG): Das Pfingsthochwasser im Mai 1999. Düsseldorf, 1999.

Grigg, N.S. and Helweg, O.J.: State-of-the-art of estimating flood damage in urban areas. *Water Resour. Bull.*, 11 (2): 379-390, 1975.

IWK (Institut für Wasserwirtschaft und Kulturtechnik): Abschätzung extremer Hochwasserabfluss-Scheitelwerte in Baden-Württemberg: Ermittlung der BHQ_1- und BHQ_2-Werte gemäß der neuen DIN 19700. IWK-Report HY 2/12, University of Karlsruhe, by order of the Landesanstalt für Umweltschutz Baden-Württemberg, 2003.

LfU (Landesanstalt für Umweltschutz Baden-Württemberg): Hochwasserabfluss-Wahrscheinlichkeiten in Baden-Württemberg. *Oberirdische Gewässer, Gewässerökologie*, Bd. 54, loose-leaf-collection (ISBN 3-88251-273-3), Karlsruhe, 1999.

LfU (Landesanstalt für Umweltschutz Baden-Württemberg): Hochwasserabfluss-Wahrscheinlichkeiten in Baden-Württemberg. *Oberirdische Gewässer, Gewässerökologie*, Bd. 69, CD-ROM (ISBN 3-88251-278-4), Karlsruhe, 2001.

MURL (Ministerium für Umwelt, Raumordnung und Landwirtschaft des Landes Nordrhein-Westfalen): Hochwasserfibel - Bauvorsorge in hochwassergefährdeten Gebieten. Düsseldorf, 2000.

NRC (National Research Council): Risk analysis and uncertainty in flood damage reduction studies. *National Academy Press*, Washington DC, 2000.

Oberle, P., Theobald, S., Nestmann, F.: GIS-gestützte Hochwassermodellierung am Beispiel des Neckars. *Wasserwirtschaft*, 90 (7-8): 368-373, 2000.

Smith, D.I.: Flood damage estimation – A review of urban stage-damage curves and loss functions. *Water SA*, 20(3): 231-238, 1994.

Wind, H.G., Nierop, T.M., de Blois, C.J. and de Kok J.L.: Analysis of flood damages from the 1993 and 1995 Meuse floods, *Water Resour Res*, 35(11): 3459-3465, 1999.

Seismic Risk Scenario for Mumbai

Ravi Sinha[1], Alok Goyal[2]
[1] Professor, Civil Engineering Department, Indian Institute of Technology Bombay, Powai, Mumbai – 400 076, India;
[2] Professor, Civil Engineering Department, Indian Institute of Technology Bombay, Powai, Mumbai – 400 076, India.

Abstract

Seismic risk assessment techniques providing scenario-like information about an urban area or a city are very useful for disaster management planning purposes. However, the development of comprehensive risk scenario requires use of data and tools that are not easily available. Simplified techniques for seismic risk assessment can be used under these situations for risk scenario development. These simplified techniques also enable one to evaluate the change in risk profile of an urban area with time. These techniques require information on the hazard, structural vulnerability and the impact of structural damage in terms of non-engineering data such as occupancy and mortality rate. This paper shows that seismic risk assessment of Mumbai using simplified techniques can provide useful information to identify the contribution of different factors to the seismic risk. The impact of mitigation measures have also been evaluated and included in the scenario.

Keywords: seismic hazard, vulnerability, seismic risk, damage scenario, Mumbai.

1 Introduction

Seismic risk assessment techniques providing scenario-like information about an urban area or a city are very useful for disaster management planning purposes. The seismic risk of a city depends on the combination of three contributory factors: (1) Hazard, (2) Vulnerability and (3) Exposure. The seismic hazard describes the likelihood of damaging earthquake affecting the city or urban area. The seismic hazard is typically assessed using information from past seismicity and the combining with seismological and geophysical data. The vulnerability describes the ability of the buildings and other structures to withstand earthquake forces of a particular intensity. Vulnerability therefore describes the inherent strength or weakness of the structure. Exposure considers the impact of earthquake damage in terms of human, economical and other losses.

The development of comprehensive risk scenario for urban areas requires use of data and tools that are not easily available. Simplified techniques for seismic risk assessment can be used under these situations for risk scenario development. These simplified techniques also enable one to evaluate the change in risk profile of an urban area with time. These risk scenario development techniques require information on the hazard, structural vulnerability and the impact of structural damage in terms of non-engineering data such as occupancy and mortality rate. This paper shows that seismic risk assessment of Mumbai, India using simplified techniques can provide useful information to identify the contribution of different factors to the seismic risk. The impact of mitigation measures have also been evaluated and included in the scenario.

Year	Month	Intensity (MMI) / Magnitude (R)
1594	--	IV
1618	May	IX[*]
1678	--	IV
1832	October	VI
1854	December	IV
1865	December	IV
1877	December	IV
1906	March	VI
1924	January	IV
1929	February	V
1933	July	V
1941	May	IV
1951	April	VIII
1963	March	IV
1965	December	IV
1966	May	V
1967	April	4.5
1967	June	4.2
1998	May	3.6

Table 1:   Some major historical earthquakes in Mumbai region ([*]There is some uncertainty about this damage being caused due to an earthquake).

Mumbai is known as the commercial capital of the India and is the centre of its economic activities. Like most major urban centres in India, Mumbai has grown tremendously in the last few decades due to unabated migration from the smaller towns and rural areas.   As a result, the city has developed in a haphazard fashion with little consideration for proper town-planning norms.   This has resulted in most areas of the city lacking basic civic amenities. In fact, over 50% of Mumbai population lives in informal houses (often illegal and of very poor quality) in slums.   Even in the non-slum areas, the basic amenities may be lacking and the structures may be of poor quality. There is, consequently, a need to be prepared against the possible natural and man-made disasters that are likely to occur in Mumbai.   For this purpose, it is essential to have realistic understanding of the consequences of likely damage in Mumbai due to different disasters. This will permit rational planning of mitigation efforts in order to minimise effects of these disasters (Sinha and Adarsh, 1997).

2   Seismic Hazard

The seismic hazard is typically determined using a combination of seismological, morphological, geological and geotechnical investigations, combined with the history of earthquakes in the region. The major lineaments that have been mapped in Maharashtra are shown in Figure 1. These lineaments do not indicate the presence of any major features very close to the Mumbai region.   However, some studies using Deep Seismic Sounding profiles and Geo-Magnetic Anomaly investigations have indicated the presence of several smaller lineaments below the Deccan Traps (Arora and Reddy, 1990) that cover the region surrounding Mumbai.   These lineaments may represent an extension of the major groups described above, or may form a part of inter-group lineaments. Recent research has also indicated that some activities such as creation of large reservoirs, extraction of hydrocarbon and extensive mining activities may trigger damaging earthquakes in relatively stable regions (Gupta, 1992 and Simpson, 1986) such as the Deccan region.

Reliable historical data for seismic activity affecting the Mumbai region is available only for the last 400 years (GOM, 1998), based on the data available from IMD, GSI, EPRI, NGRI and MERI). The most prominent recent earthquakes affecting Mumbai have been listed in Table 1. It is seen that this region has been experiencing low-intensity earthquake ground motions at frequent intervals. So far, only one record of an earthquake associated with severe damage and destruction is currently available, and that too is of dubious reliability. However, large earthquakes in the Stable Continental Regions such as the Deccan region are known to have long return periods (>500 years) (Seeber and Armbruster, 1987). The lack of information of large earthquakes may therefore be due to paucity of historical data rather than low seismic hazard.

The Indian Standard code (IS 1893:2002 (Part 1)) has placed Mumbai in Seismic Zone III. The seismic zones in the IS code are not based on analytical assessment of seismic hazard and are largely based on historical data (for example, see Kaila et al., 1972, and Rao and Rao, 1984). The seismic zones in IS 1893 are based on expected damage intensity in the event of code-level earthquake and do not denote consistent ground motion criterion such as equal peak ground acceleration levels. As per IS 1893, the design level earthquake in Zone III is expected to result in damage corresponding to MSK Intensity VII.

3 Construction Technology and Seismic Vulnerability

The seismic vulnerability of different construction practice in Mumbai has been established using the expert evaluation method (EAEE, 1995), and represents the average behavior of different types of structures. This method considers the relative strengths and weaknesses of buildings using different building materials, but does not attempt to quantify the difference in behavior of different structural forms using same building materials. GIS-based tools for evaluation of earthquake risk such as HAZUS (NIBS, 1997) use detailed analyses but cannot be applied to Indian cities due to non-availability of required data. The expert evaluation method, on the other hand, is less precise but gives "order of magnitude" information that is invaluable for development of earthquake disaster management policies.

The Mumbai region is 100% urban and the building stock exhibits a rich mix of several different building materials and technologies. The most commonly used building categories are: (1) reinforced-concrete frame buildings with partition walls; (2) brick masonry buildings (both engineered and non-engineered); and (3) buildings made of other materials such as tin sheets, thatch and other lightweight elements. In Mumbai, it has also been observed that several reinforced concrete and brick masonry buildings have been constructed without the assistance of qualified engineers. Due to this reason, these buildings are also non-engineered since they are improperly designed or constructed resulting in their lower strength.

During the 1991 census housing survey, the city was found to have a total of 2,768,910 dwellings, including residential as well as commercial and industrial establishments. Of these, 9.08% of dwellings were made of reinforced concrete, while 31.35% of dwellings were engineered masonry constructions. The remaining 59.57% of all constructions, even in Mumbai, are non-engineered. The high percentage of non-engineered constructions is due to the very large percentage of population residing in slums and informal settlements.

In this paper, the likely behavior of the different structure types have been assessed based on the authors' experience following recent Indian earthquakes and also based on the published earthquake damage reports following several earthquakes in developing countries (Sinha et al., 2001, Sinha and Goyal, 1994, Yegian et al., 1994a and 1994b, Hayes, 1996, and Hassan and Sozen, 1997). The behavior of the different building types has been quantified in terms of its damage intensity index (Sinha and Adarsh, 1999). This damage intensity index describes the probable percentage of buildings of any type that may be damaged due to an earthquake of particular strength. The vulnerability of the buildings has been determined on the basis of their construction type.

4 Earthquake Risk

The seismic hazard and vulnerability assessments have been combined to determine the seismic risk of future earthquakes affecting Mumbai. The final results of this investigation are the number of buildings that are likely to be damaged due to earthquakes of different magnitudes and the number of people who may be injured and perish. The risk scenario has been developed based on the 1991 census data for Mumbai. The effect of introduction of structural mitigation measures on the seismic risk has also been considered. The structural mitigation measures increase the inherent strength of buildings to withstand earthquakes. Since these measures can only be incorporated to those buildings constructed in future, the consequence of earthquakes several years after these measures are implemented has also been projected. This investigation has helped in the assessment of the effectiveness of mitigation measures in reducing earthquake risk.

The information on earthquake hazard and structural vulnerability has been combined to determine the risk to different building types. The procedure for determining the damage and casualty is schematically shown in Figure 2 (Murakami, 1992). In order to assess the human casualty levels due to the earthquake, the estimates of average fatality and injury levels have been used. These figures have been derived by using a mortality prediction model for different categories of structures (Coburn et al., 1992). The total number of people that may be killed due to damage of each building type can be represented by:

$$Ks_b = D_b \times \left[M1_b \times M2_b \times M3_b \times M4_b \right] \tag{1}$$

where D_b is the total number of damaged structures of building type b, $M1$ is the occupation density and $M2$ to $M4$ are conditional probability factors to modify the potential casualty figures. The factor $M1$ represents the population per tenement. For this investigation, $M1$ is taken as 5. $M2$ is the occupancy of buildings at the time of earthquake. The occupancy cycle proposed by Coburn and Spence (1992) has been used. $M3$ is the proportion of occupants who are trapped by collapse of buildings. This depends on the type of building. For all types of non-engineered buildings, this has been taken to be 20%, for engineered masonry buildings 10% and for reinforced concrete buildings 5%. These figures are derived from typical observations from damaging earthquakes. $M4$ is the proportion of injured occupants who are killed in the earthquake. It has been observed that collapsed reinforced concrete buildings lead to death of a large number of trapped occupants, while collapsed non-engineered lightweight buildings lead to death of very small number of trapped occupants. Based on the quantitative information available from several earthquakes (Coburn et al., 1992), $M4$ is taken as 0.4 for reinforced concrete buildings, 0.2 for masonry buildings and 0.1 for informal non-engineered buildings.

The Mumbai census information, earthquake hazard and vulnerability data and the mortality information have been combined to estimate the number of possible injuries

(Table 2) and the corresponding deaths (Table 3) that may occur due to earthquakes of different strengths.

Time	MSK VI	MSK VII	MSK VIII
Midnight	31,400	118,400	277,600
6 A.M.	25,000	94,600	222,100
12 Noon	18,800	71,000	166,500

Table 2. Estimated Number of Injuries Due to Different Maximum Earthquake Intensities Occurring in Mumbai.

Time	MSK VI	MSK VII	MSK VIII
Midnight	11,200	42,600	100,100
6 A.M.	9,000	34,000	80,000
12 Noon	6,700	25,500	60,100

Table 3. Estimated Number of Fatalities Due to Different Maximum Earthquake Intensities Occurring in Mumbai.

5 Mitigation Measures

The extent of damage to structures and casualty level due to an earthquake in the future can be reduced by the introduction of suitable mitigation measures. These mitigation measures can be categories as structural and/or non-structural. The structural measures are those that directly influence the performance of building stock through strengthening of code provisions and the prevalent construction practice. The vulnerability of any building type can be reduced by incorporating the appropriate structural mitigation measures. In this paper, the impact of structural measures on the future earthquake risk has been considered.

Only a small fraction of the total dwellings in Mumbai are made of reinforced concrete. Most such buildings have been designed and detailed using the prevalent design code for reinforced concrete structures. This design code is primarily intended for safe design of reinforced concrete structures for typical static loads, and does not include any ductile detailing and other special earthquake-resistant provisions. As a result, these buildings may be vulnerable to sudden and catastrophic failure if the loads exceed the carrying capacity of some critical members. The structural measures that are required for improving the seismic performance of such buildings are: (1) ductile detailing of members, (2) design for strong column and weak beam frames, and (3) rigorous construction supervision and quality control to ensure compliance with the design specifications. These measures will improve the safety margins of the building, and will also ensure that the failure causes least casualty to the occupants. The vulnerability of new buildings after inclusion of the structural mitigation measures will be similar to that expected for buildings designed according to earthquake-resistant design codes. The mitigation measures result in reduction of vulnerability by about 30%-40% (Sinha and Adarsh, 1999).

The vast majority of the buildings in Mumbai consist of engineered and non-engineered masonry buildings. Based on the evaluation of performance of typical masonry buildings, the authors feel that the inclusion of lintel band is the most effective and the only practical mitigation measure for areas with moderate seismicity. The buildings and other dwellings that are constructed from informal materials usually belong to the most economically disadvantaged people. Implementation of structural mitigation measures for these dwellings may be difficult due to their unauthorised nature and the financial resources required for these measures. It has therefore been assumed that these structures will be excluded from the implementation of structural measures, and the likely effect of mitigation measures for these structures has not been considered.

The introduction of structural mitigation measures will affect the new constructions that are designed and built in future. Due to economic reasons, it is unrealistic to expect that the existing building stock will be retrofitted to comply with the mitigation measures. It has been further assumed that only 50% of the new engineered structures shall comply with the mitigation measures. It is also assumed that 50% of the non-engineered masonry buildings shall also comply with the mitigation measures. The remaining reinforced concrete and masonry buildings and all non-engineered buildings using informal materials shall exhibit similar behaviour as the existing building stock.

6 Earthquake Risk After Mitigation

The population of Mumbai is expected to grow to approximately 12.9 million by the year 2011 (BMRDA, 1996) from the 9.9 million recorded in 1991. The morbidity model presented earlier has been used to estimate the casualty figures for a damaging earthquake in 2011 (Tables 4 and 5). The casualty levels have been estimated both when the mitigation measures discussed above have been implemented and when no structural measures have been implemented (i.e., status quo in construction practice is maintained). It can readily be seen that the implementation of mitigation measures result in a significant reduction in the number of injuries and fatalities.

Time	Mitigation					
	Yes	No	Yes	No	Yes	No
	MSK VI		MSK VII		MSK VIII	
Midnight	36,500	40,800	139,700	153,900	333,600	361,000
6 A.M.	29,200	32,600	111,800	123,100	266,900	288,700
12 Noon	21,900	24,500	83,800	92,300	200,200	216,500

Table 4: Estimated Number of Injuries Due to Different Maximum Earthquake Intensities Occurring in Mumbai in 2011.

Time	Mitigation					
	Yes	No	Yes	No	Yes	No
	MSK VI		MSK VII		MSK VIII	
Midnight	13,200	14,700	50,700	55,500	121,100	130,400
6 A.M.	10,600	11,700	40,600	44,300	96,900	104,300
12 Noon	7,900	8,800	30,400	33,300	72,700	78,200

Table 5: Estimated Number of Fatalities Due to Different Maximum Earthquake Intensities Occurring in Mumbai in 2011.

7 Conclusions

In this paper, a simple method for determination of earthquake risk to Mumbai has been presented. This method requires information on the seismicity of the region and detailed information of the city population and building stock. In case where the seismicity information is not available, approximate estimates based on the seismic zoning map given in IS 1893 can be used. The vulnerability of the building stock is determined based on observed damage of similar constructions due to past earthquakes. The information can be used for planning disaster mitigation measures based on realistic estimate of damage potential due to the postulated earthquakes.

In the analysis presented herein, the demographic and census information have been used to estimate the existing building stock and to determine the likely level of damage to the building stock. The analysis procedure has also been combined with a mortality prediction model to estimate the level of human casualty due to earthquakes of different magnitudes. The analysis has been further extended to investigate the impact of mitigation measures on the consequences of an earthquake. Only the impact of structural mitigation measures, in which the design and construction procedures are improved to make the buildings less vulnerable, have been considered. Based on this analysis, the probable casualty levels have been estimated. The corresponding casualty levels if no mitigation measures have been implemented has also been evaluated in order the estimate the impact of the mitigation measures.

It is seen that the occurrence of code-level earthquake (MSK Intensity VII) at Mumbai may lead to massive loss of life and damage of buildings. Depending on the time of the day, between 25000 and 42000 people may perish due to structural collapse and damage in the earthquake. The likely impact of an earthquake more severe and less severe than the code-level earthquake has also been presented. These give the likely range of human casualty due to earthquakes of different intensity.

The impact of implementation of structural mitigation measures has been considered in order to estimate the consequence of an earthquake in 2011. It is found that if mitigation measures are implemented in half of all the engineered buildings and in half of the masonry non-engineered buildings that are built after 1991, it will lead to over 10% reduction in the casualty levels.

References

BMRDA: *Draft Regional Plan for Bombay Metropolitan Region, 1996-2011.* Bombay Metropolitan Regional Development Authority, Mumbai, 1996.

Coburn, A. W., and Spence, R. J. S.: *Earthquake Protection.* John Wiley & Sons, Cambridge, UK, 1992.

Coburn, A. W., Spence, R. J. S., and Pomonis, A.: Factors determining human casualty levels in earthquakes: mortality prediction in building collapse. *Proceedings of the 10th World Conference on Earthquake Engineering*, pp. 5989-5994, Madrid, Spain, 1992.

European Association of Earthquake Engineering (EAEE): Report on the EAEE working group 3: vulnerability and risk analysis. *Proceedings of the 10th European Conference on Earthquake Engineering*, pp. 3049-3077, Madrid, Spain, 1995.

Government of Maharashtra (GOM): *Workshop on Disaster Management.* Government of Maharashtra Report, Mumbai, 1998.

Gupta, H.K.: Reservoir-induced earthquakes. *Current Science, Special Issue: Seismology in India - An Overview*. 62: 183-198, 1992.

Hassan, A. F., and Sozen, M. A.: Seismic vulnerability assessment of low-rise buildings in regions with infrequent earthquakes. *ACI Structural Journal*, 94: 31-39, 1997.

Kaila, K. L., Gaur, V. K., and Narain, H.: Quantitative seismicity maps of India. *Bulletin of the Seismological Society of America*, 62: 1119-1132, 1972.

Murakami, H. O.: A simulation model to estimate human loss for occupants of collapsed buildings in an earthquake. *Proceedings of the 10th World Conference on Earthquake Engineering*, pp. 5969-5974, Madrid, Spain, 1992.

National Institute of Building Sciences (NIBS): *Earthquake Loss Estimation Methodology: HAZUS97 Technical Manual*. Report to US Federal Emergency Management Agency, Washington, DC, USA, 1997.

Rao, B. R., and Rao, P. S.: Historical seismicity of peninsular India. *Bulletin of the Seismological Society of America*, 74: 2519-2533, 1984.

Seeber, L., and Armbruster, J. G.: The 1886-89 aftershocks of the Charleston, South Carolina earthquake: a widespread burst of seismicity. *Journal of Geophysical Research*, 92: 2663-2696, 1987.

Simpson, D.W.: Triggered earthquake. *Annual Review Earth Planetary Science*, 14: 21-42, 1986.

Sinha, R., and Adarsh, N.: Seismic vulnerability of urban areas. *Proceedings of CBRI Golden Jubilee Conference on Natural Hazards in Urban Habitat*, pp. 1-8, Tata McGraw Hill Limited, 1997.

Sinha, R., and Goyal, A.: Damage to buildings in Latur earthquake. *Current Science*, 67: 380-385, 1984.

Sinha, R., and Adarsh, N.: *A postulated earthquake damage scenario for Mumbai*. ISET Journal of Earthquake Technology, 36: 169-183, 1999.

Sinha, R., Shaw, R., Goyal, A., Choudhary, M. D, Jaiswal, K., Saita, J., Arai, H., Pribadi K., and Arya, A. S.: *The Bhuj Earthquake of January 26, 2001*, Indian Institute of Technology Bombay and Earthquake Disaster Mitigation Research Center, Miki, Japan, 2001.

Yegian, M. K., Ghahraman, V. G., and Gazetas, G.: 1988 Armenia earthquake. I: seismological, geotechnical, and structural observations. *Journal of Geotechnical Engineering*, 120: 1-20, 1994a.

Yegian, M. K., Ghahraman, V. G., and Gazetas, G.: 1988 Armenia earthquake. II: damage statistics and soil profile. *Journal of Geotechnical Engineering*, 120: 21-45, 1994b.

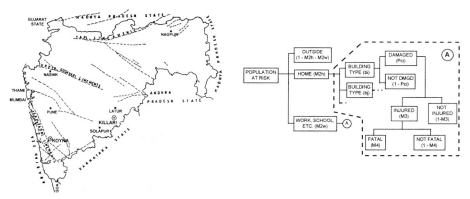

Figure 1: Map of Maharashtra showing major lineaments

Figure 2: Model for estimation of human casualty due to building damage and collapse (adapted from Murakami, 1992)

Estimation of Building Values as a Basis for a Comparative Risk Assessment

Lorenz Kleist[1], Annegret Thieken[2], Petra Köhler[2], Matthias Müller[2], Isabel Seifert[3], Ute Werner[4]

[1] *Institute for Economic Policy Research, University of Karlsruhe (TH), D-76128 Karlsruhe, Germany*
[2] *GeoForschungsZentrum Potsdam, Telegrafenberg, D-14473 Potsdam, Germany*
[3] *Postgraduate College "Natural Disasters", University of Karlsruhe (TH), D-76128 Karlsruhe, Germany*
[4] *Institute for Finance, Banking and Insurance, University of Karlsruhe (TH), D-76128 Karlsruhe, Germany*

Abstract

This paper presents the methodology and first results of a common inventory of residential buildings and their values – here defined as replacement values for the reference year 2000 - which will be used for the calculations of direct losses from various natural disasters within the CEDIM-project "Risk Map Germany". The methodology consists of two main steps. First, the values of buildings are estimated per community by means of data about the number and types of the community's buildings as well as common replacement costs. In the second step, the building assets have to be distributed spatially in the community. First results are shown for Baden-Wuerttemberg, where the replacement costs for residential buildings per inhabitant amount to EUR 50000.

Keywords: asset estimation, residential buildings.

1 Introduction

To compare risks due to different natural disasters a consistent framework is needed. Within the project "Risk Map Germany" of the Center for Disaster Management and Risk Reduction Technology (CEDIM) the term risk describes the probability that a given loss will occur. Risk encompasses three aspects: hazard, vulnerability (susceptibility) and exposed assets or people. For a comparative risk assessment a common risk indicator has to be chosen. Within CEDIM direct losses to residential buildings due to floods, storms and earthquakes will be considered first.

Whereas input data and methodologies for the hazard and vulnerability assessments vary from hazard type to hazard type, a uniform database of potentially exposed assets is essential for a consistent comparison of different risks. Therefore, a working group was established which aims to identify and assess buildings that might be exposed and especially vulnerable to hazard events. This paper presents the methodology to set up a common inventory of residential buildings in Baden-Wuerttemberg. It is shown how the building values of residential buildings - here defined as replacement values for the year 2000 - can be assessed as both, the total sum per community and the per capita value per community. As the developed methodology should be transferable to the whole of Germany the objective is to use - as far as possible - official data sets that are available for the whole country.

2 Data and Methods

The methodology to set up an inventory of residential buildings assets consists of two main steps. First, the values of buildings are estimated per community by means of data about the number and types of the community's buildings and the replacement costs of these buildings. In the second step, the building values are distributed spatially in the community.

2.1 Estimation of Values of Residential Buildings

2.1.1 Data Used

The following data were used as main sources for the calculation of replacement costs for residential buildings.

Data on buildings per community (INFAS GEOdaten GmbH, 2001)

To evaluate the assets of residential buildings several data sets of INFAS GEOdaten GmbH (2001) were used provided in two spatial units which are defined by the administrative boundaries of the communities and the postal zip codes, respectively. Both topologies do not depend on each other. This inconsistency will produce problems when transferring the results from one level to the other. Thus, the community level with the larger area per unit was used for the asset assessment. The INFAS data comprise the absolute and relative amount of buildings per community. Furthermore they are partitioned into three subclasses characterizing type, age and quality of buildings. For example, there are seven types for residential buildings, like "one-family houses", "two-families houses" or "farming houses".

The described geometric data are the basis for displaying the results of the asset estimation by maps. In figure 2 communities are visualized by polygons using a GIS Software. Their colors are related to the classified attributes.

Cost Data (BMVBW, 2001)

Since one objective is to extend the assessment approach to whole Germany, a collection of individual cost data per building was not feasible. It was decided to use mean construction costs per building type (so-called "Normalherstellungskosten" NHK) published by the German Federal Ministry of Transport, Building and Housing (BMVBW, 2001). Cost data are given in EUR per m² gross floor space. They are differentiated for a fairly high number of building types, considering the number of stories, the existence of a cellar and the type of the attic story. In addition, information on incidental construction costs, influences of the age and the quality standard of a building on the costs as well as conversion factors between living area and gross floor space are included per building type. Further differentiation can be obtained by using regional correction factors and by assumptions on the mean flat size. Owing to the objective to calculate reconstruction costs (and not e.g. market values), the provisions for the real value method (Sachwertverfahren) stipulated in the guidelines for evaluation were applied (BMVBW, 2002).

Data on total regional living area (BBR, 2003; Statistisches Bundesamt, 2004)

The calculation of costs using NHK does not lead to any results without data on the gross floor space per building type and region. Since such data do not exist, data on the living area per district – which is the regional entity above the communities – including some differentiation of building types were used. The data were provided by the Federal

Statistical Office of Germany (Statistisches Bundesamt, 2004) and the Federal Office of for Building and Regional Planning (BBR, 2003). In the asset estimation they were disaggregated to the community level and then scaled to gross floor space.

Data on the distribution of NHK-building types (Thieken et al., 2003)

Since the NHK-building types are much more detailed than the building types in the INFAS data an allocation scheme linking both classifications had to be developed. For example, the number of buildings of the INFAS type "one-family house" was split up into 12 NHK-building types. However, official data about the absolute or relative number of NHK-building types per region do not exist. Thus, results from computer-aided telephone interviews among about 1700 private households that had experienced damages during the august 2002 flood in Germany were used (Thieken et al., 2003). The survey contains detailed information about the affected buildings (number of storys, type of attic story, cellar etc.) which was used to assign a NHK-building type to each interview. A frequency analysis revealed the share of buildings among these types and the distribution was transferred to the following calculation scheme.

2.1.2 Method of Calculation

Data processing is shown in detail in figure 1. In the first stage, the number of buildings from INFAS data had to be transformed into living area. For this, the mean living area per INFAS-building type and district was calculated with the data of Statistisches Bundesamt (2004) and BBR (2003). Some simplifying assumptions concerning the number of flats per tenement (INFAS-building types with more than two flats such as apartment houses or multiple family buildings) were made before multiplying the mean living area per building type and district by the number of buildings per community to yield the total living area per INFAS-building type in the communities. The living area per building type was further divided into four quality classes according to the data on building quality.

In the second stage, the living area per INFAS-building type was further split into living area per NHK-building type. This was done on the basis of the results from the survey of Thieken et al. (2003) that show the share of NHK buildings subtypes per INFAS building type. The resulting living area per NHK-building type and community was then converted into monetary values, i.e. reconstruction costs, by multiplying it with a factor that scales living area into gross floor area and the unit NHK costs per building type and quality. Finally, after applying a regional correction factor for Baden-Wuerttemberg and aggregating the results to INFAS-building types again, the replacement costs per capita and community were calculated

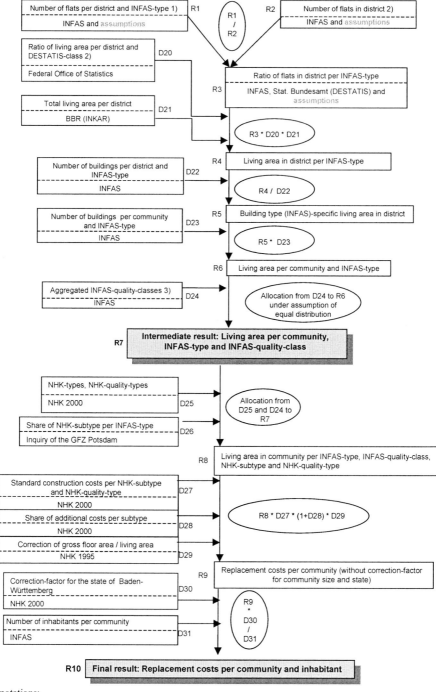

Figure 1: Data and workflow for assessing the value of residential buildings.

Annotations:
1) Calculation per INFAS-type using different assumptions for every INFAS-type.
2) DESTATIS classifies buildings in buildings with 1 flat, 2 flats and 3 and more flats.
3) INFAS uses 6 INFAS-quality-classes. First and second class, as well as third and fourth class were combined.

2.2 Spatial Distribution of Assets

Residential buildings are mainly concentrated in villages and cities. So, the asset estimation had to be combined with land use patterns. Since INFAS data give no information about the exact spatial distribution of buildings, CORINE (CoORdination of INformation on the Environment) land cover data were used (Statistisches Bundesamt, 1997) giving a European wide overview of land use in 44 categories. The data evaluation is based on satellite imagery interpretation with a defined minimum size for different areas, so CORINE land cover areas show a high degree of generalization.

To allocate the asset estimates CORINE data were intersected with INFAS community boundaries. It was further assumed that residential buildings only occur in the settlement areas of the CORINE data set with code 111 (continuous settlement area) and code 112 (discontinuous settlement area). However, even those are not completely covered by buildings. Areas for traffic and transport as well as open space (gardens, parks, undeveloped areas) had to be subtracted from the total settlement area. Moreover, different types of buildings vary in their space requirements. To get an idea of the area ratio between buildings, traffic area and open space, the distribution of buildings of community "Adelsheim" was analyzed by means of the official property register ALK (Amtliches Liegenschaftskataster) which contains single buildings and their use. Then area ratios covered by different building types and their unit values were calculated for the generalized CORINE settlement areas.

3 Results and Discussion

3.1 Estimation of Values of Residential Buildings

The results of replacement costs for the communities yield an average value of approximately EUR 50000 per inhabitant (Table 1), which seems to a be a realistic value according to estimates from experts. The replacement costs range from EUR 32029 to EUR 106627 per inhabitant (Fig. 2). Similarly, the minimum value of the living area per inhabitant ranges from 25.78 to 97.02 m² (Table 1).

Comment	Community	Inhabitants	Replacement costs per inhabitant [€]	Living area per inhabitant [m²]
	BadenWuerttemberg	10 600 906	50 480	40.29
	Stuttgart	587 152	48 115	36.30
	Karlsruhe	279 578	52 762	40.00
	Freiburg	208 294	46 136	36.10
Minimal costs per inhabitant	Schoemberg	8 592	32 029	25.78
Maximal costs per inhabitant	Buerchau	200	106 627	97.02

Table 1:   Example results for replacement costs and living area per inhabitant in Baden-Wuerttemberg.

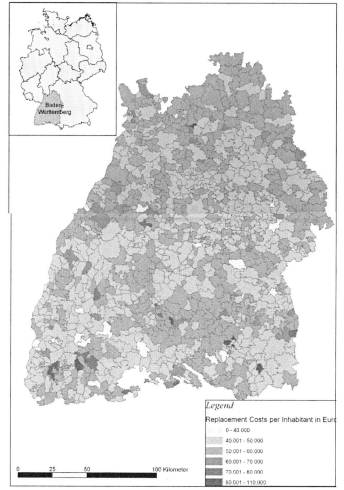

Figure 2: Asset of residential buildings per inhabitant and per community in Baden-
        Wuerttemberg.

Although variation in these values was expected and is to a certain degree in consistence
with the community data provided by INFAS, the amplitude of divergence seems to be
fairly high. Three main sources of error could explain unrealistic estimates, especially in the
smaller communities:

- INFAS data on the number of residential buildings per community may be
  faulty, thus leading to a community share of the total district living area that is
  too small or too high.

- The average number of flats per building type may vary between communities.
  Since no data on the number of flats on community level was available, this can
  also lead to flawed community estimations.

- The average flat size per building type was assumed to be constant per district,
  which is surely not the case. Especially the differences in the average household
  size (number of persons per household) per community are supposed to influence
  the average flat size.

3.2 Spatial Distribution of Buildings

The distribution of buildings was analyzed in the community Adelsheim. About 5 % of the community's area is classified as settlement area in the CORINE data set, whereas the buildings of the ALK register only cover less than 1 % of the area. Table 2 shows that buildings cover 13 to 18 % of the CORINE settlement area. 74 % of the buildings' area and 77 % of the number of buildings in the ALK register are situated in the CORINE settlement area. In a further step, the buildings in the ALK register were assigned to the building types of the INFAS data. The area per building type was calculated from the classified ALK data. By relating asset estimation for residential buildings to the area they cover unit values per building type were achieved (Table 3).

	Area [m²]	Share of CORINE settlement area	Share of the buildings' area
Total area of the community Adelsheim	43 741 978	---	---
Community's total settlement area according to CORINE	2 012 080	100.0 %	
Community's total area covered by buildings in ALK	357 939	17.8 %	100.0 %
Area covered by ALK-buildings in the CORINE settlement areas	264 743	13.2 %	74.0 %

Table 2: Comparison of CORINE settlement area and buildings from the property land register ALK for the community Adelsheim.

Building type (INFAS)	Number of buildings (INFAS)	Estimated Asset [Mill. €]	Covered area based on ALK [m²]	Area ratio in CORINE settlement area	unit-value [€/m²]
unknown type	27		21 736	1.1 %	
One/two-family houses	1176	177.2	106 835	5.3 %	1659
(semi) detached houses	300	39.0	25 151	1.2 %	1552
multiple family houses	48	35.4	9 947	0.5 %	3556
apartment buildings	15	31.3	4 573	0.2 %	6847
Farmhouses	40	6.0	7 572	0.4 %	796
administration/office building	21		27 579	1.4 %	
factories/storehouses	24		35 684	1.8 %	
Garages			30 067	1.5 %	
building for farms, barns etc.			81 987	4.1 %	
Others			6 808	0.3 %	
Total	1651	289.0	357 939	17.8 %	807

Table 3: Linking INFAS building data, asset estimation, ALK building data and CORINE land use information for the community Adelsheim to receive unit-values per building type.

4 Conclusions

The results of the asset estimation of residential buildings in Baden-Wuerttemberg confirm that the developed approach is generally suitable for investigations, when more detailed data (e.g. ATKIS, ALK) are not available. Based on the available data, the described approach can be extended to the whole country of Germany. Nevertheless, further improvements of the method can be achieved by including:

- correction factors for the village size,
- correction factors for the mean flat size,
- various designs of multiple family dwellings (two or more flats per storey) and
- the age of the buildings.

It is expected that the overall results will not change substantially, but the differences between communities could be modeled in a more detailed way. Further research is needed to also distribute buildings outside of accounted settlement areas by CORINE land use data.

Acknowledgements

We are indebted to Thomas Lützkendorf, David Lorenz (Chair of Sustainable Management of Housing & Real Estate) and Claudio Ferrara (Institute for Industrial Building Production) for providing us with first-hand knowledge on the advantages and disadvantages of various ways to assess building values.

References

BBR (Bundesamt für Bauwesen und Raumordnung): INKAR 2003 - Indikatoren und Karten zur Raumentwicklung 2003. CD-ROM, Bonn, 2003.

BMBRS (Bundesministerium für Bauwesen, Raumordnung und Städtebau): Normalherstellungskosten 1995 (NHK 1995). Bonn, 1997.

BMVBW (Bundesministerium für Verkehr, Bau und Wohnungswesen): Normalherstellungskosten 2000 (NHK 2000). Berlin, 2001.

BMVBW (Bundesministerium für Verkehr, Bau und Wohnungswesen): Wertermittlungsrichtlinien 2002 (WertR 2002). Berlin, 2002.

INFAS GEOdaten GmbH: Das DataWherehouse. Bonn. Status: December 2001.

Statistisches Bundesamt: Daten zur Bodenbedeckung für die Bundesrepublik Deutschland. No. 819 0120-97900, Wiesbaden, Status: December 1997.

Statistisches Bundesamt: GENESIS online. 3. Wohnen, Umwelt. https://www-genesis.destatis.de, Wiesbaden, Date of access: July 2004.

Thieken, Annegret, Kreibich, Heidi, Müller, Meike, Axer, Thomas and Merz, Bruno (2003): Early warning and people's reaction during the August 2002 flood – Results from a survey of private households in the Elbe and Danube region. *In Proceedings of the Second International Conference on Early Warning*, Bonn, 2003.

Monetary Evaluation of Flood Damage

Werner Buck
*Institute of Water Resources Management, Hydraulic and Rural Engineering,
University of Karlsruhe (TH), D-76128 Karlsruhe, Germany*

Abstract

Economic appraisal of flood damage is necessary for flood control planning and for informing the stakeholders. For small and medium sized areas the damage values are usually assigned to individual buildings. These values also serve as means for calibrating mesoscale estimations of flood damage derived from statistics of the national economy. Based on research and long-term experience, procedures for the assessment of reliable stage-damage relationships will be presented.

Furthermore the application of water stage – damage curves in economic appraisals of flood protection measures and risk analyses will be discussed. The problems of inappropriate flood-plain land use and residual risk are pointed out.

Keywords: flood damage, stage-damage curves, buildings, economic appraisal, residual risk, sustainable land use.

1 Introduction

Not only since the 2002 Elbe flood the question arose: What is the appropriate use of flood-endangered areas? In the course of time natural flood plains have been occupied partly or entirely for human purposes like agriculture or building development. The other side of the coin is the high damage suffered when the design level of the flood protection structures is exceeded. The trivialized term 'residual risk' counts for a rather small probability of exceedance but much higher damage than in the undeveloped flood plain.

An appropriate level of protection should be chosen according to the corresponding positive and the adverse effects. Prevented monetary damage is part and parcel of the benefits of flood control measures which can consist of various actions, e. g. structural measures and other actions such as land use regulations.

The paper deals with the monetary evaluation of flood damage to buildings, especially with the compilation of water stage – damage relationships and their use in economic appraisals of flood control alternatives (Buck, 2003).

2 Stage-Damage Relationships

In many cases the flood stage is the decisive factor for the extent of monetary flood damage. Besides there is a positive correlation between water stage and other factors like flow velocity, flood duration, sediment transport etc. in a certain flood-prone area.

Concerning the compilation of stage-damage relationships for single buildings in a so-called microscale approach (small to medium areas up to approximately 10 km^2) the following procedures can be distinguished:

- Synthetic determination for individual buildings in the project area (1a)
- Synthetic determination for typical buildings in the project area (1b)
- Derivation and application of nationwide average stage-damage curves for specific building types (2a)

- Site-specific modification of nationwide average stage-damage curves
  for specific building types   (2b)

In addition to the specified procedures, it will rarely be possible to – ideally – derive stage-damage curves in a project area from an up-to-date survey immediately after an on-site flood event. In the most unfavourable case one single (mean) damage value has to be applied to each building independently of type, for the entire project area. Such a value could be derived from a nationwide damage data base.

It has to be emphasized that damage values in principle have to reflect the current market values of the damaged items and not their replacement values (which possibly may be covered by insurance).

In the following the procedures (1a) to (2b) will be addressed.

(1a) Synthetic stage-damage relationships for individual buildings in the project area:

With high requirements for accuracy and a limited number of buildings in a project area the best approach is to compile flood damage assigned to water stages in an individual building in a synthetic way. Building appraisers from insurance companies estimate in a "what – if" manner the expected damage to the building and its contents for several assumed water stages above basement and ground floor respectively. This information is used to construct a stage-damage relationship for the whole range of potential water stages.

(1b) Synthetic stage-damage relationships for typical buildings in the project area:

With lower degree of accuracy required it can be sufficient in a not too large project area (some square kilometres) to choose one or more characteristic buildings of a certain type, such as one-family houses. For such buildings a stage-damage curve will be derived according to the procedure described in the preceding paragraph (1a).

(2a) Derivation and application of nationwide average stage-damage curves for specific building types:

The Working Committee of the German Federal States' Water Resources Administration (LAWA) stores in data base HOWAS flood damage data, which have been collected over the past two decades after several flood events in different regions of Germany. Some 3,600 cases of damage concerning buildings and other structures are grouped according to 8 sectors (e.g. dwellings, farm buildings, infrastructure, services, and industry). For example residential buildings (around 2,000 damage events) are distinguished by 4 construction periods, with – without basement, and with – without garage (IWK, 1999).

The software package HOWAS_N, developed on behalf of the Water Resources Administration, facilitates the determination of average amounts of losses, standard deviations etc. and finally, stage-damage curves. For convenient use a root function **Flood Damage D = a · (Water Stage W)$^{1/2}$** is recommended, giving plausible results. Parameter **a** denotes the damage to be expected for a water stage of one metre. Unfortunately statistical spread around the calculated curves is generally very high, despite of updating the amounts of loss by price indices. Presumably the scatter is due to continuing inhomogeneity of the samples. They contain cases of damage with different 'flood experience', flood durations, flood frequency, warning times and last but not least different 'social classes' (performance of building construction, equipment, furnishing, etc.). The two latter determinants have been taken into account in the (synthetic) stage-damage relationships for various building types, supplied for the United Kingdom for many years (Penning-Rowsell and Chatterton, 1977).

A recent evaluation of the LAWA database among others lead to the conclusion that single stage-damage curves for all residential buildings before and after 1964 should be used (IWK, 1999). No curves should be derived from random samples with sample sizes smaller than approximately 20; in every case the plausibility of the results have to be proven.

(2b) Site-specific modification of nationwide average stage-damage curves for specific building types:

To account for the immense scatter of the pair of variates in the stage-damage graph for a certain project area with its special circumstances the average curve eventually has to be shifted upward or downward accordingly. Figure 1 shows an example for the presentation of HOWAS data and gives hints for the modification of stage-damage curves. Moreover users can detect the composition of the underlying sample and can take this into account when adjusting stage-damage curves for a certain project area.

When comparing the above explained procedures for flood-damage estimation with the approaches commonly used in rainfall-runoff modelling the following parallelisms can be detected: Procedures (1a) and (1b) with site-specific evaluation of stage-damage relationships correspond to distributed rainfall-runoff models using measured discharge data for calibration and other on-site information. Whereas procedures (2a) and (2b) comply with regionalized rainfall-runoff models. They work with average statistical relationships derived from a data set including for instance data throughout the whole country. The results are less reliable than in the afore mentioned procedures.

It is understood that also mesoscale approaches for flood damage estimation in larger areas of around 10 km^2 based on statistics of the national economy have to be calibrated. For reasons of expenditure sometimes planners are obliged to restrict calibration to some subareas.

3 Annotations to Flood Protection Practice

Flood damage data are provided for the economic appraisal of flood control projects. They are also used for informing the inhabitants of flood-prone areas and the public on possible flood losses. Natural flood plains whether they are protected against rare flood events by dykes or other flood control measures or not bear a certain risk of flooding. Such areas will not be protected against the maximum possible flood event. Therefore a certain residual risk remains. Not only the damage to be prevented by a flood protection measure and the residual risk have to be taken into account in a decision by calculating expected annual monetary values (DVWK, 1985) but also the absolute damage values to be expected in case of exceedance of the design flood event.

These remarks are intended to call attention to the sword of Damocles overhanging flood plain users. Despite sometimes high flood protection degrees (design return periods) up to some hundred or even more years are provided, a certain probability of failure is remaining. Due to intensification of land use because of the better degree of flood protection the damage mostly will be much higher than in case of lower flood protection degrees. It cannot be reasonable to support intensified land use in former natural flood plains by compensating losses by the society. The risk should basically be borne by the landowners who can profit over a long time from the more fruitful land use. They should abandon intensified flood plain use and invest elsewhere, secure buildings themselves against flooding or procure insurance. An informed and rational decision of the flood plain users must be supported by comprehensive information on flood risk by the responsible agencies.

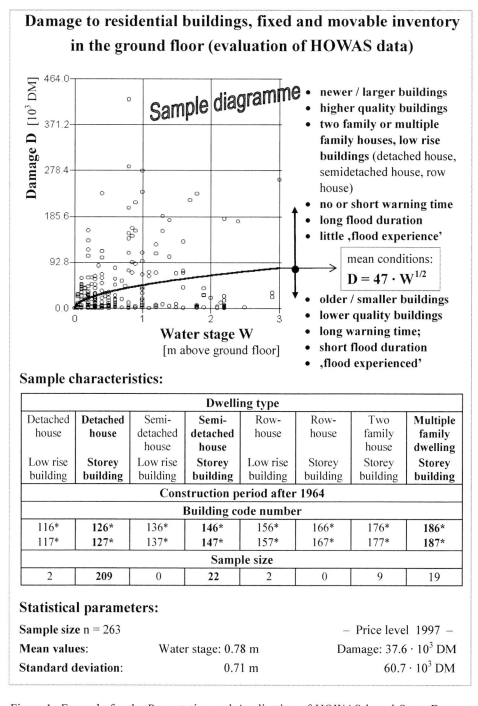

Figure 1: Example for the Presentation and Application of HOWAS based Stage-Damage Curves

Another aspect worthwhile to consider is the paradoxical situation that buildings and areas resp. suffering frequently high flood damage seemingly turn out in economic appraisals to be eligible for a higher degree of flood protection. The appropriate conclusion to be drawn would be that either exaggerated land use is not sustainable and the buildings should have been erected elsewhere (with less total costs) or there are other advantages compensating the frequent flood losses. The overall conclusion is that no further flood protection to be supplied by the society is justifiable.

Stage-damage curves have to be derived and economic project appraisals have to be applied with analytic expertise and common sense.

References

Buck, Werner: Festlegung des Hochwasserschutzgrades (Selecting the degree of flood protection). In: Seminar "Hochwasserrückhaltebecken" (Flood detention basins), Erfurt, Oktober 2003, 25 pages, ATV-DVWK, Hennef, 2003.

Buck, Werner: Auswertung einer Umfrage des BML bei den deutschen Botschaften in den EU-Mitgliedsstaaten zu Nutzen-Kosten-Untersuchungen und Hochwasserschadens-analysen/-potentialen. Zusammenfassender Bericht an das BML, Institut für Wasserwirtschaft und Kulturtechnik, Universität Karlsruhe, März/April 2000.

DVWK: Ökonomische Bewertung von Hochwasserschutzwirkungen. Arbeitsmaterialien zum methodischen Vorgehen. (Economic evaluation of flood protection measures. Working aids for the methodical approach.) Mitteilungen Heft 10, Deutscher Verband für Wasserwirtschaft und Kulturbau, Bonn, 1985.

IWK (Buck, Werner und Merkel, Ute): Auswertung der HOWAS-Datenbank (Analysis of HOWAS database). Institut für Wasserwirtschaft und Kulturtechnik, Universität Karlsruhe, by order of DVWK/LAWA, 1999.

Penning-Rowsell, Edmund C. and Chatterton, John B.: The benefits of flood alleviation: a manual of assessment techniques. Gower, 1977.

Development of a Storm Damage Risk Map of Germany – A Review of Storm Damage Functions

P. Heneka, B. Ruck

Laboratory for Building- and Environmental Aerodynamics, Institute for Hydromechanics, University of Karlsruhe (TH), D-76128 Karlsruhe, Germany

Abstract

This paper reviews activities on storm damage investigation and discusses the results with regard to an application for storm risk assessment in Germany. The major focus is on damage to buildings, especially residential buildings, which suffer most in storm events. So-called storm damage functions of various authors are presented and compared for countries exposed to tropical and extra-tropical storms. On the basis of the existing knowledge, requirements for a storm damage model for Germany are discussed.

Keywords: storm damage, storm risk assessment, vulnerability, hurricane, winter storm, buildings.

1 Introduction

For storm risk assessment both information, the storm hazard and the vulnerability of affected structures, have to be considered. An illustration of the generalized approach to wind risk assessment is presented in Figure 1. The storm hazard describes the probability of the occurrence of a certain wind speed at a specific location. Generally, the records of wind speed at meteorological weather stations serve as database for the hazard assessment. As extreme storm events are rare and associated time series very short, extreme value statistics is used to calculate local occurrence probabilities of wind speeds.

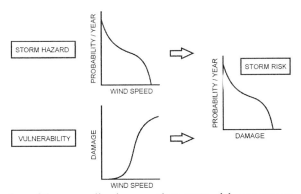

Figure 1: Illustration of the generalized approach to storm risk assessment.

Similar to other natural events, storm events are of interest only, if they cause significant damage to population or structures. Therefore, the vulnerability of the affected structures has to be taken into account. The main questions are: When does wind damage start and which loss has to be expected at a certain level of wind speed? The investigation of the vulnerability of structures leads to the development of storm damage functions, where predicted storm damage is assigned to a certain wind speed.

Currently, within the starting project of CEDIM (Center for Disaster Management and Risk Reduction Technology), a storm damage risk map is developed for Germany. For the development of storm damage functions, a literature survey was conducted. Research on this field began in the 1970s and was strongly motivated by catastrophic storm events like cyclone "Tracey" in 1974 in Australia, followed by US-Hurricanes "Andrew" and "Hugo" in 1989 and 1992, respectively, and the winter storms series of 1990 and 1999 in Europe.

2 Comparability of Literature Results

Wind speed is the most important parameter for damage estimation. Unfortunately, wind speeds are given in many different averaging time intervals reaching from seconds up to 10 min. In order to be able to compare the different results, all wind speeds are converted to a maximum gust wind speed at 10m height over ground assuming the following conversion factors: For severe winds of extra-tropical cyclones the gust factors $G_{10min} = 1.43$ and $G_{1min} = 1.22$ in urban areas are used (Schroers *et al*, 1990). For hurricanes, 1.45 in urban areas and 1.33 at coastal sides are used (Sparks, 2003). Sparks also concludes in his study that the boundary layer and turbulence characteristics in tropical cyclones appear similar to those in extra-tropical cyclones. Thus, from the point of view of wind characteristics, a comparison of damage models seems to be valid.

Also, 'damage' to buildings is expressed in various ways according to the area of interest. The mechanical damage is used to describe tendencies of vulnerability of structural parts of buildings (e.g. Buller, 1978; Sacré, 2002). As an equivalent expression for the monetary damage, the damage repair index DI (or: damage degree) is used with DI defined as percentage of repair costs to initial building costs (e.g. Leicester, 1979; Unanwa, 2000). For insurance business, the insured loss is of main interest. Some authors work with absolute values (e.g. Klawa, 2003; Dorland, 1999) while insurers prefer the usage of damage expressed as loss ratio LR (percentage of insured losses to total insured sum within a region).

The conversion between damage index and loss ratio turns out to be difficult. The latter is due to the fact that not all affected buildings are insured against wind damage and, as a consequence, are not considered in loss statistics (although they are damaged). Further on, the level of retention strongly influences the number of claims and the loss ratios of insurers (Swiss Re, 1993). Loss ratios from different insurance companies should be campared carfully when additional actuarial information is lacking (Chapter 4.1).

Another important value is the claim ratio denoting the percentage of affected insurance policies to the total number of insurance policies within a region.

3 Damage to Structures due to Extreme Winds

While this review is concerned with the vulnerability of structures in case of hazardous wind events, i.e. events with high wind speed over large areas (as it is the case in extra tropical and tropical storms), it is worthwhile mentioning that only vulnerability investigations of large numbers of structures are considered here. The vulnerability of individual buildings with specific properties is not considered (e.g. Natasha-A, 1996). Also, tornado damage is not included in this review, because tornados cause damages different to tropical and extra-tropical storms due to their extremely high wind speeds.

The physical storm damage is the consequence of wind loads that are stronger than the resistance of the structure and affects parts of buildings, like roofs, envelopes, and openings. Sacré (2002) described storm damage in France caused by winter storm "Lothar" in 1999 and gave tendencies of the reported damage with respect to the surface roughness type of the environment. More severe and structural damage were observed on buildings located near the sea and in open country as well as in front of open fields. Buildings in urban areas suffered more damage to openings.

A survey of storm damage to buildings in the United Kingdom was conducted by Buller (1978) based on newspaper reports for the period 1970-1976. It is shown that 40% of all incidents suffered damage to roofing, although the removal of complete roofs was rarer and confined to older buildings. Further notable damage occurred to chimneystacks, to gable and sides walls, and breakage of windows. Concerning the financial side, an annual damage of at least 13 million British Pounds (in 1970 values) was observed which led to an average domestic building damage of some 20 pounds. This low average was justified with the very large numbers of small incidents.

Sparks (1994) gives a detailed study of the insured damage caused by the Hurricanes "Hugo", in 1989, and "Andrew", in 1992. Most of the loss (60%) took place to housing in urban non-coastal areas. Ocean-front communities had losses due to storm surge and wind forces of only 20% of the total loss, although they provided the most spectacular failures. The envelope of buildings, especially the roof, wall envelopes and openings, suffered most wind damage. Very few buildings collapsed completely, serious damage was mainly caused by rain entering the building. The average insured loss to residential buildings from Hurricane Hugo in South Carolina was 5,550$ and 44,350$ from Hurricane Andrew in Florida (Sparks, 2003). According to the author, the average loss has strongly been influenced by a few poorly constructed buildings experiencing extensive loss.

Investigations of Munich Re (1993, 2001) of the 1990 and 1999 storm series showed that the main damage was also to roofs, facades, scaffolds, forests and power lines. Millions of properties were affected with an average insured damage to residential buildings of some 400€ (1990) and 1300€ (1999).

Reference	Observed area + year	Damage due to	Average loss per building
Buller (1978)	UK 1970-1976	Winter storms	20 BP (1970)
Sparks (1994)	US 1989 and 1992	Hurricanes	5,550$ (1989) 44,350$ (1992)
Munich Re	Europe 1990, 1999	Winter storms	400€ (1990) 1300€ (1999)

Table 1: Summary of storm damage investigations in different countries.

The results show similarities in the observed damage to buildings. Roofs and building envelopes (facades, openings) are most endangered by severe winds in the investigated countries. Large differences can be seen with regard to the average insured damage per building for hurricanes and winter storms. Although the sample size of the investigated events is surely not sufficient to be representative, the trend indicates a higher vulnerability to storm damage in the US. As we will see later, Figure 2 points up this assumption: The upper curves of loss ratio are assigned to hurricane loss in the United States, the lower curve to winter storm loss in Europe.

4 Storm Damage Functions

The motivation of modeling storm damage is the need to assess expected damage in the future. On the one hand, this holds for insurance companies, but on the other hand also for the society as the knowledge of risk provides helpful information for loss mitigation and management strategies. As mentioned before, the published models give an assessment of storm damage behaviour of large units, from postal-code areas to districts. All models, except the theoretical approach of Unanwa (2000), are empiric functions and adjusted to loss data provided by insurers.

Huang (2001) studied insurance loss data of the Hurricanes "Hugo" and "Andrew" in South Carolina and Florida. The expected loss ratio *LR(v)* in percent corresponding to a surface 10-min mean wind speed v in 10m heights is given as

$$LR(v) = \exp(0.252v - 5.823), \quad v \le 41.1$$
$$LR(v) = 100, \quad v > 41.1 \tag{1}$$

The loss ratio curve adjusted to the gust wind speed is shown in Figure 2. Also a claim ratio function is given in Figure 3; however, the underlying equation was not published.

Dorland (1999) investigated the vulnerability to storm damage in the Netherlands by analysing insurance data of 5 storm events between 1987 and 1992. An exponential storm damage function applicable for 2-digit postal code areas including three parameters is suggested:

$$\ln TDO = \alpha \ln A + \beta \ln O + \chi \ln v_{max} + C \tag{2}$$

where *TDO* is the total loss for objects (houses or business), O the number of objects, A the postal code area, and v_{max} the maximum gust speed. α, β, χ, and C are regression coefficients. Although the model works with absolute figures instead of ratios, which raises problems for a reasonable comparison, the behaviour of storm damage seems to be quite similar to the study of Huang (2001).

Munich Re (1993, 2001) suggests a power law model to describe storm losses related to wind speed. Analyses of insurance data of the winter storm series in 1990 (Gales "Daria", "Vivian", "Wiebke") and 1999 (Gales "Anatol", "Lothar", "Martin") show that the loss ratio increases with maximum wind speed to the power of α=2.7, for the 1990 data, and of α =4 to 5, for the 1999 data. The equation writes simply as

$$LR(v) = LR(160)\left(\frac{v}{160}\right)^{\alpha} \tag{3}$$

where *LR(v)* is the loss ratio at maximum wind speed v, *LR*(160) is the calculated loss ratio at 160 km/h based on "Lothar"-Data and α the power coefficient.

Further investigations of storm damage in Germany were conducted by Klawa and Ulbrich (2003) in order to quantify a model for the estimation of annual insured storm losses. Input wind data are the daily maximum wind speeds exceeding the local 98% percentile at meteorological stations. Assuming a cubic relationship between loss and maximum wind speed, the model is adjusted to fit the annual insurance loss data published by the Germany Insurance Union (GDV). The equation writes as

$$loss = c \sum_i pop(i) \left(\frac{v_{max}(i)}{v_{98}(i)} - 1 \right)^3 \qquad \text{for} \quad v_{max} > v_{98} \qquad (4)$$

where c is a regression coefficient, *pop* the population number, v_{max} the maximum wind speed and v_{98} the 98%-percentile of the wind speed for every district i. It has to be noted that this model considers the local wind climate because the maximum wind speed is related to the local 98% percentile. Unanwa (2000) proposed a new approach to hurricane damage prediction using the concept of wind damage bands. The damage band methodology considers component cost factors, component fragility, and location parameters to assess upper and lower bounds to building damage thresholds. The lower bound of the residential band is show in Figure 2.

Swiss Re (1993) found a strong relationship between loss and storm duration as well as precipitation. In Figure 3 two claim ratio curves for storm durations of 6h and 24h, respectively, are given.

Reference	Damage definition	Wind speed	Wind damage relationship	Further model Parameters
Huang 2001	Loss ratio	10-min mean speed	Exponential	
Dorland 1999	Total absolute loss	Max. gust speed	Exponential	Number of objects, size of area
Munich Re 1993, 2001	Loss ratio, claim frequency	Max. wind speed	Power law v^α, $\alpha = 2.7$ (1990), $\alpha = 4\text{-}5$ (1999)	
Swiss Re 1993	Claim ratio	Max. wind speed		Storm duration, insurance conditions
Klawa 2001	Absolute loss	Max. wind speed	Power law v^α, $\alpha = 3$	
Unanwa 2000	Damage degree	1-min mean speed	Wind damage bands	

Table 3: Summary of the characteristics of published storm damage model

4.1 Further Parameters Influencing Insured Wind Damage

As insurance companies provide the most complete information about storm events and damage, it is necessary to highlight further actuarial parameters influencing the insured storm damage. Additionally to meteorological and structural parameters of the storm damage functions, the storm loss of insurers is highly variable with the extent of cover and with the level of the retention (Swiss Re, 1993). The former determines, whether storm damage is insured and consequently registered in databases and the latter affects strongly the amount of loss and claims. A ratio of 0.2 of level of retention to average damage leads already to a reduction of losses of about 20%.

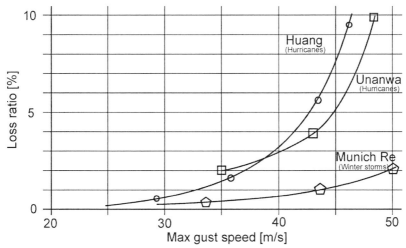

Figure 2: Loss ratio functions of Huang (2001), Munich Re (2001) and Unanwa (2000).
Note that no adjustment of the functions could be executed concerning different
insurance conditions of the models.

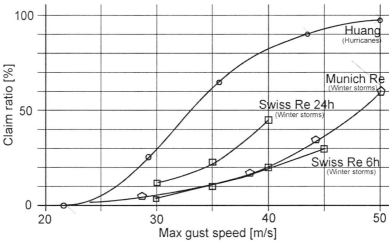

Figure 3: Claim ratio functions of Huang (2001), Munich Re (2001) and Swiss Re (1993).
Note that no adjustment of the functions could be executed concerning different
insurance conditions of the models.

5 Discussion

In the previous chapter all published storm damage functions were presented. However,
only few models are described in detail, especially publications of insurance companies
often do not contain detailed information about the models' applicability. This discussion
shall point out the requirements of a storm damage model for Germany and whether existing
models can be used to be implemented into the model or for the validation of the model.

Which damage definition is reasonable? Since the total damage to structures should be
calculated, i.e. not only the insured loss, the damage repair index DI should be used
preferably. In Germany, however, only loss data of insurers are available, no information of
damage repair costs has been registered so far. Thus, in order to calibrate a DI model with

loss ratio data of insurers, a conversion technique has to be developed including actuarial information (Chapter 4.1).

Does vulnerability to wind loading of structures vary over the country? In Germany, the north is more exposed to higher wind speeds than the south. Also, from the point of view of construction requirements, building resistance of existing structures cannot be considered as homogeneous over the country. Klawa (2003) uses wind speeds relative to the local wind climate with the result that the same wind speed would produce more damage in Southern than in Northern Germany. So, it is likely to assume that building resistance is somehow adapted to wind climate. This assumption is also reflected in the Munich Re (2001) model as it is calibrated with real local damage data.

Which relationship between wind speed and damage shall be used? Huang (2001) and Dorland (1999) proposed exponential functions, whereas Munich Re (1993, 2001) suggested potential functions later on also applied by Klawa (2003). Both types where derived to fit statistical damage data, thus, both may be valid for the observed regions and wind speeds. However, Munich Re revised their 1990-prediction of the power coefficient from $\alpha=2.7$ to $\alpha=4\text{-}5$ for the 1999 storm events. It was noted that storm loss increases much faster with higher wind speeds. Thus, as this characteristic is already the nature of the exponential functions, one might favor an exponential function as relationship between wind speed and damage.

Finally, it should be marked that the proposed functions given by the authors in their publications represent the "best-fit" curves while the data themselves are highly variant. As a consequence, storm damage is not described accurately by simple models with wind speed as the only parameter. Therefore, existing storm damage functions can reflect tendencies only for wind speed ranges based on a great number of data, however, this implies that it might be difficult to apply these functions to extreme wind speeds with a low probability of occurrence.

References

Buller, P.S.J.: Wind damage to buildings in the United Kingdom 1970-1976. *Building Research Establishment Current Paper*. CP42/78, 1978.

Dorland, C., Tol, R.S.J. and Palutikof, J.P.: Vulnerability to the Netherlands and Northwest Europe to storm damage under climate change. *Climatic Change* 43: 513-535, 1999.

Huang, Z., Rosowsky, D.V. and Sparks, P.R.: Hurricane simulation techniques for the evaluation of wind-speeds and expected insurance losses. *J. Wind Eng. Ind. Aerodyn.* 89: 605-617, 2001.

Klawa, M. and Ulbrich, U.: A model for the estimation of storm losses and the identification of severe winter storms in Germany. *Natural Hazards and Earth System Sciences* 3: 725-732, 2003.

Leicester, R.H, Bubb, C.T.J., Dorman, C. and Beresford, F.D..: An assessment of potential cyclone damage to dwellings in Australia. *Proc. of the 5th Int. Conf. on Wind Engineering*: 23-36, 1979.

Munich Re: Winterstürme in Europa – Schadensanalyse 1990 und Schadenspotentiale. *Münchner Rückversicherungs-Gesellschaft*, 1993.

Munich Re: Winterstürme in Europa (II) – Schadensanalyse 1999 und Schadenspotentiale. *Münchner Rückversicherungs-Gesellschaft*, 2001.

Nateghi-A, F.: Assessment of wind speeds that damage buildings. *Natural Hazards* 14: 73-84, 1996.

Sacré, C.: Extreme wind speed in France: the '99 storms and their consequences. *J. Wind Eng. Ind. Aerodyn.* 90: 1163-1171, 2002.

Schroers, H., Lösslein, H. and Zilch, K.: Untersuchung der Windstruktur bei Starkwind und Sturm. *Meteorol. Rdsch.* 42: 202-212, 1990.

Sparks, P.R., Schiff, S.D. and Reinhold, T.A.: Wind damage to envelopes of houses and consequent insurance losses. *J. Wind Eng. Ind. Aerodyn.* 53: 145-155, 1994.

Sparks, P.R.: Wind speeds in tropical cyclones and associated insurance loss. *J. Wind Eng. Ind. Aerodyn.* 91: 1731-1751, 2003.

Swiss Re, Stürme über Europa – Schäden und Szenarien. *Schweizer Rückversicherungs-Gesellschaft*, 1993.

Unanwa, C.O., McDonald, J.R., Metha, K.C. and Smith, D.A.: The development of wind damage bands for buildings. *J. Wind Eng. Ind. Aerodyn.* 84: 119-149, 2000.

Drought and Famine Disasters in China: From Risk Assessment Based on Historical Records to Contingency Planning

Fengxian Bu[1,2], Hendrik J.Bruins[1]
[1] Ben-Gurion University of the Negev, Jacob Blaustein Institute for Desert Research, Department Man in the Desert, Sede Boker Campus, 84990, Israel
[2] Northwest Sci-Tech University of Agriculture and Forestry, Yangling, Shaanxi, 712100, P.R.China

Abstract

Ancient Chinese manuscripts contain abundant information about disasters in the past. The recording style is generally brief and concise, but relevant information is conveyed. However, the denotation of ancient disasters is not always clear, as exceptional phenomena are sometimes mentioned. A methodology for evaluating the historical materials of famine and drought disaster is presented in this study, based on the principles of disaster science, famine theory, as well as the character of historical materials. Our approach enables identification of disaster category, the time of its occurrence, the region affected, the resulting impact, as well as disaster relief measures. Based on the historical materials we present a long-term time series of the occurrence of drought and famine. There were 2547 cases of drought and 3192 cases of famine in ancient China during the period 1765 B.C to A.D 1911. Differences are present in the frequency of famine and drought among different dynasties. Our results show that drought was the predominant cause for famine, being responsible for about 70% of all cases. Significant drought years in modern times are distinguished by us on the basis of the size of the affected agricultural lands: 1959, 1960, 1961, 1972, 1978, 1980, 1981, 1986, 1987, 1988, 1989, 1992, 1994, 1997, 1999, 2000, 2001, 2002, and 2003. The largest loss of food production due to drought, almost 60 million ton, occurred in 2000. Given the low level of annual world food reserves, we recommend the formation of significant national food reserves as the principal component in contingency planning to reduce drought impact and famine risk.

Keywords: drought disasters, famine, chinese history, risk assessment, contingency planning.

1 Introduction

The record of famine and disasters in ancient China can be traced back 3000 years ago to the Shang dynasty. The oracle inscriptions of the Shang dynasty, written on bones or tortoise shells, recorded the actions of the emperor to pray for rainfall. The total number of such oracle inscriptions about prayers for rainfall numbers about 151 pieces (Hu, 1990). Since the Shang Dynasty, many other documents related to disasters and famine was collected in various ancient history books. This collection of written records constitutes a more or less continuous historical time series about disasters, which is unique in the world.

The famine records of ancient China appear in a variety of sources, which amount to 10 different types, including government archives, local documents, inscriptions on stones and archaeological objects (Zhang, 1992). Each of the written sources can be linked to time within the respective dynasty. China suffered serious famine problems in the past, caused by three principal phenomena according to the observations of ancient Chinese scholars. These three main causes for famine disasters were drought, floods and locusts (Xu Guang-qi).

The past may be the key to the future in terms of modern hazard assessment. Our approach is to evaluate and analyse the historical disaster record in a thorough and comprehensive manner in order to establish the frequency of occurrence of drought and famine. Such information can be used in hazard and impact assessment for modern times, in order to develop contingency planning. Previous studies by scholars in the 20th century (Chen, 1934; Yao, 1942; Deng, 1958; Zhu, 1979; Chen, 1986) about the number of different disasters in ancient China led to different conclusions (Bu, 1998). The reasons for these differences are related to the following two aspects: (1) Use of difference source materials. (2) Different methodologies to deal with the historical records.

2 Data Sources and Methodology

This study of drought and famine disasters is based on two main data sources: 1) The compilation of much historical literary data concerning agricultural natural disasters in China into one book by four modern scholars (Zhang, B., Zhang, L., Li, H. B., Feng, F., 1994). This book has collected almost all of the famine related history materials from official ancient history texts. 2) Another book has compiled a table of natural disasters and human-made catastrophes that occurred in past Chinese dynasties, which is based on the ancient sources mentioned above, but also on other important ancient documents, such as Zizhi Tongjian, Xu Zizhi Tongjian, Qing Dynasty History Draft, Qing Inspection, Gujin Tushu Jicheng.

Therefore, most of the ancient history materials are at our disposal for research, based on these Chinese books. The historical sources of disasters and famine include three aspects: (1) the date of the disaster, (2) the region of the disaster, and (3) the severity and impact of the disaster. Concerning the third aspect, the historical information about the disaster impact can be evaluated from descriptions about the number of people who perished and also the reduction in crop yields. The ancient documents may also contain important information about the development of the famine disaster through time, as a kind of domino effect: beginning of famine, food shortages and price rises, disaster relief activities, and even cannibalism.

The ancient literary data about historical drought can be transformed into statistical form, because the information is concise and reliable. There are several Chinese words that were used to describe or indicate drought conditions in ancient China, such as direct terms for drought, severe drought, or indirect terms such as no rainfall, no snow in winter, pray for rainfall and rivers become dry. We have systematically placed all this information into a statistical database concerning drought.

The famine record in ancient texts is not as clear as the information about drought, described above. Therefore, we developed new criteria to distinguish the terms that indicate famine in the ancient sources. The key indicator to make a correct judgement about past famine in our research is based on the degree of food shortages. Three aspects are involved in this textual indication to judge famine conditions: information about (1) society, (2) economy and (3) calamity victims. Concerning social aspects, descriptions of measures taken by the central government to provide famine relief, including the movement of people suffering from malnutrition to other areas, indicate that food shortages occurred in the region. Regarding economic aspects, information about very high food prices suggest food shortages in the region. Finally, descriptions about calamity victims that people had no food to eat, or were eating tree bark and other substitute food, or even mention of cannibalism, also indicate famine conditions in the region.

The severity of the famine has two aspects, relating to the degree of malnutrition and to the size of the region affected. A mild famine usually did not occur in more than a few counties, up to one Zhou-Fu. The imperial term for land administration in ancient China is Zhou-Fu or only Fu, which includes several counties. A "common" famine usually covered a larger region, having the size of one to ten Zhou-Fu and the victims needed to be given food to survive. A severe famine usually affected more than several provinces, involving more than about 10 Zhou-Fu. The impact of a severe famine is characterised by the death of many people, the sale of children or people, even for cannibalism. The word hungry in the historical sources is synonymous with famine, so the mention of severe hunger means in fact severe famine conditions..

3 The Frequencies of Drought and Famine in Ancient China: A New Risk Assessment

According to the criteria described above, we evaluated all the literary historical data from the beginning of the Shang Dynasty in 1765 BC to the end of the Qing Dynasty in 1911 AD. We compiled and analysed both the frequencies of drought and famine for this unique period of 3,675 years (Table 1).

Chinese Dynasties	Drought Occurrence	Relative Frequency of Drought per Dynasty (%)	Drought Frequency per Century	Famine Cases	Relative Frequency of Famine per Dynasty (%)	Famine Frequency per Century
Shang 1765-1122 BC	6	0.2	0.9	0	0	0
Zhou 1121-249 BC	35	1.4	4.0	10	0.3	1.1
Qin 248-207 BC	1	0.04	2.4	4	0.1	9.8
Han 206 BC-220 AD	112	4.4	26.3	76	2.4	17.8
Wei-Jin 220-580 AD	192	7.5	53.3	195	6.1	54.2
Sui 581-618 AD	11	0.4	29.7	13	0.4	35.1
Tang 618-960 AD	232	9.1	67.8	150	4.7	43.9
Song 960-1279 AD	388	15.2	121.6	386	12.1	121.0
Yuan 1279-1367 AD	212	8.3	241.0	533	16.7	605.7
Ming 1368-1644 AD	328	12.9	118.8	437	13.7	158.3
Qing 1644-1911 AD	1030	40.4	385.8	1388	43.5	519.9
Total	2547	100		3192	100	

Table 1:   Frequency of Drought and Famine in Ancient China.

The total number of droughts for this period is 2547 and the total number of famine cases is 3192. Our conclusions in both cases represent new numerical results, which are quite different from those presented by previous scholars (Mallory, 1928; Chen, 1934; Deng, 1958; Chen 1986). The large number of drought and famine cases in China during the past 3,700 years underlines the very serious risk of both hazards in this country, which currently has the largest population of any nation in the world, *i.e.* 1,300 million or 1.3 billion people.

Our results presented in Table 1 show the very large differences in drought and famine frequency among the different dynasties. The last one, the Qing Dynasty has by far the highest number of drought and famine cases. It is obvious that the number of cases recorded in historical documents may not necessarily give a fully comprehensive account of all cases. The older the period, the more likely that ancient document got lost and that the number of drought and famine cases will be lower as a result. This problem has been discussed by scholars in the first half of the 20th century who pointed out that the more recent Chinese documents are usually more detailed and comprehensive than the older ones, but there are some exceptions (Ting, 1935). Therefore, the relative frequency of drought and famine per dynasty or per century may partly be an artefact of the number of documents and data that have survived until modern times. Moreover, the criteria to interpret the historical records will also affect the numerical compilation and statistical conclusions of famine and drought frequency. Another possible factor is that the more recent dynasties may have lowered their own criteria, as compared to older periods, and even included rather mild cases of drought and famine into historical books, which were perhaps not recorded in ancient dynasties. On the other hand, the total size of the agricultural area used for farming during the Qing Dynasty exceeded that of all previous periods. Thus the increase in farming land may also have led to an increase in the cases of drought occurrence and famine in terms of administration as the number of recorded cases collected into historical books.

Famine also occurred in China in the 20th century, in the modern period not covered by the above table. Severe famine affected the country in 1920-21, 1928-30, 1958-61. Drought was the principal natural cause in all three cases, but socio-economic causes, related to vulnerability, also played important roles. An American scholar, W. H. Mallory, who was secretary of the China International Famine Relief Commission during the great famine in 1920-21, wrote a book entitled *"China, Land of Famine"*. The author had personally witnessed the tragic life of Chinese peasants during the above famine (Mallory, 1928). However, the food relief mentioned above proved very important to mitigate the impact of the drought and the resulting famine disaster. Nevertheless, about 500,000 people perished during this famine. The 1928-30 famine affected more or less the same region, but due to the internal war conditions it was much more difficult to supply food relief into the countryside. The number of famine victims was much higher as a result, numbering 2,500,000 people. The third severe famine in China in the 20th century, during a three-year period from 1958 to 1961 had a terrible impact. The area affected was much larger than in the two previous cases and about 30,000,000 people died as a result of the famine. Drought played again an important role, besides socio-economic and political causes.

4 The Causes of Famine

Famine was usually defined in simple terms of food shortages in the past, but the issue can also be approached in a more sophisticated way according to the food availability entitlement theory, developed in the 1980's by Sen (1981). However, both approaches are valid and represent two aspects of famine, as both food shortages and entitlement to food play a role in most famine cases. The basic reason for famine can often be linked directly to a decline in food production, which may be caused by drought, floods and other disasters.

The cause of famine was also recorded in the Chinese historical documents, mentioning one of the above causes, as well as others. Occasionally the text is ambiguous concerning the reasons for a particular famine. We made an evaluation of all famine causes in Chinese history and we have presented this compilation in Table 2.

Drought, floods and locusts were already for a long time considered to be the dominant natural hazards in China, which may lead to famine disasters. Our evaluation of the ancient texts suggests that, in unambiguous terms, floods appear to be the most frequent cause of famine. However, Chinese historical writers and scholars often linked famine together with drought, without mentioning drought specifically. Indeed drought can be considered as a non-dramatic passive phenomenon. Unlike a flood, hurricane or locust plague, a drought does not occur because something happens, but because the rains fail to come (Gillette, 1950). Therefore, drought may not have been mentioned explicitly in the historical records of many famine cases. Yet it seems very likely that drought was the common cause of famine in most cases, described in ambiguous terms, being so obvious that it was perhaps not worth mentioning, also due to its passive nature. On the other hand, if the famine was caused by clear, active "happenings" then those causes were recorded specifically by the ancient scholars. Indeed, all famine cases with ambiguous causes in a textual sense were, in accordance with this approach, placed into the drought category by Chen (1986).

Cause of Famine	Number of Causes	Relative Frequency (%)
Drought most likely (text less specific)	1691	49.0
Floods	877	25.4
Drought (text clear)	503	14.6
Hail	131	3.8
Locusts	67	1.9
Strong Winds	51	1.5
Frost	44	1.3
Snow	28	0.8
Excessive or Intensive Rainfall	25	0.7
Crop Pests	19	0.6
War	14	0.4
TOTAL Number of Famine Causes	3450	100
TOTAL Number of Famine Cases (see table 1)	3192	

Table 2: The Causes of Famine in Ancient China.

Note: One famine case may result from several causes; therefore, the number of causes is larger than the number of cases.

Accepting this principle, drought becomes the predominant cause of famine in ancient China, representing almost 70% of all famines cases recorded in Chinese history. Drought was and is, given the experience of the 20th century, the most formidable hazard that may cause a severe decline in food production, which may lead to famine.

5 Drought in Modern Times

Our above conclusions are confirmed by modern studies of drought in China since the 1950s, in which the hazard of drought is clearly found to be the most serious threat to agricultural losses (Li and Lin, 1993). Although drought may occur in any climatic belt, its impact is most dangerous in the semi-arid and sub-humid region, in which rainfed agriculture is widespread. The Huang-Huai-Hai plain region, which includes the North

China Plain, the Loess Plateau and the Inner Mongolia Plateau, is the most vulnerable region to drought in China. Wheat is here the main food crop. Severe droughts were observed in this region in 13 years out of 40 years in the period 1951-1990 (Li and Lin, 1993).

China is such a large country that drought will occur somewhere each year. Our criterion to classify a year as a significant drought year on a national scale is based on the size of the drought impacted area of agricultural land. If the area affected in a certain year is above 10 million hectare, we consider it a significant agricultural drought year (Table 3). Thus the following years can be singled out as drought years in the period 1950-2003: 1959, 1960, 1961, 1972, 1978, 1980, 1981, 1986, 1987, 1988, 1989, 1992, 1994, 1997, 1999, 2000, 2001, 2002 and 2003 (Table 3). It seems clear that the occurrence of agricultural drought in China was limited in the 1950s, 1960s and 1970s. However, a dramatic increase in the frequency of drought took place in the 1980s with six significant agricultural drought years. This trend continued more or less in the 1990s with four drought years. Since 2000 every year so far showed a large area of agricultural land impacted by drought. The years 2000 and 2001 stand out as the worst years since 1950 in terms of the size of agricultural lands affected by drought.

Drought Year	Population (million)	Total Cultivated land (ha)	Drought impacted area (ha)	Lost Food Production due to Drought (10^3 ton)	Total Food Production (10^3 ton)	Food/ Capita (kg)
1959	672	104,579,300	11,173,000	10,800	169,680	252
1960	662	104,861,300	16,177,000	11,300	143,850	217
1961	659	103,310,700	18,654,000	13,200	136,500	207
1972	872	100,614,700	13,605,000	13,673	240,480	276
1978	963	99,389,500	17,970,000	20,046	304,770	317
1980	987	99,305,200	14,174,000	14,539	320,560	325
1981	1001	99,035,100	12,134,000	18,548	325,020	325
1986	1075	96,229,900	14,765,000	25,434	391,510	364
1987	1093	95,888,700	13,033,000	20,955	404,730	370
1988	1110	95,721,800	15,303,000	31,169	394,080	355
1989	1127	95,656,000	15,262,000	28,362	407,550	362
1992	1172	95,425,800	17,049,000	20,900	442,658	378
1994	1199	94,906,600	17,049,000	26,200	445,101	371
1997	1236	129,933,000	20,250,000	42,600	494,171	400
1999	1258	129,205,361	16,614,000	33,300	508,386	404
2000	1267	128,233,100	26,784,000	59,900	462,175	365
2001	1276	127,615,800	23,698,000	54,800	452,637	355
2002	1285	125,930,000	13,267,000	31,300	457,100	356
2003	1292	123,392,200	14,467,000	30,800	430,670	333

Table 3:   Food production and food losses during significant drought years.

Sources: Population data based on the National Bureau of Statistics of China. Data about total area of cultivated land (column 3) and drought impacted area (column 4) before 1997 are from the database of the Chinese Agriculture website; from 1997 data are based on the official website of the Ministry of Land and Resources (column 3) and the National Bureau of Statistics of China (column 4). Data about lost food production due to drought (column 5) are based on Fan Baojun (1999:576) for the period 1950-1990, on the Water Conservation Yearbook for the years 1991-2001, on the Water Resource Statistical Bulletin of the Ministry of Water Conservation for 2002 and on an official report (Suo, 2004) for 2003.

6   Risk of Food Shortages and the Need for Contingency Planning

The risk of drought and famine in China led the authorities to develop drought mitigating strategies, which include improvement of agricultural technology, the building of a water conservation and water transportation infrastructure for irrigation to increase food production. Although the Chinese authorities have made great progress in food production and disaster reduction, the rather low amount of food grains available on the world market is a cause for concern (Postel, 1998; Bruins, 2000; Bruins et al., 2003). Moreover, groundwater used for irrigation farming and food production in the drylands of northern China is declining at an alarming rate (Postel, 1998), which could lead to regional water shortages and a severe decline in food production in this area.

However, the past record of drought shows the very serious risk of extreme drought that may occur in the near future. Then much higher food imports may be required. Such potential quantities may surpass the annual availability of food on the world market. In such a scenario, severe food shortages and famine cannot be prevented, unless the country develops large internal food reserves as the main component in contingency planning to avert massive famine. The relevance of conventional planning, proactive planning and disaster contingency planning is discussed by Bruins and Lithwick (1998) in relation to the risk of drought.

Acknowledgements

This research was carried with the support of the Jacob Blaustein Center for Scientific Cooperation (Jacob Blaustein Institute for Desert Research, Ben-Gurion University of the Negev) from whom Dr. Fengxian Bu received a Postdoctoral Fellowship. We also thank the Ben-Gurion University for providing travel grants that enabled us to participate in the International Conference "Disasters and Society: From Hazard Assessment to Risk Reduction" (University of Karlsruhe).

References

Bu, Feng-xian: Some problems in the research of agricultural disasters history. *Agriculture Archaeology*, 18(1): 280-284, 1998(in Chinese).

Bruins, H.J.: Proactive contingency planning vis-à-vis declining water security in the 21st century. *Journal of Contingencies and Crisis Management* 8:63-72. 2000

Bruins, H.J., Akong'a, J.J., Rutten, M.M.E.M., Kressel, G.M.: *Drought Planning and Rainwater Harvesting for Arid-Zone Pastoralists: the Turkana and Maasai (Kenya) and the Negev Bedouin (Israel)*. NIRP Research for Policy Series 17, The Hague, KIT Publishers, Amsterdam (ISBN 90-6832-682-1), 2003.

Bruins, H.J. and Lithwick, H., edtors: *The Arid Frontier - Interactive Management of Environment and Development.* Kluwer Academic Publishers, Dordrecht / Boston / London (ISBN 0-7923-4227-5), 1998.

Chen, Gao-yong: *The Table of Natural Disasters and Man-made Catastrophe in Chinese Past Dynasties*, Shanghai Bookstore Press, Shanghai, 1986 (in Chinese).

Chen, Ta: *Population Problem*, Business Press, Shanghai, 1934 (in Chinese).

Chinese Agriculture Website: http://www.sannong.gov.cn/tjsj/lssj/1/default.htm

Deng, Yun-te: *Famine Relief History of China*, Sanlian bookstore, Shanghai, 1958 (in Chinese).

Fan, Baojun: *Natural Disasters in Modern China*. Modern China Press, Beijing, 1999

Hu, Hou-xuan: Study on the climate variation in the Shang Dynasty, in: H.X. Hu (Ed.), *A Collection of the History of the Shang Dynasty in Inscriptions on Bones or Tortoise Shells*, pp. 1-6, Shanghai Press, Shanghai, 1990 (in Chinese).

Li, K. and Lin, X.: Drought in China: Present Impacts and Future Needs. *In* D.A. Wilhite (ed.) *Drought Assessment, Management and Planning: Theory and Case Studies.* Chapter 15, pp. 263-289, Kluwer Academic Publishers, London, 1993.

Mallory, W.H. (1928), *China: Land of Famine*, American Geographical Society, New York, 1928.

Ministry of Land and Resources of China: http://www.mlr.gov.cn/query/gtzygk/index.htm

Ministry of Water Resources of China: *China's Water Conservation Yearbook*, 1991-2001, China's Water Conservation and Water Electricity Press, Beijing, 1991-2001.

Ministry of Water Resources of China: Water Resource Statistical Bulletin 2001-2003, at: http://ghjh.mwr.gov.cn/document/index.asp?LMark=H3&category=3

National Bureau of Statistics of China: http://www.stats.gov.cn/tjsj/ndsj/index.htm

Postel, S.: Dividing the Waters: Food Security, Ecosystem Health, and the New Politics of Scarcity, *Worldwatch Papers 132*, Washington, D.C., 1998.

Sen, Amartya, *Poverty and Famine: An Essay on Entitlement and Deprivation*. Oxford University Press, Oxford, 1981.

Suo, L.S.: Special Attention to the Effect of Water Resources Problems in the Development Process of Agriculture and Ecology, *China's Market of Water Resource and Water Electricity*, Compact Disc Version, 4, 2004.

Ting, V.K.: Notes on the Records of Droughts and Floods in Shensi and the Supposed Desiccation of N.W. China, Geografiska Annaler, 17 (Issue Supplement): 453-462, 1935.

Walford, Cornelius. The Famines of the World: Past and Present, Journal of the Statistical Society of London, 41(3): 433~535, 1878.

Xu, Guang-qi, *Complete Treatise on Agricultural Administration*, compiled in Ming Dynasty, the modern version was edited by Shi Shenghan, *Collated and Annotated Version of Complete Treatise on Agricultural Administration*, Shanghai ancient books press, Shanghai, 1979(in Chinese).

Yao, Shan-you: The Chronological and seasonal distribution of floods and droughts in Chinese History: 206B.C.-A.D.1911, Harvard Journal of Asiatic Studies, 6 (3), (4): 273-312, 1942.

Zhang, B., Zhang, L., Li, Hong-bing and Feng, F.: Aspects of historical data about natural disasters in China, China Historical Materials of Science and Technology, 13(3): 9-13, 1992(in Chinese).

Zhang, B., Zhang, L., Li, Hong-bing and Feng, F.: *Historical Materials Collection of China's Agricultural Natural Disasters*, Shaanxi Science & Technology Press, Xi-an, 1994(in Chinese).

Zhu, Ke-zhen (1979), the climate changes in Chinese history, in *The Article Collections of Zhu Kezhen*, pp. 58-68, Science Press, Beijing, 1979(in Chinese).

Earthquake Risk Assessment of Existing Structures

Marcel Urban

International Graduate College 802 "Risk Management of Natural and Civilisation Hazards on the Built Environment", Technical University of Braunschweig, D-38106 Braunschweig, Germany;

Abstract

Historical structures, especially churches and vaulted structures are very vulnerable with respect to ground motions such as earthquake shaking. The risk analysis and the risk assessment of these structures are of great importance, especially with regard to the cultural value. Due-to the peculiarities in structure and design, existing concepts of vulnerability assessment cannot be easily applied. In addition, the evaluation of the existing risk is becoming even more complicated with respect to the high degree of uncertainties involved in the modelling of loads and structures. Starting with a description of typical damages and the major points of modelling, the scheme of analysing the risk by performing a nonlinear dynamic analysis is described. The paper concludes with some ideas to evaluate the loss of cultural properties.

Keywords: earthquake, risk assessment, historical structures, churches.

1 Introduction

1.1 Motivation

Experience shows that, although a structure is already standing for a long period, the earthquake hazard is a serious problem. Noticeable is, that after every european earthquake the effects of ground vibrations show mainly on historical buildings. Especially churches and vaulted structures are damaged profoundly. This is of great importance since the damage potential and the overall risk of churches concerning earthquakes are extremely hard to determine. Particular characteristics have to be regarded because of the cultural value and the high degree of uncertainties involved in the models. In addition, the number of endangered churches is high enough to constitute a significant public risk, even in areas with a small seismic hazard. In cooperation with researchers from the Università degli Studi di Firenze and Università degli Studi di Genova in Italy, a concept is being developed to assess the risk inherent in historical structures. Research is at first concentrated on churches.

1.2 The concept of Risk Management in Civil Engineering

The assessment of the risk is a complicated task regarding several aspects of which not all do have their origin in engineering. The following figure shows the flow of assignments to fulfil.

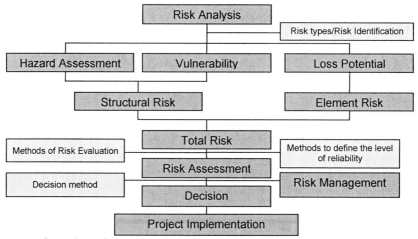

Figure 1: A flow chart of the tasks within the risk management process.

Although the main aspect for engineers is the combination of hazard and vulnerability, several other aspects have to be accounted for, to interface with other disciplines or to allow decision makers to implement the results of the assessment in the overall risk assessment scheme and decision procedure. Especially for historical buildings the additional loss potential to items which are not related to the structure itself has to be assessed. For this reason the process differentiates the evaluation of the structural risk related to the building itself and the additional loss potential.

1.3 State-of-Art of Vulnerability Assessment

Furthermore, despite calling it risk assessment what is usually done is an assessment of the structural safety only. Common approaches for this are empiric formulas, vulnerability curves or the capacity spectrum method. They are based on typical building styles and used for a rather fast assessment of regional earthquake risks. For monumental buildings these concepts cannot be applied due to the irregular architecture or unpredictable behaviour of large masonry structures. Instead, usually a detailed FEM study is performed, in which mostly deterministic values of load properties, effective geometry and material parameters are assumed. Another innovative approach was developed by italian researchers depending on possible ways for a structure to collapse (Lagomarsino 1998). Although this approach is very practical, only few studies on its reliability have been performed. Figure 2 shows some examples with cracks and the behaviour of collapse.

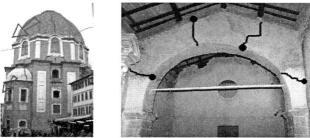

Figure 2: Collapse behaviour in Domes (Capella di Medici, Florence) and arches (chapel in Norcia, Italy). Note: cracks were drawn, circles show possible hinge mechanisms.

2 A Procedure for the Assessment of Risk

2.1 Remarks

There are several ways for the assessment of risk and vulnerability of structures with respect to ground motions. Most common are the response spectra method, pushover curves, collapse mechanisms and time history analyses. With respect to the complexity and uniqueness of historical buildings the most precise way is the application of natural or artificial accelerograms. The additional effort seems to be acceptable with respect to the high cultural value of these buildings. Therefore, during this paper emphasis is laid on nonlinear dynamic analyses with accelerograms applied as loadings.

2.2 Analysis of Impact

In contrast to the normal design, which is governed by safety factors, the loads for the risk assessment have to be characterised more thoroughly to insure the correct computation of the existing risk. The goal should be not to apply the highest possible load or to concentrate on the strongest event, but to describe the probability and properties of certain events accurately. Next to the precise evaluation of loadings, impacts of different strength have to be regarded, to reach a more distinguished assessment of the risk.

To gain valuable information about the seismic hazard, the Gutenberg-Richter relationship is applied. In Figure 3 this was done for the Lower Rhine Embayment using data of Hinzen (Hinzen, Pelzing, Reamer, Mackedanz, 2000) with the seismic constants $a=$ 2.00 and $b = 0.85$. Alternatively these constants can be determined by examination of earthquake catalogues (Leydecker, 1986).

$$\log_{10}(N/a_T) = a - b \cdot M_w \qquad (1)$$

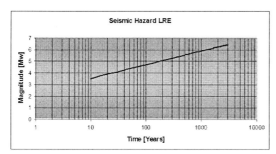

Gutenberg-Richter variables:
N = Number of events
a_T = Time period
a,b = seismic constants
    for given region
M_w = Moment Magnitude

Figure 3: Seismic hazard in terms of Magnitude-Return period relationship.

The design events given in the codes with a return period of 475 years would correspond to a Magnitude of 5.5 for example. It has to be noted, that this information does not include distances between structures and epicentres. Effects of distance and soil can be taken into account by the use of attenuation function as the one proposed by Ambraseys (1999).

As mentioned above, to perform a more qualified assessment of risk several events have to be considered. Thus it is useful to apply standards of Performance Based Seismic Engineering (PBSE) for the assessment of the seismic risk especially with regard to the long life time of historical structures. An event with a smaller magnitude should also be included to evaluate previous damages and their effects on the structural safety. For the four given different return periods in the codes for PBSE (FEMA, 1997) the corresponding magnitude in the Lower Rhine Embayment is shown in Table 1.

Earthquake description	Probability	Return period	Magnitude in LRE
Frequent earthquake	50% in 50 years	72 years	4.53
Occasional earthquake	20% in 50 years	225 years	5.12
Rare earthquake	10% in 50 years	475 years	5.50
Very rare earthquake	2% in 50 years	2475 years	6.34

Table 1:   Magnitude values for return periods of PBSE in the Lower Rhine Embayment.

Knowing the magnitudes and return periods of events, the loading input may be implemented in several methods. Although the use of response spectra is widely used in design, the focus here lies on the more precise calculation with time-history records. Since for Germany no recorded measurements for events around a magnitude of 6.0 exist, accelerograms have to be produced to be used as the loading. For these records either natural earthquake recordings should be applied – in the best case modified, so that the strongest frequency in the response spectra of the earthquake corresponds to the first eigenfrequency of the structure – otherwise, artificial earthquake motions may be generated, which can be fitted easily to the conditions at the site. For this problem several algorithms exist of which in Germany one of the most widespread is the one developed by Meskouris (Meskouris and Hinzen, 2003) to fit target spectra. Further algorithms were developed by Armouti (Armouti, 2003) and Sabetta (Sabetta and Pugliese, 1996). Tests have shown that the results of algorithms differ sincerely from natural earthquake recordings. Some of the comparisons between several earthquakes for a magnitude of 6.0 for stiff soils are shown in Figure 4.

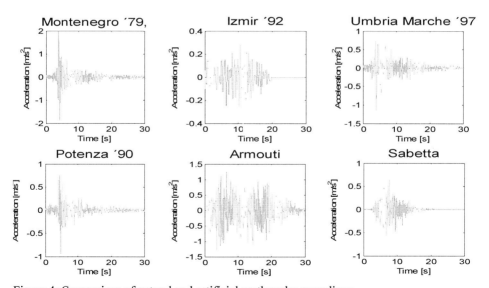

Figure 4: Comparison of natural and artificial earthquake recordings.

The accelerograms show four natural recordings of earthquakes with a magnitude of 6.0 in a distance between 29 to 31 kilometres from the epicentre on stiff soil as well as two artificial recordings (Armouti and Sabetta) generated with the same parameters. On the following figure the energy release rates of these quakes and six events on soft soils are shown. It can be seen that there are two remarkable differences between artificially generated and natural earthquakes. At first, artificial earthquakes overestimate the energy of earthquakes, especially those fitted to target spectra, because they contain all frequencies given in the target spectra, if taken from codes. Furthermore the generated earthquakes cannot properly reflect the behaviour over time. The algorithm of Armouti for example produces long earthquakes with a steady energy release rate. Of the three algorithms regarded, the approach of Sabetta was found to produce results which reflect properties of natural events most closely.

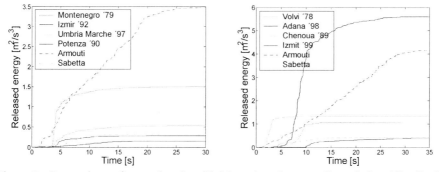

Figure 5: Comparison of natural and artificial earthquake recordings (left: stiff soil, right: soft soil, M_w around 6.0, distance from epicentre around 30 km)

It can be concluded that if comparable recordings cannot be found, the nonlinear approach from Sabetta is most useful for the generation of loads. This approach may be fitted to the situation at the given site with several return periods and distances to the hypocentre. Otherwise the application of similar recordings is recommended.

2.3 Structural Diagnosis

The next step in the assessment of risk is the modelling of the structure. It is important to realise that construction of historical churches lasted for years. During this time, several different materials were used, some parts collapsed, the originally planned design was changed. Sometimes, parts of the structure were added centuries after. In consequence masonry and structure show a very inhomogeneous behaviour affected by many uncertainties. Due-to these uncertainties, modelling of masonry is a very complicated task. Apart from the uncertainties in the loading, which were described in the previous chapter, uncertainties of material properties, structural properties (i.e. damping, nonlinear behaviour) and geometrical measures have to be assessed.

Thus, a proper campaign for the risk assessment should always be accompanied by a larger monitoring campaign to reduce the influence of uncertainties. Most important tests for the analysis are flat jack tests to detect the stress state within a structure and to measure elastic moduli, core drillings for the evaluation of material properties and vibration measurements to detect eigenfrequencies. Further tests as ultrasonic tests, control of the ground water level and settlements as well as crack and displacement monitoring complete the information used in modelling.

After data on material properties and geometry were gathered, the numerical model has to be developed. The model strongly depends on the calculation technique chosen. Although in this text the nonlinear dynamic analysis with application of time histories of events is promoted, benefits of the response spectra method have to be stated. Already applied to the Aachen Cathedral (Kuhlmann, Butenweg, Meskouris, 2003), high stressed areas can be easily identified. Calculating the eigenfrequencies is an important task, since it shows the major movements of the building, which are also those mostly affected by random vibrations, with a reasonable amount of computation time. The following figure shows some results of the modal analysis of the Church St. Martinus, Linnich placed in the region of the Lower Rhine Embayment. The damage sustained in reality during the Roermond 1992 earthquake was located mainly in the abside. It can be seen that there is a coherence between the model shapes and the occurred damage. Colours show the total displacement of the nodes, from dark (none) to light (maximum). The abside, which is excited in the first and second eigenmode, is also the part of the structure where the highest damage in reality occurred.

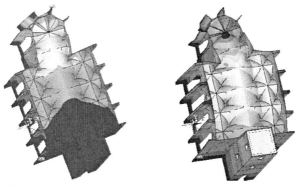

Figure 6: First (left: 2.8 Hz) and Second (right: 3.5 Hz) eigenmode of St. Martinus, Linnich.

Furthermore the modal analysis demonstrates why churches are very vulnerable. Following the combination between the high mass and low stiffness, the eigenfrequencies of masonry churches lie around 1-3 Hz. This is of great importance, since the main energy of earthquakes is released within these frequencies.

Despite the existence of this simplified technique, also more sophisticated calculations should be made for the risk assessment of singular structures. Since computer speed is increasing and material models gain in reliability the complexity of models and calculation techniques can increase. Still, there should remain the possibility to cross check with experimental data or to review the reliability of the results. The main efforts are to include the time dependency of loadings and the nonlinear material behaviour.

In these studies we used the material model for masonry developed by Gambarotta and Lagomarsino (Gambarotta and Lagomarsino, 1997). This model is capable of representing the damage in the masonry by introducing damage factors for the mortar and for the bricks which result in degrading stiffness with increasing damage. It is also possible to represent the brittle behaviour of masonry. The damage factors may furthermore be used as a measure for the structural vulnerability. For additional information the reader is referred to the source mentioned above. Most uncertainties in the model can be covered with this description of the material and a sensitivity analysis with a variation of parameters. Some sample analyses are shown in figure 7.

With the implementation of the proper material model and the application of loads described in Chapter 2.1 including the four different return periods defined by the regulations of PBSE and nonlinear dynamic calculations a close description of damage and risk is possible.

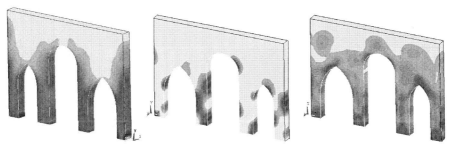

Figure 7: Distribution of damage in the mortar joints (left), position and size of cracks (middle) and distribution of damage in the bricks (right) of a triumphal arch of St. Giovanni a Mare, Naples after an earthquake from none (light) to maximum (dark).

2.4 Analysis of the Structural Risk

In this part of the process the data from the calculations has to be transferred to economic values. After the hazard was evaluated and several earthquake records with different return periods were simulated it is possible to express the damage which occurred in different parameters of which the most frequent are:

- Damage index (Lagomarsino,1998) $i_d = \frac{1}{3 \cdot N} \cdot \sum_{n=1}^{16} d_k$ (2)

- Damage parameter of calculations   -> referred to calculations

- Displacement ductility $\mu = \frac{|u|_{max}}{u_y}$ (3)

- Normalised Hysteretic Energy $NHE = \frac{\sum_{1}^{N} (\oint R_u \, du)}{R_y \cdot u_y}$ (4)

The variables in detail denote: i_d = Damage index, N = Number of mechanisms potentially activated, d_k = damage in the k-th mechanism (from 0 to 3), μ = Displacement Ductility, $|u|_{max}$ = maximum displacement, u_y = Yield Displacement, NHE = Normalised Hysteretic Energy, N = number of cycles, R_u = Force over time, R_y = Yield strength.

While the displacement ductility is more suited for assessment of steel structures the Normalised Hysteric Energy is a measure that can also be applied in the case of monumental buildings. This value is related to the damage parameter of the calculations used in the material model of Gambarotta and Lagomarsino. Currently it is evaluated how this measure can be transformed easily into monetary values.

3 Outlook

3.1 Evaluation of the Overall Risk

This chapter has to be an outlook, since the question of performing a complete assessment of the inherent risk has still to be solved. As it was seen in figure 1, the calculation of the risk is governed by three major points. The description of the hazard, which was dealt with in Section 2.1 and the vulnerability of the building, which may be evaluated by the proceedings described in Section 2.2. But still the loss potential of the building has to be assessed. Work is under way to receive at least an estimate by reviewing information on the following factors:

- Vulnerability of the building
- Vulnerability or hazard to the surrounding area
- Importance of items of art
- Touristic importance
- Data concerning the municipal and religious community

A database is being setup, whose data will be used to implement the aspects mentioned above into the assessment of risk as well as for the definition of risk classes and target reliabilities. Afterwards criteria are developed to connect these data to risk analysis and risk optimisation criteria.

Acknowledgements

I would like to thank Professor Bartoli from the Universita degli Studi di Firenze for taking me along to perform testing campaigns at the Capella di Medici in Florence and for his knowledge in historical structures which he was willing to share with me at all times. Also I have to express deepest gratitude to Professor Lagomarsino and his co-workers Ms. Chiara Calderini and Ms. Sonia Giovinazzi for their support in implementing their material model into my research.

References

Ambraseys, N.N., Simpson, K.A. and Bommer, J.J.: *Prediction of horizontal response spectra in Europe*, Earthquake Engineering and Structural Dynamics, Vol. 25, pp.371-400,1996.

Armouti, N.S.: *Response of Structures to synthetical earthquake,* Proceedings 9th Arab Structural Engineering Conference, Vol.1, pp.331-339, Abu Dhabi, UAE, 2003.

Hinzen, K.-G., Pelzing, R., Reamer, S. and Mackedanz, J.: *Seismisches Risiko in der Niederrheinischen Bucht*, DGG Annual Meeting, Frankfurt, 2001, Available electronically at http://www.seismo.uni-koeln.de/projects/risk/poster_1.htm, 1997.

Kuhlmann, W., Butenweg, C., Meskouris, K.: *Baudynamische Untersuchung des Aachener Doms unter Erdbebenbelastungen*, Bautechnik 2003

Lagomarsino, S.: *A new methodology for the post-earthquake investigation of ancient churches*. 11th European Conference on Earthquake Engineering, Balkema, Rotterdam, 1998.

Gambarotta, L., Lagomarsino, S.: *Damage Models for the seismic response of brick masonry shear walls. Part I and II*. Earthquake Engineering and Structural Dynamics, Vol.26, pp. 423-462, 1997

Leydecker, G.: *Erdbebenkatalog für die Bundesrepublik Deutschland mit Randgebieten für die Jahre 1000-1981*, Geologisches Jahrbuch Reihe E, Band E 36, 1986, Available electronically at http://www.bgr.de/quakecat/ger/printref.htm

Meskouris, K. and Hinzen, K.-G.: *Bauwerke und Erdbeben*, Vieweg, Wiesbaden, 2003

Sabetta, F., Pugliese A.: *Estimation of Response Spectra and Simulation of Nonstationary Earthquake Ground Motions*, Bulletin of the Seismological Society of America, Vol86. No.2, pp.337-352, April 1996

FEMA 273: *NEHRP Guidelines for the Seismic Rehabilitation of Building*, 1997

South African National Disaster Hazard and Vulnerability ATLAS

Dusan Sakulski
United Nations University Institute for Environment and Human Security, D-53113 Bonn, Germany

Abstract

The increase in the frequency of disasters and their associated damages globally is part of a worldwide trend, which results from growing vulnerability and may reflect changing climate patterns. Global risks seem to be increasing. These trends make it all the more necessary for the South African National Disaster Management Centre (NDMC) to initiate the development and implementation of the National Disaster Hazard and Vulnerability ATLAS. The main idea was to design and develop database-driven, web-enabled interactive "virtual book" (ATLAS). It consists of various "chapters", such as drought, flood, cyclones, storms, It enables users, using just web browser, to search and select various data, images, maps, graphs, to perform different analysis, to run selected models on-the-fly, and copy-paste results to the local computer and to print "their own page of the ATLAS".

1 Introduction

Natural disasters can be defined as temporary events triggered by natural hazards that overwhelm local response capacity and seriously affect the social and economic development worldwide. The sources of risk in developing countries are both natural and man-made. Because of the various geographical conditions, regions are prone to natural events of severe intensity. But the large economic and human cost associated with these natural events is mainly the result of extreme vulnerability. This vulnerability stems from the pattern of socio-economic development in the region as well as inadequate risk management policies.

Despite preventative efforts at national and regional levels, as well as globally, the risk associated with natural events has not decreased. Economic costs can be expected to increase, as economic assets accumulate and economic interdependence reaches new levels. While the human toll taken by disasters has remained more or less stable, it is unlikely to decrease because of the persistence of widespread poverty, continuing demographic growth and migration towards urbanized areas. Finally, preliminary evidence regarding climate change seems to indicate that the probability of occurrence of severe weather events will rise in the region.

Natural disasters jeopardize sustainable development. Inaction today regarding natural hazards compromises safety, economic growth, and environmental quality for generations to come. Furthermore, the burden of natural disasters falls disproportionately on the poor. However, forward-looking decision making today regarding land use, the direction and nature of economic development, and needed investment in societal infrastructure and capital facilities can improve the prospects and opportunities afforded to future generations.

A closer analysis of what transforms a natural event into a human and economic disaster reveals that the fundamental problems of development that the world faces are the very same problems that contribute to its vulnerability to the catastrophic effects of natural hazards.

The principal causes of vulnerability in the region include, very often, rapid and uncontrolled urbanization, the persistence of widespread urban and rural poverty, the degradation of the global environment resulting from the mismanagement of natural resources, inefficient public policies, and lagging and misguided investments in infrastructure. Development and disaster-related policies have largely focused on emergency response, leaving a serious under investment in natural hazard prevention and mitigation.

2 Why ATLAS?

The increase in the frequency of disasters and their associated damages globally is part of a worldwide trend, which results from growing vulnerability and may reflect changing climate patterns. Global risks seem to be increasing.

These trends make it all the more necessary for the NDMC to initiate the development and implementation of the National Disaster Hazard and Vulnerability ATLAS. The main idea was to design and develop database-driven, web-enabled interactive "virtual book" (ATLAS). It consists of various "chapters", such as drought, flood, cyclones, storms, etc.

It enables users, using web browser, to search and select various data, images, maps, and graphs, to perform different calculations, to run certain model on-the-fly, and copy-paste results to the local computer and to print "their own page of the ATLAS".

Figure 1 shows the main page of the South African National Disaster Hazard and Vulnerability Atlas. Using Internet browser Atlas can be accessed using the following URL: http://sandmc.pwv.gov.za/atlas/.

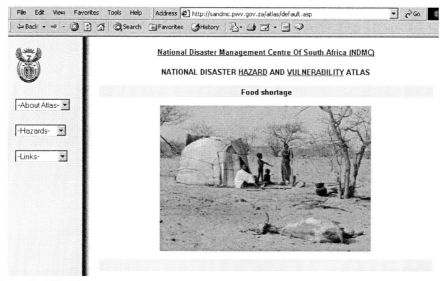

Figure 1: ATLAS Home Page.

Key ATLAS tasks are:

- Development of the disaster related hazard, vulnerability and risk assessment tools, to be able to report periodically on the global, regional as well as national exposure to natural hazards, patterns and trends or changes in the exposure and to guide priorities in natural disaster vulnerability reduction efforts.

- Development of an integrated global, regional as well as national disaster hazard and vulnerability information network to provide the tools needed by various levels of government, the private sector, and the general public. The network will also facilitate much-needed augmentation of education and training.
- Augmentations of the comprehensive, hazard specific programs.

Some of the main elements of the ATLAS are:

- Developing a comprehensive database to identify hazard, vulnerability and risk-prone areas.
- Understanding and addressing risk.
- Assimilating and disseminating information.

In particular, Atlas attempts to:

- Carry out research on factors contributing to disaster hazard and vulnerability and measures to alleviate this vulnerability.
- Develop methodologies for the analysis of disaster related hazard, vulnerability and risk indicators, and to improve disaster management.
- Disseminate the research results and methodologies through national, regional as well as global disaster information network and other channels to promote increased awareness and preparedness to natural and man-made hazards.

3 Rainfall Dynamics Chapter of ATLAS

Since mid 1980s frequency of tropical cyclones occurrences above the Indian Ocean, between Australia and Madagascar, has been increasing. In average during rainfall season, between October and April, 8 to 12 tropical cyclones are born and heading towards African continent (east-to-west). Figure 2 shows two tropical cyclones (twins) approaching Mauritius and Madagascar, in February 2003.

Figure 2: Twins tropical cyclones (NOAA).

Still in our memory is the year 2000/2001. 16 tropical cyclones were born in that region between December 2000 and March 2001 (Figure 3, from University of Hawaii).

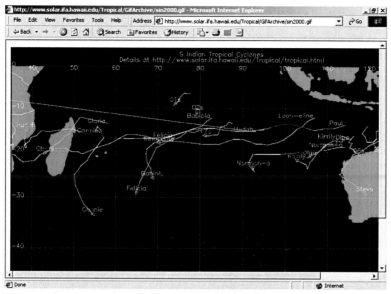

Figure 3: Tropical cyclones above Indian Ocean, 2000/2001 season.

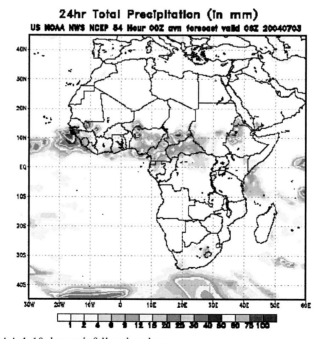

Figure 4: NOAA 1-10 days rainfall estimation.

Sub-Saharan Africa is very well known as a region of sudden heavy rain. It is nothing unusual to have 70 – 150 mm of rain in 24-hour period. Disastrous impact of such rain is on the area of the informal settlements (informal housing, rural communities). People at the NDMC are maximizing effort to have that kind of information as early as possible. Very useful web site is NOAA satellite 24-hours rainfall estimation for 1-10 days in advance (Figure 4).

South African National Weather Services (http://www.weathersa.co.za) is the biggest national rainfall data collector. Apart from the ground based gauging rainfall measurement, they operate the network of 11 ground radars (Figure 5).

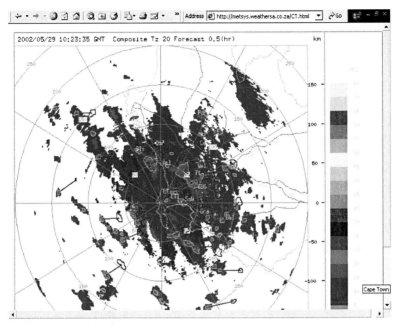

Figure 5:   SAWS radar image.

From the disaster management point of view, two kinds of data / information are extremely important:

- Close-to-real-time storm development and 30 min estimation.
- Accumulated hourly rainfall estimation.

4 Drought Chapter of ATLAS

Drought should not be viewed as merely a physical phenomenon or natural event. Its impacts on society result from the interplay between a natural event (less precipitation than expected resulting from natural climatic variability) and the demand people place on water supply. Human beings often exacerbate the impact of drought. Recent droughts in both developing and developed countries and the resulting economic and environmental impacts and personal hardships have underscored the vulnerability of all societies to this "natural" hazard.

Basic for drought analysis is a relatively new index called the Standardized Precipitation Index (SPI). Index was developed by McKee et al., with the intention to give a better representation of wetness and dryness than other indexes.

In accordance with sequences of drought impact, for selected rain station or selected area (i.e. quaternary catchment or quad), SPI is calculated according to five different time scales of water deficit (Fig. 6):

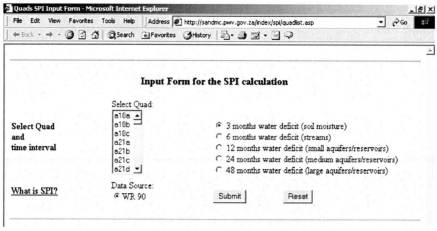

Figure 6: Input screen for the SPI on-the-fly analysis.

- 3 months water deficit (soil moisture)
- 6 months water deficit (streams)
- 12 months water deficit (small aquifers / reservoirs)
- 24 months water deficit (medium aquifers / reservoirs)
- 48 months water deficit (large aquifers / reservoirs)

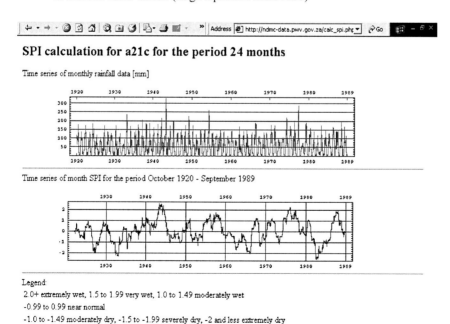

Figure 7: On-the-fly SPI analysis result screen.

As a result SPI time series has been calculated and two graphs are returned to the user's browser window: Time series graph of the monthly rainfall (pink line) and SPI time series graph (blue line) for the chosen quad and the 70 year period (Fig. 7).

For example, the calculation of the SPI for all quaternary catchments, for the period October 1920 – September 1990, shows four major drought periods (Fig. 8):

Figure 8: SPI for South Africa for the period October 1920 – September 1990.

- mid-to-end 1920s
- mid 1940s
- beginning-to-mid 1960s
- end of 1970s–to-beginning of 1980s

This type of graph (Fig. 8) represents the temporal dimension of the SPI. But, from that type of graph it is not possible to find out what part(s) of the country were affected by drought, for example in mid 1940s, or in mid 2002?

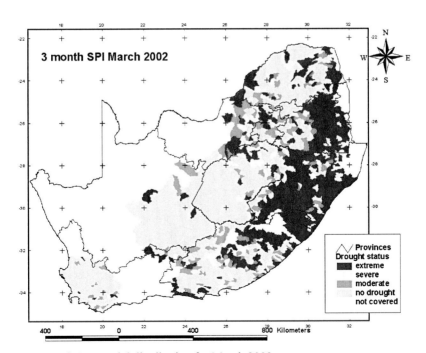

Figure 9: 3 month SPI spatial distribution for March 2002.

Advantage of the SPI is in its multi-dimensionality, and can be seen on Figure 9 representing spatial distribution of the 3 month drought for March 2002.

Acknowledgements

Author would like to acknowledge and give credits to the following institutions / organisations:

- National Disaster Management Centre, Republic of South Africa (http://sandmc.pwv.gov.za).
- Department of Water Affairs and Forestry, Republic of South Africa (http://www.dwaf.gov.za).
- South African Weather Service (http://www.weathersa.co.za).

References

McKee, T. B., N. J. Doesken, and J. Kleist. The relationship of drought frequency and duration to time scales. *Preprints, 8th Conference on Applied Climatology*, 17-22 January, Anaheim, CA, pp. 179-184, 1993.

Palmer. Meteorological Drought. *Research Paper No. 45*, U.S. Department of Commerce Weather Bureau, Washington, D.C, 1965.

National Disaster Management Centre (NDMC), South Africa. (http://sandmc.pwv.gov.za).

National Oceanographic and Atmospheric Administration (NOAA), USA. Geostationary Satellites Monitoring Server (GOES) (http://www.goes.noaa.gov).

University of Hawaii, Institute for Astronomy, USA. Tropical Storms Monitoring Worldwide (http://www.solar.ifa.hawaii.edu/Tropical/).

National Oceanographic and Atmospheric Administration (NOAA), National Weather Service, Climate Prediction Center, African Desk, USA (http://www.cpc.ncep.noaa.gov/products/african_desk/mrf_fcst/mrf_QPF24_96.gif).

METSYS, South African Weather Service, (http://metsys.weathersa.co.za).

Forecasting and Early Warning

Real-Time Prediction of Earthquake Casualties

Max Wyss
World Agency of Planetary Monitoring and Earthquake Risk Reduction, 36A Route de Malagnou, CH-1208 Geneva, Switzerland.

Abstract

Quantitative estimates of the number of casualties are necessary immediately after an earthquake for organizing an appropriate response. Reliable earthquake parameters become available within about two hours of an earthquake occurrence, worldwide, and faster in countries with advanced seismograph networks. Losses are estimated, based on a dataset of population and condition of buildings in 0.8 million settlements worldwide. The intensity of shaking is calculated, assuming standard transmission of seismic energy. Major earthquakes with insignificant consequences are identified as such in 93% of the cases, and 71% of the major disasters are recognized as such within two hours.

Keywords: earthquake risk, earthquake loss estimates in real-time.

1 Introduction

Predicting how many people may be trapped beneath collapsed buildings must be attempted to help rescue agencies and disaster managers to implement a response appropriate to the extent of the disaster. If an earthquake has likely caused no serious disaster, in spite of its significant magnitude, local organizations can cope and international rescuers need not waste a mobilization effort. On the other hand, it is essential to know if a major disaster is likely to have happened, as soon as possible after an earthquake because the chance of rescue decreases rapidly with time.

For estimating the probable degree of destruction, accurate hypocenters and magnitudes are necessary. A difference in depth or coordinates of only 10 km can affect the numbers of casualties many fold. By 'casualties' we refer to the injured plus fatalities.

The quality of the built environment is crucial in earthquake losses, because collapsing buildings are the chief cause of fatalities and injuries. Dwellings' resistance to shaking and their potential to injure varies enormously globally as a function of the degree of development of the affected society. The most primitive dwellings, bamboo huts or slum cabins, are so light that they cannot kill inhabitants and barely injure them, when they collapse. Well built wood-frame single story homes, as is customary in the US, together with buildings engineered, based on modern building codes, are also fairly safe, whereas shoddily built five to ten story apartment buildings are among the most dangerous.

Detailed surveys by engineers, of the type and quality of all buildings would be desirable, but are impossible in developing countries, because they would exceed resources, by far. Therefore, the approach taken here is to estimate the distribution of buildings into classes with specific fragility curves, and then to adjust these assumptions, based on calibration, using the losses sustained in past earthquakes. This means that losses can be estimated with more confidence in countries experiencing earthquakes frequently, than in those where no calibration events are available.

2 Method

The intensity of the ground motion is proportional to the magnitude of the earthquake and decreases as a function of distance from the source. The rate of decrease is not the same everywhere, but its variation influences the loss estimates by only about 10%. Therefore, it is reasonable to use an average attenuation curve, unless the specific values for the region in question are known.

Local amplification of the shaking, due to the top rock layers on which a building is resting, can be very significant. For important cities in developed countries, microzonation studies map the soil conditions. In that case, these effects can be taken into account in the loss estimates. However, for the vast majority of cities in zones of earthquake hazard, soil conditions are unknown. Thus, most of the loss estimates have to be carried out in ignorance of local amplification factors.

The most basic information about the population at risk is the number of inhabitants in each of about 0.8 million cities, worldwide. Without knowing the number of buildings in these settlements, one may calculate the percentage of injured and fatalities, based on the percentage of buildings collapsed and otherwise damaged. The loss to the built environment is also expressed in percent. In this case, it is the average percentage of the building value that will be required for restoration of the built environment to its pre-earthquake state. From this, and an estimate of the total value of the built environment, one can then derive the cost to rebuild.

For calculating the effect of the shaking on buildings, the latter are divided into five classes, each with an assumed fragility curve. The fragility curve specifies the probability that a building sustains a given damage state (e. g. collapse) as a function of the shaking intensity. Yet another set of equations connects the probability of a given damage level to the probability that an injury or fatality results. Summing the results over all the building classes gives then the most likely percentage of the population that will be injured and killed, with formal error limits.

Additional errors are introduced through inaccuracies in depth, coordinates and magnitude of earthquakes. In order to estimate the variability of the loss calculations due to earthquake parameter errors, we evaluate the losses for several hundred combinations of positions and magnitude within the error limits of the parameters. As a result, the loss estimates often range over an order of magnitude.

The hour of day is also taken into account. At night, most people are at home and at risk, whereas many of them are out of doors during daytime, especially during the early morning hours.

These calculations are performed by QUAKELOSS, a computer code and database that runs under the operating system of windows and developed by personnel of the Extreme Situations Research Center, Moscow. This program has grown out of an earlier program and data base EXTREMUM (Larionov et al., 2000; Larionov, 1999a; Larionov, 1999b; Shakhramanjyan et al., 2000; Shakhramanjyan et al., 2001; Shojgu et al., 1992).

3 Performance Test

The performance of the computer code QUAKELOSS was tested by comparing the number of fatalities reported, with that calculated (Wyss, 2004) for historic earthquakes. This test was based on a data set composed of events for which users expressed an interest. It included earthquakes that had caused losses and were part of the calibration data set, as well as events that had not caused losses, but had attracted the interest of rescue agencies. The total number of events used was 544.

The number of fatalities had to be used as the key test-parameter because the observed values are published for many events. Only rarely are numbers of injured, or dollar losses, listed.

In this performance test, it became evident that the success rate was a function of the size of the disaster. Cases of insignificant losses were correctly identified with a 92% success rate. The success rate for intermediate sized disaster (200 to 1000 fatalities) was lowest, 58%. Disasters exceeding 1000 fatalities were correctly identified in 71% of the cases (Wyss, 2004).

4 Real-Time Predictions

Real-time predictions of losses were started in October 2002, in collaboration with the Swiss Seismological Service (SSS). At first, loss estimates were distributed to a few people only, but later the distribution was formalized. The chief recipient was the Humanitarian Aid and Swiss Humanitarian Aid Unit (SKH), which used the loss estimates, together with other information, for their decisions concerning offering aid in case of significant disasters. In addition, the members of WAPMERR's advisory board and other interested individuals were informed.

In the procedure worked out with the SSS, alerts are transmitted automatically from the SSS to WAPMERR, when an earthquake above a threshold magnitude occurs. This minimum magnitude varies as a function of position on the globe (Europe: lowest, Pacific Ocean: highest) and reflects the needs of the SKH. WAPMERR then scans all sources available on the internet for earthquake parameters. The first automatic solution becomes available typically within about 0.5 hours, but is not accurate enough to allow loss calculations. Thanks to its worldwide network of seismographs, the US Geological Survey (USGS) is able to announce reliable solutions within about two hours. These solutions have an epicenter error of typically 20 km, and furnish magnitudes derived from surface waves or moment magnitudes. These improvements over the inaccurate automatic solutions that are distributed earlier are essential, especially because early magnitudes are based on body waves, which are unreliable for large events. Once the USGS parameters are available, WAPMERR calculates losses and distributes these estimates by email within 10 minutes. When serious disasters are suspected, the possible errors and other sources of information are discussed by telephone between WAPMERR, SKH and SSS.

The 60 real-time loss estimates carried out to date are listed in Table 1, giving the minimum and maximum of the formal error calculations. In addition to the source parameters of the earthquakes in question, the last column gives the time the loss estimate was sent by email. A few of the messages were sent with many hours delay (emboldened in the table) because they were not initiated by WAPMERR, but were made in response to a delayed request for evaluation by an agency or an individual. The hours emboldened in Table 1 are not used to calculate the average response time between earthquake occurrence and email distribution, which is two hours.

The rules for judging success or failure of the estimates are based on the needs of the rescue agencies (Wyss, 2004). Successes in Figure 1 are those for which the reported number of deaths lies between the minimum and maximum estimate, or within 100 from these limits. Acceptable estimates are defined as within 200 fatalities, or within a factor of two from a minimum or maximum. In two cases, the number of fatalities is not known.

Table 1: Loss Estimates by QUAKELOSS in Real Time

No	Date	H	Min	Lon.	Lat.	Z	M	Deaths	Deaths	Deaths	Location	Delay
				Deg	deg	km		obsvd	est min	est max		hours
1	2002 10 31	10	2	14.9	41.8	10	5.7	29	13	42	Italy	
2	2002 11 02	1	26	96.2	3	20	7.5	30	16	55	Sumatra	
3	2002 11 20	21	32	74.52	35.4	20	6	19	7	12	Kashmir	
4	2003 01 22	2	6	104.1	18.5	20	7.3	29	42	130	Mexico	
5	2003 05 21	18	44	3.78	36.9	10	6.7	2217	1,690	3,660	Algeria	
6	2003 05 26	9	24	141.5	38.9	53	6.9	0	0	30	Japan	
7	2003 07 06	19	10	26.1	40.4	10	5.7	0	0	0	Turkey	
8	2003 07 10	17	6	54.2	28.3	10	5.7	1	6	19	Iran	
9	2003 07 21	15	16	101.2	26	14	6	16	95	260	China	
10	2003 07 25	22	13	141	38.4	15	6.1	0	29	100	Japan	
11	2003 07 26	23	18	92.31	22.9	10	5.6	2	10	28	Banglad.	
12	2003 08 14	5	14	20.74	39.2	10	6.1	0	70	230	Greece	
13	2003 08 21	12	12	167.2	45.1	20	7.4	0	0	0	N. Zeal.	
14	2003 09 01	23	16	75.21	38.6	10	5.8	0	0	1	China	
15	2003 09 21	18	16	95.72	19.9	10	6.7	?	80	200	Myanmar	
16	2003 09 22	4	45	-70.67	19.7	20	6.5	3	130	350	Dominic.	
17	2003 09 25	19	50	143.8	41.9	33	8	0	29	160	Hokkaido	
18	2003 09 25	21	8	143.5	41.8	33	7	0	0	6	Hokkaido	
19	2003 09 27	11	33	87.73	50.1	18	7.5	?	540	1,230	S. Sibiria	3.0

20	2003 10 09	22	19	120	13.8	16	6.1	0	0	10	Mindoro	12.8
21	2003 11 05	0	58	-77.75	4.97	20	5.8	0	0	0	Colombia	6.3
22	2003 11 09	19	23	127.3	1.6	15	6.1	0	0	0	Halmahera	2.5
23	2003 11 12	8	26	137	33.6	391	6.4	0	0	0	Japan	0.8
24	2003 11 14	18	49	141	36.4	33	5.5	0	0	0	Honshu	0.6
25	2003 11 18	17	14	125.4	12	37	6.5	1	0	1	Philippines	0.7
26	2003 12 01	1	38	80.65	42.9	10	5.7	11	2	7	China	1.1
27	2003 12 03	14	11	144.6	42.5	23	5.7	0	0	0	Hokkaido	1.4
28	2003 12 10	4	38	121.4	23.1	23	6.6	0	130	330	Taiwan	1.0
29	2003 12 10	15	51	120.7	17.8	10	5.8	0	15	48	Luzon	2.0
30	2003 12 19	0	11	95.74	19.9	10	5.7	?	2	5	Myanmar	8.6
31	2003 12 22	19	15	-121.1	35.71	7	6.5	2	0	3	California	0.6
32	2003 12 25	7	11	-82.82	8.43	25	6.3	2	11	45	Panama	1.3
33	2003 12 26	1	57	58.27	29.1	20	6.3	26,271	410	1010	Iran	1.4
33a	2003 12 26	1	57	58.35	29.1	10	6.7	26,271	6090	11,810	Iran	9.1
34	2003 12 28	1	32	144.6	42.6	10	5.6	0	0	0	Hokkaido	0.9
35	2004 01 01	23	32	-101.17	17.57	14	6	0	0	10	Oaxaca	10.1
36	2004 01 09	22	35	149.36	-6	34	6.3	0	0	0	Papua	8.9
37	2004 01 14	16	58	52.3	27.7	10	5.4	0	0	100	Iran	1.0
38	2004 01 28	9	6	57.8	27.1	33	5.8	0	0	1000	Iran	1.2
39	2004 02 04	11	59	82.95	8.42	29	5.8	0	0	100	Panama	2.8
40	2004 02 05	21	5	135.52	-3.6	10	6.8	37	0	8	Indones.	9.4
41	2004 02 07	2	41	134.9	-3.8	10	7.3	0	0	10	Indones.	1.8

42	2004 02 22	6	46	100.4	-1.9	10	6.5	0	0	0	Sumatra	1.0
43	2004 02 24	2	27	4	35.2	10	6.2	628	0	2700	Morocco	0.7
44	2004 03 02	3	47	86.91	11.67	99	5.5	0	0	0	Nicarag.	2.2
45	2004 03 07	13	29	91.21	31.57	10	5.5	0	0	50	China	4.1
46	2004 03 17	3	21	-65.57	21.13	288	6.1	0	0	0	Bolivia	0.8
47	2004 03 17	5	21	23.54	34.65	40	5.7	0	0	0	Crete	1.5
48	2004 03 24	1	54	118.2	48.2	10	5.7	0	0	12	China	1.4
49	2004 03 25	19	31	40.9	39.9	10	5.5	10	0	12	Turkey	1.9
50	2004 03 26	15	20	144.09	41.91	39	5.5	0	0	0	Hokkaido	1.4
51	2004 03 27	18	47	89.25	34.03	10	5.8	0	0	0	Tibet	2.3
52	2004 04 05	21	24	71.04	36.47	196	6.6	3	0	0	Hindu Ku.	1.6
53	2004 04 11	18	8	144.82	42.91	40	6.1	0	0	0	Hokkaido	2.6
54	2004 04 23	1	50	122.82	-9.44	78	6.4	0	0	0	Savu Sea	1.6
55	2004 04 29	0	57	-85.95	10.7	21	6.1	0	0	1	Costa Ric.	3.1
56	2004 05 03	4	36	73.2	-37.6	25	6.6	0	170	450	Chile	1.5
57	2004 05 05	13	40	14.9	38.7	249	5.7	0	0	13	Sicily	0.6
59	2004 05 08	8	3	121.5	22	12	5.7	0	0	0	Taiwan	1.8
60	2004 05 28	12	38	51.57	36.27	22	6.2	35	250	680	Iran	1.8

Table 1: Minimum and maximum estimated numbers of deaths compared to reported numbers (from the USGS list of significant earthquakes). The parameters of the earthquakes are those available at the time of the real-time calculations, not the ultimately correct values. Last column gives the time difference between the earthquake occurrence and the posting of the loss estimate by email.

The correct plus acceptable results sum up to 93%. However, this result is dominated by cases with zero, or very few, fatalities. The five outstanding disasters during the study period were: (5) 21 May 2003, M6.7, Algeria; (19) 27 September 2003, M7.5, Southern Siberia; (33) 26 December 2003, M6.7, Iran; (43) 24 February 2004, M6.2, Morocco; (60) 28 May 2004, M6.2, Iran (numbers in parentheses refer to Table 1). For (19) the number of deaths is not accurately known, but the relief effort of the Russian federation was massive, thus the alarm issued in this case is scored, together with cases (5) and (43), as a success. These three successes are somewhat offset by the two Iranian cases, numbers (33) and (6), which are classified as failures.

The number of casualties in the Bam earthquake (33) was at first underestimated because the available hypocenter was at a greater distance and greater depth from the city of Bam than the true energy release of this earthquake (Wyss et al., 2004). However, Bam was identified immediately as the crucial city. SKH was informed that it was important to find out what had happened in Bam, and that a major disaster could have happened there, although the calculation did not reflect this. Thus, this case was a success, in practice, but a failure according to our formal definitions.

The number of casualties in the second Iranian earthquake (60) was overestimated because the hypocentral depths broadcast and assumed (22 km) were shallower than the depth ultimately calculated (29 km). Also, in real-time, the possibility that the depth was as shallow as in the Bam event (4 km) had to be considered. Thus, the two Iranian cases are failures in the opposite sense. The first of these (Bam) was not a failure from the point of view of rescue agencies, because WAPMERR recommended mobilization of the rescue team. However, in the second Iranian earthquake (number 60) WAPMERR also recommended mobilization, unnecessarily, as it turned out.

5 Discussion

Predicting the number of casualties in real-time after an earthquake is not easy. Nevertheless, one should attempt to make these estimates on a quantitative basis. Because some of the error sources cannot be excluded, sometimes the calculated real-time loss estimates are not much better than expert opinions of specialists familiar with the earthquake source region. However, the over all score, 93% acceptable estimates, is excellent. Because this applies mostly to small earthquakes, this chiefly benefits rescue agencies, which do not have to mobilize unnecessarily.

For major disasters, the acceptable loss estimates still outnumber the incorrect ones, in real-time as well as in performance testing after the fact (Wyss, 2004), but errors do occur and cannot be avoided, unless more accurate information on all necessary input parameters (earthquake source, transmission path, built environment) become available.

The ways in which real-time loss estimates can be improved include the following. (1) Deepen the database on population and building stock. (2) Refine the earthquake source model used for strong motion calculations to allow extended sources of a length appropriate for the announced moment magnitude. (3) Reduce the uncertainty in hypocentral depth estimates by a catalog of regional most likely depths for the entire globe. (4) Educate local disaster managers on the benefits and limitations of international loss estimates in real time.

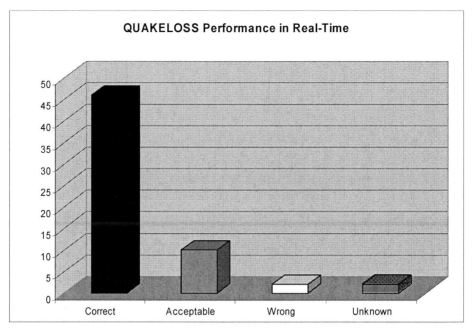

Figure 1: Number of real-time loss estimates as a function of performance: correct, acceptable, wrong and unknown, as defined in the text.

This latter point is important because local managers, along with news media, regularly insist for hours and days that no disaster, or only a minor one, had occurred, when in reality the real-time loss estimate was correct in predicting one or two orders of magnitude more casualties. This means that the best efforts to work toward efficient rescue efforts are of no use to the injured trapped beneath rubble, if their regional disaster managers do not accept help, because they do not recognize the value of international real-time loss estimates.

It is concluded that the real-time loss estimate technique used in QUAKELOSS has matured to the point that it can be used worldwide. Ways in which uncertainties can be reduced and usefulness of real-time loss estimates can be increased have been identified and should be implemented.

Acknowledgements

This work was in part supported by the Humanitarian Aid and Swiss Humanitarian Aid Unit. Thanks go to the S. Suschev, V. Larionov, and N. Frolova, staff members of the Extreme Situation Research Center, Moscow, who supplied WAPMERR with the computer code and data base QUAKELOSS.

References

Larionov, V., N. Frolova, and A. Ugarov: Approaches to vulnerability evaluation and their application for operative forecast of earthquake consequences, in *All-Russian conference "Risk-2000"*. edited by A. Ragozin, pp. 132-135, ANKIL, Moscow, 2000.

Larionov, V. I.: Basis of Theory of Efficiency. PC Application for Solving the Tasks of Civil Defence and Emergency Response, 277- 428, in Russian pp., Military Engineering University, Moscow, 1999a.

Larionov, V. I.: Forecast of Emergencies, Mechanics of Destruction, in *Theoretical basis of response to emergency situations*, pp. 1-276, in Russian, Military Engineering University, Moscow, 1999b.

Shakhramanjyan, M. A., G. M. Nigmetov, V. I. Larionov, N. I. Frolova, S. P. Suchshev, and A. N. Ugarov: Seismic and complex risk assessment and management for the Kamchatka region, in *XII World Conference on Earthquake Engineering*, Auckland, New Zealand, 2000.

Shakhramanjyan, M. A., G. M. Nigmetov, V. I. Larionov, A. V. Nikolaev, N. I. Frolova, S. P. Sushchev, and A. N. Ugarov: Advanced procedures for risk assessment and management in Russia. *International Journal of Risk Assessment and Management*, 2 (3/4), 303-318, 2001.

Shojgu, S. K., M. A. Shakhramanjyan, G. L. Koff, E. T. Kenzhebaev, V. I. Larionov, and G. M. Nigmetov: Seismic risk analysis, population rescue and life support during catastrophic earthquakes (seismological, methodological and systematic aspects), 295, in Russian pp., State Committee of Russian Federation on Civil Defense and Emergency Situations, Moscow, 1992.

Wyss, M.: Estimates of losses due to earthquakes worldwide. *Bull. Seismol. Soc. Am.*, 94, in press, 2004.

Wyss, M., R. Wang, J. Zschau, and Y. Xia: Instability of Loss Estimates After Earthquakes: Bam, 26 December 2003. *Bull. Seismol. Soc. Am.*, submitted, 2004.

Near Real-Time Estimation of Ground Shaking for Earthquake Early Warning in Istanbul

Maren Böse[1], Mustafa Erdik[2] and Friedemann Wenzel[1]

[1] *Institute of Geophysics, Karlsruhe University (TH), D-76187 Karlsruhe, Germany; maren.boese@gpi.uni-karlsruhe.de*
[2] *Kandilli Observatory and Earthquake Engineering, Bogazici University, 81220 Cengelkoy, Istanbul, Turkey*

Abstract

Earthquake early warning systems (EWS) —either embedded in real-time earthquake information systems (see Wenzel *et al.*, in these proceedings) or operated as stand-alone—can help to significantly reduce losses due to major earthquakes. Advanced earthquake early warning systems allow a prognosis of ground shaking within an endangered area a short period of time before the seismic waves arrive. If the shaking is classified as critical, automatisms are triggered to minimize damage. Short warning-times in case of earthquake catastrophes prevent the determination of magnitude and source location using standard methods. Therewith, the applicability of conventional attenuation laws in EWS is mostly impossible. Using the example of Istanbul (Turkey) we utilize information that is obtained in real-time from the low amplitude compression waves (P-waves) recorded by a network of seismic sensors in order to make estimates of high amplitude shear and surface waves. Multi-layer feed-forward neural networks are suitable for mapping input information onto ground motion parameters such as horizontal peak ground acceleration (PGA). The consideration of only two seismic stations can already give good estimations of ground shaking in Istanbul and its vicinity before the seismic waves arrive.

1 Earthquake Early Warning

Key to earthquake early warning is a continuous communication link between seismic sensors, ideally deployed close to the seismic source, and a central processing facility. If the incoming ground motion data at the processing unit is classified as critical, the system issues a warning to endangered people or facilities. Thereby one takes advantage of the fact that seismic waves propagate slower (3-6 km/s) than electromagnetic waves that are used for transmission of the warning. Generally, the warning time is in the range of a few up to 70 seconds, which certainly depends on the distances between the seismic source, the recording instruments, and the endangered area or facility.

The response to an alert given by the earthquake early warning system generally proceeds fully automatically:

- High-speed trains and metros are slowed down or stopped to avoid derailments.
- Pipelines and gas lines are shutdown to minimize fire hazards.
- Manufacturing operations are shutdown to inhibit likely damage to the equipment.
- Vital computer information is saved and disk heads are retracted away from the disk surface.

Further examples are, e.g., given by Goltz (2002) and Harben (1991). The benefit of audio and visual alarms to people is subject to controversial discussions, since it requires proper training of people and reasonable long warning times. At the time being, only Mexico City has experience with public earthquake early warning (Espinosa Aranda et al., 1995).

Configurations in which seismic sensors (accelerometers) are installed between a known source area and the user are called *front-detection* systems. However, if the location of the seismic source is unknown in advance, the realization of the early warning system requires the installation of an expanded network of seismic instruments. Typical examples for *front-detection* systems are installed in Mexico (Espinosa Aranda et al., 1995) and Romania (Wenzel et al., 1999), while large and complex fault systems like the San Andreas Fault zone in California demand for *complex-detection* systems.

Important progress in data transfers and real-time data processing in the last years lift previous restrictions of the realization of early warning systems. However, there are still unsolved problems concerning, e.g., the reliability of alerts and the question of liabilities of the institution that issues the warning (Goltz, 2002).

While the basic idea of earthquake early warning is now about 135 years old (Cooper, 1868), it took nearly one century until the first system was realized. In the mid 1950s the Japanese National Railways started the installation of simple alarm seismometers–strictly speaking this was no EWS–along their railway lines. Later, with the operation of the "bullet" trains Tokaido Shinkansen (1964) and Tohoku Shinkansen (1982), this front alarm system was improved and alarm seismometers have been installed every 20 km along the lines. The stations issue a warning whenever a preset level of horizontal ground acceleration (40 gal) is exceeded (Nakamura, 1989). Trains in the vicinity to the respective alarm station are automatically slowed down or stopped in order to avoid derailments. The contemporary Japanese Urgent Earthquake Detection and Alarm System (UrEDAS) consists of 30 strong motion instruments capable of determining the area of sustain damage using information about magnitude and hypocenter obtained from P-wave observations ("P-wave alarm"). Ordinary alarm seismometers and UrEDAS are combined into the Japanese "Compact UrEDAS".

The most prominent earthquake early warning system is the Mexican Seismic Alert System (SAS), since it is the only public alarm system in the world (Espinosa Aranda et al., 1995). The warning time of 60-70 seconds before the high amplitude shear and surface waves arrive at the capital Mexico City in a distance of about 320 km from the Guerrero subduction zone provides an almost perfect background for earthquake early warning. More recently, earthquake early warning systems have been developed for Taiwan (Wu and Teng, 2002), California (Allen and Kanamori, 2003), Romania (Wenzel et al., 1999) and Turkey (Erdik et al., 2003b).

2 Seismic Hazard in Istanbul

Seismic hazard for the Turkish megacity Istanbul is extremely high: the Main Marmara Fault, which is the western continuation of the North Anatolian Fault zone (NAFZ), runs in close vicinity to the city through the Sea of Marmara. The favorable geographical position between Europe and Asia and the advantageous routes of transport have led to a high concentration of industrial facilities in and around Istanbul. Today, Istanbul has more than 10 million inhabitants. Disobedience of building regulations as well as a high number of informal settlements aggravate the cities's vulnerability to a possible earthquake catastrophe.

A historic earthquake catalogue for the past 2000 years (Ambrayses, 2002) reveals a statistical recurrence of one destructive earthquake hitting Istanbul each century. Erdik *et al.* (2003a) estimate that a major earthquake of $M_w7.5$ in the Marmara region might cost 40-50.000 lives and cause economic losses of more than US$ 11 billions. Observations of past earthquakes in the region show a westward migration of epicenters along the NAFZ. The enormous seismic hazard of the NAFZ was felt in northwestern Turkey and partly also in Istanbul when the two strong earthquakes in 1999 in Kocaeli ($M_w7.4$) and Duzce ($M_w7.1$) occurred. Approximately 1000 people in the Istanbul suburb Avcilar were killed and damage of buildings and structures was rather serious, though the epicenter of the Kocaeli earthquake was more than 110 km away (Özel *et al.*, 2002).

These events have raised the public awareness of seismic risk within the Marmara region and Istanbul in particular (e.g., Hubert-Ferrari *et al.*, 2000; Parsons *et al.*, 2000). In cooperation with governmental institutions the Kandilli Observatory of the Bogazici University in Istanbul has designed and installed the earthquake information system IERREWS (Istanbul Earthquake Rapid Response and Early Warning System). Besides the earthquake early warning component (Fig. 1), IERREWS comprises a so-called rapid response system, whose purpose is the rapid processing and broadcasting of information on the distribution of ground motion parameters relevant to structural damage (e.g., horizontal peak ground acceleration (PGA), seismic intensity or spectral displacement). This *shake map* is supplemented by a tool for damage projection (Erdik *et al.*, 2003b).

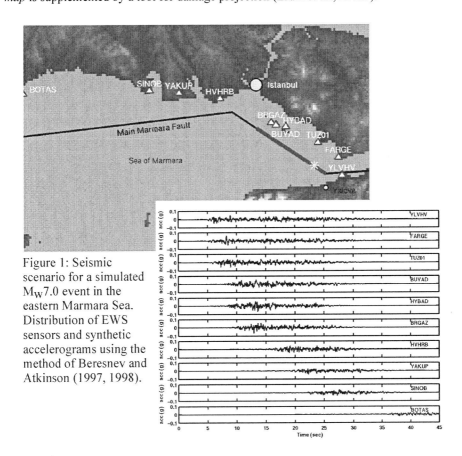

Figure 1: Seismic scenario for a simulated $M_W7.0$ event in the eastern Marmara Sea. Distribution of EWS sensors and synthetic accelerograms using the method of Beresnev and Atkinson (1997, 1998).

The earthquake early warning system itself consists of 10 tri-axial accelerometers, which are installed along the coast of the Marmara Sea, close to the Main Marmara Fault (Fig. 1). A radio link guarantees a continuous real-time communication between the accelerometers and a central processing facility at the Kandilli Observatory in Istanbul. The present system is based on thresholds of ground acceleration, whose exceedance leads to an alarm. There are three distinct alarm levels, associated with thresholds of 0.02g, 0.05g, and 0.1g. Since these thresholds can be exceeded by the low amplitude P-waves at the seismic stations, the maximum ground acceleration in Istanbul might be significantly higher than the threshold values. To increase reliability confirmation of at least one further station is required before a final alert is given. The system is now in a test stage.

3 Estimation of Ground Shaking using Empirical Attenuation Laws

The level of ground motion at a given point of observation depends on the strength and radiation of the earthquake source and on the attenuation of the signal with distance. Besides, local site conditions at the point of observation play an important role. Empirical attenuation laws derived from local and regional earthquakes in the past allow a rather good estimation of ground shaking at a specified location if magnitude and source location are given. Frequently, the integration of knowledge about the respective local underground, characterized, e.g., by the average shear wave velocity of the uppermost 30 meters, can significantly improve these estimations.

A notable number of attenuation laws for different tectonic settings has been published over the last decades. Normally, these laws have in common that nonlinear relations between input parameters (magnitude, distance, local site condition, source mechanism) and wanted ground motion parameter are considered. These nonlinearities are usually accounted for during the regression analysis in an iterative procedure by Taylor-expansion considering only the linear term (e.g., Joyner and Boore, 1993). An alternative approach is the gradient descent method, in which the error is minimized following the negative gradient of the error function.

Artificial neural networks (ANN) are able to approximate any linear or nonlinear relationship with a reasonable degree of uncertainty (e.g., Zell, 1994). In this work we make use of so-called multi-layer feed-forward networks, which are trained by means of the backpropagation-learning algorithm. Backpropagation is a realization of the gradient descent method mentioned above, in which the free parameters of the neural net (*weights*) are determined using a set of training examples of the respective relation. The error between target and network output is iteratively decreased by small changes of the weights according to fractions of the negative gradient of the error function (*learning rate*). Generally, the error is defined by the sum of quadratic distances between target and network outputs. Once the neural network is trained, new data, i.e., data that were not within the training dataset, can be processed successfully. Thereby, the ANN is in most cases capable to interpolate, while extrapolation works only to a small extent.

Using the example of a nonlinear attenuation relationship for PGA, Fig. 2 demonstrates the ability of neural networks to nonlinear mapping. However, due to the black-box character of ANNs, it is generally advisable to make use of known functional dependencies based on physical background knowledge if given. On the other hand, in many cases relationships are complex and cannot be formulated in form of simple mapping functions, particularly if they are nonlinear. ANNs can give a suitable strategy for solving such problems.

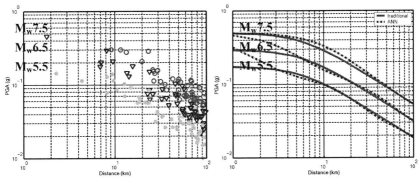

Figure 2: Comparison of regression results for synthetic PGA values using standard regression techniques (Joyner and Boore, 1993; solid line) and backpropagation for artificial neural networks (dashed line) using data shown in the figure on the left ($\sigma_{\ln(\text{PGA})}$=0.25).

4 Artificial Neural Networks for Estimation of Ground Shaking from P-wave Observations

Short pre-warning times of a few up to about 20 seconds for Istanbul prevents the exact localization of earthquake sources and the determination of magnitudes for earthquake early warning using conventional techniques. Therefore, attenuation laws based on these very parameters as described in the previous section cannot be applied within the IERREWS. However, as the distribution of ground shaking is important for emergency response to major earthquakes, we make use of the faster P-wave recorded at the EWS stations in order to derive information that allow a rough estimation of ground shaking in Istanbul and neighboring areas before the seismic waves arrive. With ongoing time, these prognoses are automatically updated with more data available from the stations.

Because of the limited database of moderate and strong earthquake records for the Marmara region, synthetic time series of ground acceleration are used for the following analyses (stochastic point source method expanded to a finite source after Beresnev and Atkinson, 1997, 1998). The adaption of modeling parameters to the tectonic and geophysical situation within the Marmara region rely on empirical attenuation laws for shallow earthquakes in active tectonic regions (Boore *et al.*, 1997) and on specific local knowledge and estimations of parameters obtained from literature (Q, density, seismic velocity, fmax, ...). An example for ground motion modeling is shown in Fig. 1.

Using artificial neural networks we realize a mapping of input parameters extracted from the (synthetic) P-waves recorded at the EWS stations of IERREWS onto the horizontal peak ground acceleration (PGA) for different locations in Istanbul and its vicinity. We distinguish between event-specific information and site-specific information for the ANN input. Event-specific information is information about a detected earthquake that is obtained from the EWS stations. From the time a station is triggered by the arrival of the P-wave, each station starts the computation of the cumulative absolute velocity (CAV) (Benjamin *et al.*, 1988). From the time differences of triggering of different seismic stations the location of the earthquake epicenter can be approximated using the hyperbola method proposed by Mohorovicic (1915). If less than three stations are involved in the prediction process, which is the minimum number for the hyperbola method, it is assumed, that the epicenter is in a certain area around the firstly triggered station (Voronoi cell). In general, the uncertainties of these estimates do not pose a problem to the neural networks since the information is supplemented by the additional CAV values obtained from the different stations (*strong motion centroid* after Kanamori, 1993).

Site-specific information involves (a priori) knowledge on the specific site where PGA is to be estimated. That is, e.g., evidence of soil class, shortest distance to the fault system, and distance and azimuth to the estimated epicenter.

5 Artificial Neural Networks for Earthquake Early Warning

For the following example we simulate a dataset of 70 records of earthquakes at different locations, of variable magnitudes ($3.5 \leq M_w \leq 7.5$) and varying directions of rupture propagation (unidirectional and bidirectional cases), whereby the ruptures run parallel to the Main Marmara Fault. The size of the rupturing area is determined from empirical relations as a function of magnitude. For the consideration of amplifications by local soil deposits we utilize a map of soil classification for the Istanbul area (Erdik *et al.*, 2004). The target map– that is, the map of PGA values–is defined by the application of empirical attenuation laws after Boore *et al.* (1997). After the training the ANNs are able to estimate the PGA distribution for earthquakes not within the training dataset.

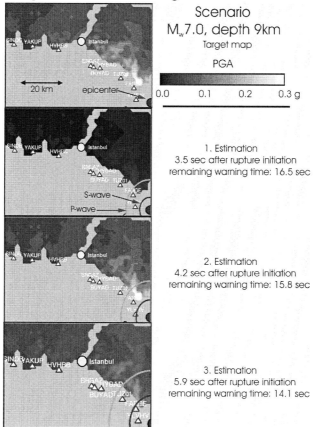

Figure 3: Scenario earthquake in the Marmara region ($M_w7.0$, 9 km depth, rupture length: 30 km, rupture direction: northwest, unidirectional). Top: PGA map determined from attenuation laws after Boore *et al.* (1997) (target map) without consideration of directivity effects. Below: estimates of PGA distribution by ANNs at three time steps after rupture initiation using P-wave information.

Fig. 3 shows an example of dynamic PGA prediction in three time steps after rupture initiation for a scenario earthquake in the Izmit Bay of $M_w7.0$. The target map is shown on the top of the figure. The remaining warning time until the arrival of high amplitude shear-waves in downtown Istanbul is specified for each prognosis. The prediction error decreases clearly from the first to the second estimation considering information from one or two stations, respectively. The third estimation–which by means of the used approach is completely independent from the other two prognoses–is somewhat inferior but still satisfying. The consideration of an increasing number of stations in the procedure requires a growing set of training data due to the rise in dimensions of the ANNs.

However, due to short warning times the consideration of more than three stations seems useless for early warning. Yet it is possible to update the map by recorded PGA values obtained from the 140 dial-up stations of IERREWS at a later time (a few minutes after the catastrophe). These shake maps reflect the actual–not estimated–distribution of ground shaking and are fundamental for disaster management during and after an earthquake catastrophe.

References

Allen, R.M. and Kanamori, H.: The potential for earthquake early warning in southern California, *Science*, 300, 786-789, 2003.

Ambraseys, N.: The seismic activity of the Marmara Sea region over the last 2000 years, *Bull. Seism. Soc. Am.*, 92, 1, 1-18, 2002.

Benjamin, J.R. and Associates: *A criterion for determining exceedance of the operating basis earthquake*, EPRI-report NP-5930, Electric Power Research Institute, Palo Alto, California, 1988.

Beresnev, I.A. and Atkinson, G.M.: Modeling finite-fault radiation from the w**n spectrum, *Bull. Seism. Soc. Am.*, 87, 1, 67-84, 1997.

Beresnev, I.A. and Atkinson, G.M.: FINSIM - a FORTRAN program for simulating stochastic acceleration time histories from finite faults, *Seism. Res. Lett.*, 69,1, 27-32, 1998.

Boore, D.M., Joyner, W.B., Fumal, T.E.: Equations for estimating horizontal response spectra and peak acceleration from western north American earthquakes: A summary of recent work, *Seism. Res. Lett.*, 68, 1, 128-153, 1997.

Cooper, M.D.: Editorial in San Francisco Daily Evening Bulletin, November 3, 1868.

Erdik, M., Aydinoglu, N., Fahjan, Y., Sesetyan, K., Demircioglu, M., Siyahi, B., Durukal, E., Ozbey, C., Biro, Y., Akman, H., Yuzugullu, O.: Earthquake risk assessment for Istanbul metropolitan area, *Earthquake Engineering and Engineering Vibration*, 2, 1, 1-23, 2003a.

Erdik, M., Fahjan, Y., Ozel, O., Alcik, H., Mert, A., Gul, M.: Istanbul Earthquake Rapid Response and the Early Warning System, *Bulletin of Earthquake Engineering*, 157-163, 1, Kluwer Academic Publishers, Netherlands, 2003b.

Erdik *et al.*: *Earthquake Risk Assessment for the Istanbul Metropolitan Area*, Report prepared by B. U. Kandilli Obs. And Earthquake Research Ins., Dept. of Earthquake Engineering, Bogazici University Press, (Red Cross Project) 300 pages, 2004.

Espinosa Aranda, J.M., Jimenez, A., Ibarrola, G., Alcantar, F., Aguilar, A., Inostroza, M., Maldonado, S.: Mexico City Seismic Alert System, *Seism. Res. Lett.*, 66, 6, 42-53, 1995.

Goltz, J.D.: *Introducing earthquake early warning in California: A summary of social science and public policy issues*, Caltech Seismological Laboratory, Disaster Assistance Division, A report to OES and the Operational Areas, 2002.

Harben, P.E.: *Earthquake alert system feasibility study*, Lawrence Livermore National Laboratory, Livermore, CA, UCRL-LR-109625, 1991.

Hubert-Ferrari, A., Barka, A., Jacques, E., Nalbant S.S., Meyer, B., Armijo R., Tapponnier, P., King, G.C.P.: Seismic hazard in the Marmara Sea following the 17 August 1999 Izmit earthquake, *Nature*, 404, 269-273, 2000.

Joyner, W.B. and Boore, D.M.: Method for regression analysis of strong motion data, *Bull. Seism. Soc. Am.*, 83, 2, 469-487, 1993.

Kanamori, H.: Locating earthquakes with amplitude; application to real-time seismology, *Bull. Seism. Soc. Am.*, 83, 1, 264-268, 1993.

Mohorovicic, A.: Die Bestimmung des Epizentrums eines Erdbebens, *Gerl. Beitr. z. Geophys.*, 14, 199-205, 1915.

Nakamura, Y.: Earthquake alarm system for Japan Railways, *Japanese Railway Engineering*, 28,4, 3-7, 1989.

Özel, O., Cranswick, E., Meremonte, M., Erdik, M., Safak, E.: Site effects in Avcilar, West of Istanbul, Turkey, from Strong- and Weak-Motion data, *Bull. Seism. Soc. Am.*, 92, 1, 499-508, 2002.

Parsons, T., Toda, S., Stein, R.S., Barka, A., Dieterich, J.H.: Heightened odds of large earthquakes near Istanbul: an interaction-based probability calculation, *Science*, 288, 661-665, 2000.

Wenzel, F., Oncescu, M.C., Baur, M., Fiedrich, F.: An early warning system for Bucharest, *Seism. Res. Lett.*, 70, 2, 161-169, 1999.

Wu, Y.-M. and Teng, T.-l.: A virtual subnetwork approach to earthquake early warning, *Bull. Seism. Soc. Am.*, 92, 5, 2002.

Zell, A.: *Simulation Neuronaler Netze*, Addison-Wesley, (in German), 1994.

Monsoon Rainfall Prediction using a GCM and its Application on Farming Risk Reduction in Anas Catchment India

Anupam K. Singh[1], Erwin Zehe[2] , Andras Bardossy[3] , Franz Nestmann[1]
[1] *Institute for Water Resources Management, Hydraulics and Rural Engineering, University of Karlsruhe (TH), D- 76128 Karlsruhe, Germany;*
[2] *Institute for Geo-ecology, University of Potsdam, D-14476 Golm/ Potsdam, Germany*
[3] *Institute for Hydraulics Engineering, University of Stuttgart, D-70550 Stuttgart, Germany*

Abstract

The inter-annual and inter-seasonal variation of monsoon rainfall has a large impact on agriculture as 63% of the cultivated land is under rainfed farming which supports approximately 400 million people in India. Since 1960 the India in general and Anas catchment in particular has been affected by recurrence droughts during 1962, 1965, 1975, 1985 and 1997. Low performance of monsoon for any year in rainfed area would have a direct influence, not only on agricultural yield and gross domestic product, but may also widen regional disparities and rural out-migration. Thus, the importance of issuing a reliable monsoon rainfall predictions on the likelihood of good or bad rainfall season is indispensable for farming risk reduction. In this paper, the potential of surface based approach for rainfall prediction using data from general circulation model (GCM) has been explored using a statistical downscaling method. The prediction methodology has been applied in a highly varying seasonal semi-arid environment of Anas catchment in India. An automatic circulation pattern (CP) classification method in which daily rainfall occurrence has been conditioned on CP-type using a fuzzy-rule based process. The predicted mean monthly rainfall shows a good-fit with observed rainfall within a mean variability range of 1-15% at different time scale which may be considered to be more plausible than the raw GCM data. A good correlation between 0.85 to 0.99 for mean observed and simulated rainfall amount for various stations at varying elevations have been obtained. The rainfall downscaling model also predicted a good correlation coefficient of 0.89 for simulated and observed mean number of rain days per month. (wc262)
Keywords: monsoon prediction, statistical downscaling, risk reduction, India

1 Introduction

The Indian subcontinent is extremely sensitive to the water resources problems, drought has been a periodic phenomena for arid and semi-arid region of India which effects around 400 million people. Time of monsoon arrival, rainfall amount and distribution of rainfall has large impact on agriculture as 63% of the cultivated land falls under rainfed farming in arid and semi-arid agro-climatic zone. Indian agriculture sector contribute about 30% of the national gross domestic product (GDP) and employ 67% of its workforce during the year ending in 2000. For rainfed farming, weather has been a key determinant for the productivity of the crop grown and seasonal agricultural yields. Low performance of monsoon for any given year would have a direct influence not only for agricultural yield but may also widen regional disparities and finally rural out-migration. After observing the annual GDP statistics released by the Central Statistical Institute reports low annual growth rate (almost 4%) for several drought years during 1980-2003.

Since the agricultural yields for rainfed areas tend to fluctuate year to year, the farmers decision on whether to invest in water conservation measures, supplementary irrigation, fertilisers, quality of seeds depends upon the net benefits. Given the low levels of resources availability, farmers tend to minimise additional benefits. Thus pre-forecasting of monsoon rainfall has to be seen as one of the principal management strategy. In a research study conducted by (Siddiq, 1999) for rice growing areas, has been of the opinion that enhancement of food production will result from tailoring of weather related management strategies rather than changes in crop varieties for the complex rainfed ecologies. Thus the importance of issuing a reliable monsoon rainfall prediction on the likelihood of the good, poor or bad rainfall season and development of coupled crop model are indispensable on drought risk reduction. The scope of this research paper has been limited to the issuing reliable monsoon rainfall predictions and their possible application for farming risk reduction.

In recent years substantial progress has been made for the development of the long-range (seasonal to monthly) and short-range (weekly to daily) climatic prediction models. Indian Meteorological Department (IMD) have developed a regional scale power regression model for rainfall forecasting (Gowariker et al., 1991) based on time domain approach for various climatic regions in India. This model has been based on sixteen parameters namely SST, air temperature, geoptential heights at 500hPa, ocean SST index etc. and almost 40 years of rainfall data. Later (Rajeevan et al., 2004) used an eight and ten parameters model for rainfall forecasting and found a mean error of 7% and 6% respectively. The above meteorological models predicts rainfall at regional scale limited to just 34 regions in India. Rajeevan et al., 2004 reported that attempts to forecast monsoon rainfall over a geographical area of 2500 km^2 for ten days time scale become unsuccessful. But these regional models can neither be used appropriately for hydrological modelling studies nor for farming decision making at local scale.

The growing demand for climate scenarios for future hydrological impact assessment, farming risk studies and climate change signals have created a need for statistical downscaling methods. The statistical downscaling models have been applied in several research studies (Bardossy et al., 1992), (Wilby, 2001), (Uvo et al., 2001), (Stehlik et al., 2002) for rainfall forecasting and climate change modelling. Majority of the statistical models have been conditional stochastic models based on fuzzy rule based to canonical correlation analysis. The downscaling model used under this research study, for prediction of seasonal monsoon rainfall is based on downscaling model proposed by (Bardossy et al., 2003) and (Stehlik et al., 2002). It is postulated that the statistical model prediction could support probability based planning, and provide farming community, government department and disaster preparedness agencies time to respond.

2 Study Area and Database

Anas catchment is a head watershed of Mahi basin which falls under semi-arid agro-climatic zone in western India. The catchment covers a geographical area of 1750 km^2 having a mean altitude between 280m to 560m at downstream and upstream respectively. The daily rainfall data records for 10 stations between 1985-2000 have been collected from State Water Data Centre (SWDC) at Bhopal. The location of rainfall stations and their distribution has been given in Figure 1. The mean seasonal rainfall differ from dry to wet year between 350mm to 1300mm while 75-80% of the total rainfall pore during monsoon season (typically during June-September) from 30-40 rainfall events. More than 50% of the cultivated land dependents on the direct rainfall as rainfed farming. Almost 90% of the main workforce are engaged in agriculture and primary activities in the Anas catchment. As per the study conducted by (NCHSE, 1993) during the poor rainfall season as much as 57% of the total households out-migrate to nearby urban centres in search of job.

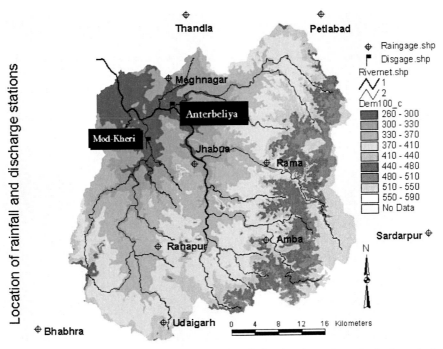

Figure 1: A comparison between mean observed and simulated number of rain days during monsoon season of 1985-94.

The large scale daily atmospheric circulation re-analysis data of mean geo-potential heights of 500hPa at 5°X5° grid size have been obtained from National Centre for Environmental Predictions (NCEP/NCAR) for a period of 1962-94. A large scale atmospheric circulation window having geographical coordinates between 05°N40°E and 35°N95°E over Indian continent has been selected for downscaling. A common dataset period both for rainfall and atmospheric circulation series between January 1985 to December 1994 have been chosen for model calibration. The model validation have been performed for station Jhabua during the period between 1962-1994. Since 80-90% of the rainfall falls during monsoon season which is significant for hydrological studies and runoff generation, only the monsoon season downscaling analysis has been carried out.

3 Results and Discussion

The rainfall downscaling results are simulated for two objective functions namely conditional rainfall probability and conditional rainfall amount which are obtained from wet and dry circulation types. Earlier the downscaling model sensitivity analysis has been assessed in terms of performance of the objective functions, a more detail description on downscaling methodology and sensitivity analysis has been given in (Singh, 2004). The statistics of monsoon season mean observed and simulated rainfall totals for various stations are given in Table 1. It is clear that mean value shows a best compromise with an error of 1.2% for Rama to 2.3% for Bhabhra and 10.6% for Udaigarh stations respectively. This shows very good model simulation performance given the stochastic nature of the rainfall. The monthly cycle of rainfall totals for observed and simulated rainfall have been found to be highly correlated. The order of Pearson-type II correlation for mean monthly observed and simulated rainfall vary between 0.85 for station Petlabad to 0.99 for stations Amba and Meghnagar.

Stations	Rainfall amount [mm]				Wet days [No.]	
	Mean 1985-94		Standard deviation			
	Obs.	Sim.	Obs.	Sim.	Obs.	Sim.
Jhabua	787.0	715.8	109.9	97.7	47.0	54.6
Ranapur	780.1	760.7	113.7	115.3	35.6	48.7
Udaigarh	817.3	729.8	98.2	99.9	52.0	53.7
Amba	852.3	873.9	129.9	120.1	39.2	55.3
Rama	934.3	922.8	127.5	110.0	39.4	58.3
Meghnagar	728.3	705.6	107.3	99.6	40.1	51.2
Thandla	881.1	844.1	116.5	115.6	48.3	55.7
Bhabhra	795.9	777.9	99.5	97.0	47.9	52.4
Sardarpur	769.2	775.4	93.2	79.5	38.7	51.9
Petlabad	1016.2	983.0	127.5	125.3	54.1	62.0
Obs. = Observed rainfall and Sim.= Simulated rainfall						

Table 1:   Statistics on observed and simulated seasonal rainfall totals for 1985-94 for various stations in Anas catchment.

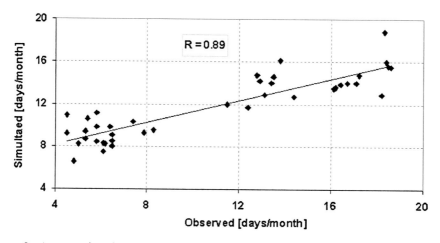

Figure 2:   A comparison between mean observed and simulated number of rain days during monsoon season of 1985-94.

The downscaling model also simulated mean number of rain days during monsoon season as good as 1day to 19days. It has been observed that model simulated more number of rain days but still a mean correlation coefficient of 0.80 for observed and simulated rain days. A comparison between observed and simulated number of rain days has been given in Figure 2 above.

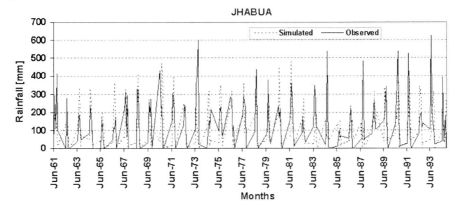

Figure 3: Observed and simulated monthly rainfall time series at station Jhabua during monsoon season of 1961-94.

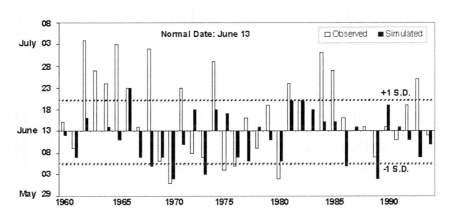

Figure 4: Dates of onset of observed and simulated monsoon over Jhabua during 1961-94.

The discrete-continuous model parameters have been obtained from daily rainfall time series of 1985-94 just limited to 10years period for predicting the long term rainfall at station Jhabua for a period between 1962-94 (refer Figure 3). The overall monthly correlation coefficient of 0.40 have been found for observed and simulated rainfall during 1962-94 period at Jhabua.

There is a fair amount of consistency for the dates of monsoon arrival of over Anas catchment which has been practical for farming community since decades. It has been found that 13th June as mean date of monsoon arrival with a standard deviation of 8days can be considered for Anas catchment as suggested by IMD. Figure 4 shows onset of monsoon over station Jhabua during 1961-94 period derived from large scale atmospheric circulation. The standard deviation (S.D.) of 8days ahead (+1) and later (-1) from mean date of arrival have been marked with dotted lines. Almost for two-third cases the rainfall downscaling model could explain the tentative dates of monsoon arrival but could not take account for some one-third cases.

4 Application for Farming Risk Reduction

As explained earlier under section 1, some of the important factors for sustainable agriculture in rainfed regions have been variations in the seasonal rainfall amount, distribution during rainfall season and types of crops. In semi-arid regions, soil moisture prior to monsoon is negligible thus crop planting is restricted to the beginning to the monsoon season after first showers. The crops may also suffer due to the late onset of monsoon as well as extended breaks during monsoon season. Thus agriculture planning is specially critical and minimising the risk of crop failure for small and marginal farmers is important. Predicting the seasonal rainfall amounts, distribution of rainfall (number of rain days) and onset of monsoon season are important for farming risk reduction.

In a study by Singh et al. 1994, on seasonal rainfall variability and agriculture yield found a correlation coefficient of 0.64 for a rainfed region in south India. The crop yield depends upon seasonal rainfall totals and also on the distribution of rainfall within the season. Thus predicting reliable seasonal rainfall amount directly affects the farm level decision making. Timing and choice of the sowing and harvest depends on reasonable prediction of onset of monsoon. However a wrong prediction can lead to a crop failure and restrict the investment in farming system operation and water conservation activities.

Rao et al. 2000 in their research have found a strong relationship between agricultural yield and time for crop sowing. Early and late crop sowing between 2-3 weeks resulted just 20% of agricultural yield. This was primarily due to the fact that probability of dry week is very low during given sowing period. If the time of sowing is so chosen that the dry spells occur at cropping stages which are less sensitive then the yield is high. Thus the rainfall and coupled crop modelling reduces the farming risk.

Use of rainfall forecasts at seasonal, monthly, weekly scale for water allocation is another field of application for crop risk reduction and agricultural planning. The farmers risk can be estimated based on expected rainfall during growing season and permitted water allocated. Hence the land planted under agriculture can be defined based on the amount of available water.

5 Conclusions

This research study demonstrates the benefit of using statistical downscaling model based on GCM for rainfall prediction in Anas catchment. The model has clearly depicted the importance of rainfall variability in addressing the farming risk reduction and benefit for the farming society. The optimum strategy in rainfed situation are associated with tailoring to the rainfall variability and crop sowing duration.

The rainfall model calibration parameters have been obtained from a daily rainfall time series just limited for 10 years period between 1985-94 which has been a typical case for developing regions where poor database exists. These model parameters have been applied for prediction of long term rainfall during 1961-94 at station Jhabua. In general, the model gave sound explanation on inter-seasonal rainfall variability and distribution, development of monsoon system and onset of monsoon. The model has been successful in predicting number of rain days per month with higher correlation coefficient. The model results have been very promising for averaging long time series, having correlation coefficients in the range of 0.85 to 0.99 for all stations. Further efforts have been directed for rainfall prediction even at smaller time scale as the case with onset of monsoon.

If the aim of rainfall downscaling model has been for agriculture planning and crop re-insurance application then model performance need to be improved. This can be done easily by using additional GCM predictor such as air moisture indices or sea surface temperature

(SST). Webester et al. 1999 in their research findings described the association of inter annual variability in the monsoon system to the ENSO and Indian Ocean Zonal Mode (IOZM). Thus the rainfall model prediction performance need to be assessed for new predictors. This direct application of this research study can be in terms of crop re-insurance. Since 1999 the General Insurance Corporation (GIC) of India has been providing re-insurance to farmers, thus rainfall downscaling model can be considered as an interesting tool for risk reduction and uncertainty assessment.

The level of uncertainty in the final prediction can be estimated from the probabilistic information contained in the GCM output. It is postulated that rainfall prediction which is one of the assessment tool will be helpful for farming risk reduction.

Acknowledgements

The lead author wish to thank State Ministry of Science and Culture (IMK), Government of Baden-Württemberg(Germany) for financial support and State Water Data Centre (SWDC), Government of Madhya Pradesh (India) for providing necessary rainfall data for Anas catchment.

References

Bardossy A., Plate E.J.: Space-time model of daily rainfall using atmospheric circulation patterns. Water Resources Research 28(5): 1247-1259, 1992.

Bardossy A., Stehlik J., Caspary H-J.: Automated objective classification of daily circulation patterns for rainfall and temperature downscaling based on optimised fuzzy-rule. Climate Research 23: 11-22, 2003.

Challinor A.J., Slingo J.M., Wheeler T.R., Craufurd P.Q., Grimes D.I.F.: Towards a combined seasonal weather and crop productivity forecasting system: Determination of the working spatial scale. Journal of Applied Meteorology 42: 175-192, 2003

Gowarikar V., Thapliyal V., Kulshrestha S.M., Mandal G.S., Sen N., Sikka D.R.: A power regression model for long range forecast of south west monsoon rainfall over India. Mausam 42: 125-130, 1991.

NCHSE: Sustainable utilisation of natural resources in Jhabua district. National Centre for Human Settlements and Environment (NCHSE) Bhopal, India, 298p, 1993.

Rajeevan M., Pai D.S., Dikshit S.K., Kelkar R.R.: IMD's new operational models for long-range forecasts of SW monsoon rainfall over India and their verification for 2003. Current Science 86(3): 422-431, 2004.

Rao K.N., Gadgil S., Rao P.R., Savithri K.: Tailoring strategies to rainfall variability- The choice of showing window. Current Science 78(10): 1216-1230, 2000.

Siddiq E.A.: Rainfall prediction for rice growing areas. In : Abrol Y.P. and Gadgil S. (Eds.), Rice in a variable climate, APC Publication Delhi, pp.107-123, 1999.

Singh A.K.: Towards decision support models for an un-gauged catchment in India. Institut für Wasserwirtschaft u- Kulturtechnik, Universität Karlsruhe (TH), Heft 225, 2004.

Singh P., Boote K.J., Rao A.Y., Iruthayraj M.R., Sheikh A.M., Hundal S.S., Narang R.S., Singh P.: Evaluation of groundnut model PNUTGRO for crop response to water availability, sowing dates and seasons. Field Crop Research 39(2-3):147-162, 1994.

Stehlik J., Bardossy A.: Multivariate stochastic downscaling model for generating daily rainfall series based on atmospheric circulation. Journal of Hydrology 256: 120-141, 2002.

Uvo C.B., Olsson J., Morita O., Jinno K., Kawamura A., Nishiyama K., Koreeda N. Nakashima T.: Statistical atmospheric downscaling for rainfall estimation in Kyushu Island Japan. Hydrology and Earth System Sciences (B) 5(2): 259-271, 2001.

Webster P.J., Moore A.M., Loschnig J.P., Leben R.R.: Coupled ocean-atmosphere dynamics in the Indian Ocean during 1997-98. Nature 401: 356-360, 1999.

Wilby R.L.: Downscaling summer rainfall in the UK from North Atlantic Ocean temperatures. Hydrology and Earth System Sciences 5(2): 245-257, 2001.

Enhancing the Operational Forecasting System of the River Rhine

Silke Rademacher, Mailin Eberle, Peter Krahe, Klaus Wilke
Federal Institute of Hydrology (Bundesanstalt für Gewässerkunde), Mainzer Tor 1, D-56068 Koblenz

Abstract

 Early flood warning and flood forecasting is the basis for efficient flood management to reduce damage and losses.

At the Flood Warning Centre (HMZ) Rhine at Mainz the forecasting system WAVOS is in operational use. The system is developed by the Federal Institute of Hydrology (BfG). WAVOS Rhine is built up by hydrodynamic models for the River Rhine from Karlsruhe/Maxau down to the German/Dutch border and for the two main tributaries, River Main and River Moselle. For the inflow it can take into account forecasts from other forecasting centres.

To increase the warning leadtimes for the River Rhine downstream of Karlsruhe, a precipitation-runoff model has been built up at the BfG to calculate the inflow of the tributaries. The model is taking into account short and middle range numerical weather predictions of the German National Meteorological Service (DWD).

In this paper first results from the test stage are presented. In addition, the potential of improving the hydrological forecasts with respect to the meteorological input data is presented including ensemble predictions.

Keywords: flood warning, flood forecasting, ensemble predictions

1 Introduction

The Flood Warning Centre (HMZ) Rhine in Mainz is a joint centre of the Federal State of Rhineland-Palatinate and the Federal Waterways and Shipping Administration. As a part of an international chain of forecasting centres (Wilke and Rademacher, 2002) it is responsible for the dissemination of flood warnings along the German reach of the Rhine downstream of Karlsruhe. The HMZ computes water level forecasts up to 36 hours ahead for the major Rhine gauges by the use of the forecasing system WAVOS Rhine. The system, developed by the the Federal Institute of Hydrology (BfG), is in operational use since 1998.

Conversely, inland navigation is negatively affected during sustained low water periods. To optimize loading-rates an operational water level forecast is highly recommended also in these cases. For this reason the BfG makes use of WAVOS Rhine for daily forecasts up to 48 hours at low flow conditions.

The International Commission for the Protection of the Rhine (ICPR) initiated the compilation of an inventory of flood reporting systems and proposals for improved flood forecasting in the Rhine basin (ICPR, 1997). On the basis of this report, the following objectives were included in the ICPR Flood Action Plan of 1998 (ICPR, 1998):

- Improving the reporting systems

- Prolonging the forecasting leadtime by 50% until the end of the year 2000 (related to 24 hours in Germany and 48 hours in The Netherlands)

- Prolonging the forecasting leadtime by 100% until the end of the year 2005

This means that the HMZ has to compute flood forecasts with a leadtime up to 48 hours till the end of 2005. Also the inland water navigation has a substantial interest in a prolongation of the forecast length.

To satisfy the demand of increasing the leadtime, the BfG has built up a precipitation-runoff (PR-) model for the Rhine catchment downstream of Rheinfelden as a further part of WAVOS Rhine. The PR-model is in a pre-operational test stage at the BfG.

2 Forecasting System WAVOS Rhine

The water level forecasting system WAVOS Rhine has a modular set-up and consists at present of four different coupled model components:

(a) hydrodynamic model of the River Rhine between the gauges Karlsruhe-Maxau and Emmerich and of the River Moselle downstream of Trier (685 km);

(b) hydrodynamic model of the River Main downstream of Trunstadt (378 km);

(c) hydrodynamic model of the River Moselle downstream of the French gauge Custines, the River Saar downstream of Fremersdorf, and the River Sauer downstream of Bollen-dorf (426 km);

(d) statistical multi-channel filter model MKF (Wilke, 1984) for the gauges Maxau/Rhine, Rockenau/ Neckar and Fremersdorf/Saar.

The core of WAVOS Rhine is (a), the hydrodynamic model of the River Rhine. The modelling approach is based on the complete Saint-Venant equations and some modelling

Figure 1: Course of the River Rhine and its tributaries with WAVOS input and forecast (cap-ital letters) gauges

extensions (Steinebach, 1998). As shown in Figure 1 there are 13 tributary streams, which are considered. For all of these tributaries forecasts of the discharge at the most downstream gauge are necessary as an input for the Rhine model. In WAVOS Rhine external forecasts from other authorities for these gauging stations can be included. If they are not available the other models (b) - (d) are used for producing the required input forecasts. For tributaries, where neither an external nor an internal WAVOS model exists, the forecasts has to be estimated by the user.

If forecasts for the inflow of tributaries should be made over longer leadtimes, PR-models must be applied. Therefore, the BfG has developed a PR-model for the whole catchment downstream Rheinfelden to obtain a consistent forecast for all tributaries, which is described in the next chapter.

Because operational flood forecasting is highly dependent upon reliable and timely data, furthermore WAVOS offers numerous interfaces for automatic data collection.

2.1 Precipitation-Runoff Model

For extending the leadtime of reliable flood forecasts, it is necessary to make use of precipitation measurements and forecasts. Based on these, discharge can be simulated by PR-modelling.

As part of a cooperation between the Dutch Rijksinstituut voor Integraal Zoetwaterbeheer en Afvalwaterbehandeling (RIZA) and BfG, an HBV PR-model (Bergström, 1995) for the River Rhine basin downstream of Basel has been built up on an hourly basis. This model, or if necessary an enhanced version, is meant to be used in the operational forecasting systems

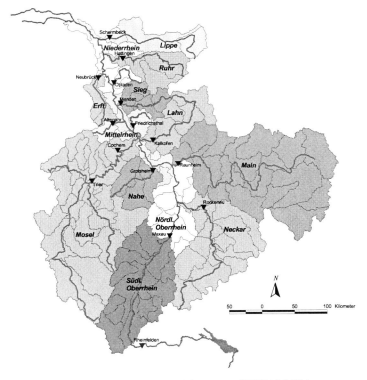

Figure 2: Spatial distribution of the HBV model as part of WAVOS Rhine

of both institutions. More details on the PR-model can be found in Eberle et al. (2001), for the application at RIZA see Sprokkereef (2002).

HBV is a conceptual semi-distributed PR-model. In this case the commercial version IHMS-HBV was applied. For modelling, the River Rhine basin downstream of Basel, with an area of more than 120,000 km^2, is divided into 117 subbasins (see Figure 2). Most of the subbasins cover between 500 and 2000 km^2. Inside the subbasins some processes are simulated separately for different elevation zones and forested and non-forested areas.

Calibration of the model parameters has been carried out manually by comparison of observed and computed hydrographs and statistical criteria. For some parameters, for example the parameter representing the maximum water storage in the soil, values are estimated from catchment characteristics. The calibration period is 1990 to 1999, for the River Moselle 1990 to 1998. One criterion, that is commonly used, to appraise the model effiency is the Nash-Sutcliffe coefficient R^2 (Nash and Sutcliffe, 1970):

$$R^2 = 1 - \frac{\sum\limits_{i=1}^{n}(Q_{sim_i} - Q_{obs_i})^2}{\sum\limits_{i=1}^{n}(Q_{obs_i} - \overline{Q_{obs_i}})^2} \tag{1}$$

where n is the number of discharge values of the selected hydrographs, Q_{sim} and Q_{obs} are the simulated and the observed values, and $\overline{Q_{obs}}$ is the average of observed values. R^2 approaches 1 for small deviations.

For the large tributaries calibration results are satisfactory. The values of the Nash-Sutcliffe coeffizient R^2 exceed 0.75, except for the River Erft where discharge dynamics are dominated by technical measures related to brown coal mining. Results tend to be best for the Rivers Ruhr, Moselle and Lahn, where R^2 is above 0.9. Concerning the River Ruhr this is rather surprising because the large reservoirs in the River Ruhr basin have not been taken into account explicitly.

Necessary input data for the model are precipitation and temperature data as well as discharge values at the input gauge Rheinfelden. In addition, water stage or discharge data are needed initially for calibration and during forecasts for adapting the simulation to the last known measured discharge values (updating).

Phase	Data	Parameter	Spatial Resolution	Time Step	Leadtime
Initialisation	Measured TTRR station data	P, T	45 stations	1 h	
	Measured Synop station data	P, T	40 stations	6 /12 h	
	Measured water level	W	15 gauges	≤ 1 h	
Forecast	DWD weather forecast Local Model (LM)	P, T	∼ 7 km	1 h	48 h
	DWD weather forecast Global Model (GME)	P, T	∼ 55 km	3 h	174 h
	BWG external forecast	W	gauge Rheinfelden	1 h	67 h

Table 1: Input data for the precipitation-runoff model in forecast mode

For the calibration period, hourly areal precipitation values for the subbasins have been calculated by combining grid based daily data and hourly station values. However, the grid data are not available operationally and can, thus, not be used in a forecasting system. The forecast input data that are interpolated to areal values for the subbasins and used in the test phase are shown in table 1.

Discharge data are provided by different state authorities, the major part of meteorological data by the German National Meteorological Service (DWD). The input forecast for Rheinfelden is prepared by the Swiss Bundesamt für Wasser und Geologie (BWG) (Bürgi, 2002).

As mentioned before, the original HBV simulations are adapted to the last known measured discharges during forecasts. At the moment this is done by an output error correction implemented by WL|Delft Hydraulics (Weerts and van der Klis, 2003) that is fitted during the initialisation phase and applied to the forecast simulation afterwards.

3 Example of the Preoperational Use

At the begin of 2004 a minor flood event took place in the River Rhine catchment. Figure 3 shows the 48h-forecasts from 8 to 17 January 2004 at the gauge Cologne calculated with WAVOS Rhine. The results based upon the above described progression of GME numerical weather forecasts as input for the PR-model and subsequent PR-model forecasts as input for the hydrodynamic model.

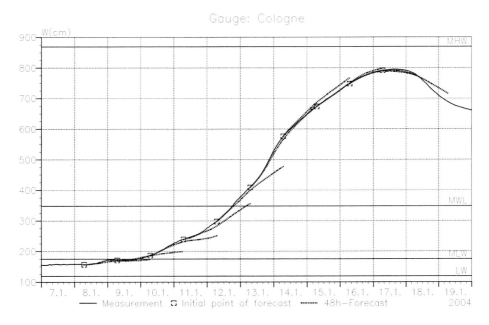

Figure 3: WAVOS results of 48h-forecasts of waterlevel at gauge Cologne 8.-17.01.2004

In this case the 48h-forecasts are surely acceptable and the claimed prolongation of the 36h-forecast currently disseminated would be possible. However, analysing other periods and further gauging stations reveals that considerable over- and underestimation occurs from time to time. Moreover the discharge simulation based on operational point measured precipitation seems to be less good than based on offline available grid based data.

Before forecast results can be released to the public operationally their quality must be analysed in more detail. Furthermore, all possible options for improving the system have to be investigated. Particularly with regard to extend the leadtime of hydrological forecast to a couple of days, i. e. use WAVOS Rhine as a flood alert system, these aspects are highly true.

4 Capability of Improvement

In general it can be stated that in order to prolong the leadtime of the hydrological forecasts hydro-meteorological data have to be considered. Reliable precipitation estimates may reduce the discharge forecast error in the case if the forecast leadtime exceeds the memory of the hydrological system.

There are many efforts e.g. the use of advanced devices for precipitation measurements such as radar on the ground (Griffith et al., 2001) together with a dense rain gauge network to estimate the temporal and spatial behaviour of rainfall fields for the observed period. The appropriate system RADOLAN (radar online adjustment, 1 km^2, hourly) of DWD (Weigl) is so far in the testing phase and its products will be implemented within the hydrological forecast system when they become operational in the near future.

Concerning forecasts, the use of deterministic meteorological models for prediction of the near future rainfall (quantitative precipitation forecasts, QPF) and other meteorological variables becomes more and more common. By application of these data in operational hydrological forecasting systems it becomes clear that the confidence of the hydrological forecasts is strongly limited to the meteorological forecasts.

To overcome this obstacle, the use of meteorological Ensemble Prediction Systems (EPS) may be a solution. Within the project EFFS (European Flood Forecasting System) the ensemble predictions of the European Centre for Medium Range Weather Forecasts (ECMWF) were applied for several river basins spread over Europe (Bates et al., 2003). For the River Rhine the above described HBV model was applied in combination with results of the ensemble prediction system of ECMWF. The basis of this system (Molteni et al., 1996) is the Atmospheric General Circulation Model (AGCM).

Even the application of coarse meteorological ensemble forecast products for flood forecasting achieved encouraging results (Bates et al., 2003). However, the appropriate use of meteorological ensemble predictions in order to produce reliable hydrological ensemble forecasts that can be used by the emergency management and water resources sectors are just in an early stage. But, together with the further development of deterministic meteorological models it is conceivable that the uncertainty which is inherent to meteorological forecasts will be reduced or at least ratable.

Cooperative research and development of a coupled weather and hydrologic ensemble forecast system is needed. These research issues form the basis for an international research programme called Hydrological Ensemble Prediction Experiment (HEPEX) which is just in the phase of initialisation . The expected progress in ensemble techniques encouraged by this programme will also affect the progress of flood forecasting for River Rhine.

It is worthwhile to mention that the hydrological forecast system itself has to be adjusted and improved permanently. The hydrodynamic models of WAVOS Rhine are presently revised, e.g. to take into account the filling of polders and to offer a better approach to transfuse the river geometry, and the PR-model is still being adapted to revised data sets of DWD.

Concerning PR-modelling, the first option for improving results that will be tested is the use of enhanced measured precipitation data or an upgraded interpolation of the data used so far. Regarding the PR-model itself, the parameterisation, the spatial structure or even the conceptual equations are potential options for improvement. This topic is investigated currently at BfG within the project FLOODMAN. The further analysis of the test phase forecasts will show for which tributaries upgrading of the PR-model is needed most.

5 Summary

In this paper we have presented the forecasting system WAVOS used at the Flood Warning Centre HMZ Rhine in Mainz. To yield the claimed extension of leadtime WAVOS was enhanced by developing a PR-model for the whole catchment downstream Rheinfelden. The advantage of one model for the entire catchment is to obtain a consistent forecast for all tributaries.

The first results of the testing phase are promising. However, there is still a major need for detailed analysis of the results and for identifying options for further improvement. But the flood forecasting is becoming more and more an essential part of effective flood management and emergency planning.

References

Bates, Paul D., Ad P.J. De Roo, Ben Gouweleeuw, Jutta Thielen, Jens Bartholmes, Paolina Bongioannini-Cerlini, Ezio Todini, Matt Horritt, Neil Hunter, Keith Beven, Florian Pappenberger, Erdmann Heise, Gdaly Rivin, Michael Hils, Anthony Hollingsworth, Bo Holst, Jaap Kwadijk, Paolo Reggiani, Marc Van Dijk, Kai Sattler, and Eric Sprokkereef. Development of a european flood forecasting system. *International Journal of River Basin Management*, 1(1), 2003.

Bergström, Sten. The HBV model. In Singh V.P., editor, *Computer Models of Watershed Hydrology*. Water Resources Publications, Highlands Ranch, Colorado, 1995.

Bürgi, Therese . Wasserstands- und Abflussvorhersagen für den Rhein. *Wasser Energie Luft*, 94(7/8), 2002.

Eberle, Mailin , Eric Sprokkereef, Klaus Wilke, and Peter Krahe. Hydrological modelling in the river rhine basin, Part II. Technical Report 1338, Bundesanstalt für Gewässerkunde, Koblenz, 2001.

FLOODMAN. http://projects.itek.norut.no/floodman/. Online in Internet, 2004/06/24.

Griffith, R.J., I.D. Cluckie, G.L. Austin, and D. Han, editors. *Radar Hydrology for Real Time Flood Forecasting*. Proceedings of an Advanced Study Course, University of Bristol, 24 June to 3 July 1998. EUR 19888. European Commission, 2001.

HEPEX. http://www.ecmwf.int/newsevents/meetings/workshops/2004/HEPEX/. Online in Internet, 2004/06/24.

ICPR (International Commission for the Protection of the Rhine). Bestandsaufnahme der Meldesysteme und Vorschläge zur Verbesserung der Hochwasservorhersage im Rheineinzugsgebiet. Koblenz, 1997.

IPCR (International Commission for the Protection of the Rhine). Aktionsplan Hochwasser. Koblenz, 1998.

Molteni, F. , R. Buizza, T.N. Palmer, and T. Petroliagis. The ECMWF ensemble prediction system: methodology and validation. *Quarterly Journal of the Royal Meteorological Society*, 122(529), 1996.

Nash, J.E. and J.V. Sutcliffe. River flow forecasting through conceptual models. Part I: a discussion of principles. *Journal of Hydology*, 10(3), April 1970.

Sprokkereef, Eric. Extension of the flood forecasting model FloRIJN - executive summary of IRMA-SPONGE project no.12. Technical report, Netherlands Centre for River Studies (NCR), Delft, 2002.

Steinebach, Gerd. Using hydrodynamic models in forecast systems for large rivers. In K.P. Holz, W. Bechteler, S. S. Y. Wang, and M. Kawahara, editors, *Advances in Hydro-Science and -Engineering*, 1998.

Weerts, Albrecht and Hanneke van der Klis. Data assimilation methods for Delft FEWS. Technical report, WL|Delft Hydraulics Research Report, Delft, 2003.

Weigl, Elmar. Online-angeeichte Radarniederschlagsprodukte als zukünftige Komponente für die Hochwasservorhersage. http://www.dwd.de/de/wir/Geschaeftsfelder /Hydrometeorologie/a_href_pages/RADOLAN/Aneichung_Radardaten.pdf. Online in Internet, 2004/06/24.

Wilke, Klaus. *Kurzfristige Wasserstands- und Abflussvorhersage am Rhein unter Anwendung ausgewählter mathematischer Verfahren*. Number 65 in DVWK-Schriften. Verlag Paul Parey, 1984.

Wilke, Klaus and Silke Rademacher. Operationelle Wasserstands- und Durchflussvorhersagen im Rheingebiet. *Österreichische Wasser- und Abfallwirtschaft*, 34(9-10), September/Oktober 2002.

Real-Time Earthquake Information Systems

Friedemann Wenzel[1], Klaus Bonjer[1], Frank Fiedrich[2], Dan Lungu[3], George Marmureanu[4] , Wolfgang Wirth[1] and Maren Böse[1]
[1]*Geophysical Institute, Karlsruhe University, Germany, E-mail: friedemann.wenzel@gpi.uni-karlsruhe.de*
[2]*Institute for Technology and Management in Construction, Karlsruhe University, Germany*
[3]*Technical University of Civil Engineering / National Institute for Building Research, Bucharest, Romania*
[4]*National Institute for Earth Physics, Bucharest, Romania*

Abstract

Shortcomings of information before, during and after strong earthquakes frequently aggravate the extent of catastrophes significantly. Real-time information systems can help to overcome this lack by providing rapid information for disaster management. In terms of a temporal hierarchy this starts with literal early warning a few seconds before the disaster strikes. Then shake maps provide near real-time information on the level of ground motion within minutes. As a next step damage estimates are provided which are based on previously developed models of vulnerability. These projections are continuously updated by observations from the field (air borne photos, ground reports, etc.). Key to real-time information systems is seismological instrumentation and real-time communication. Despite of short warning times a number of potential applications of earthquake early warning can be specified to reduce losses before the seismic waves stimulated by a distant earthquake source arrive at some critical point. Shake maps provide rapid information on ground shaking parameters in a specified area within minutes after an earthquake. The area of a shake map can be national, regional and urban. The parameters that quantify ground shaking are intensity, horizontal peak ground acceleration (PGA) and -velocity (PGV), and spectral values of acceleration at specified periods. As site effects play a critical role in the spatial distribution of ground shaking, their understanding and quantification appears to be critical for the design of a shake map. A further step consists in the projection of damage due to ground shaking. This can be done with damage estimation tools, usually requesting a database of buildings and infrastructure and associated vulnerability functions. The feasibility of a potential real-time earthquake information system is demonstrated for the Romanian capital Bucharest.

1 Introduction

In the case of earthquakes, warning times are fairly small, ranging from seconds to a maximum of about one minute. However, even this small time window can provide opportunities to automatically trigger measures, such as the shutdown of computers, the rerouting of electrical power, the shutdown of disk drives, the shutdown of high precision facilities, the shutdown of airport operations, the shutdown of manufacturing facilities, the stoppage of trains, the shutdown of high energy facilities, the shutdown of gas distribution, the alerting of hospital operating rooms, the opening of fire station doors, the starting of emergency generators, the stoppage of elevators in a safe position, the shutoff of oil pipelines, the issuing of audio alarms, the shutdown of refineries, the shutdown of nuclear power plants, the shutoff of water pipelines, and the change to a safe state in nuclear facilities (Harben, 1991). Some of these measures have been implemented or are under consideration in Japan, Mexico, Taiwan, California, Romania, Turkey and other locations. For design principles of earthquake early warning systems (EWS) see, e.g., Böse *et al.* in these proceedings.

2 Earthquake Early Warning for Bucharest

Recently, the National Institute of Earth Physics (Bucharest) and the Geophysical Institute Karlsruhe, Germany, started to design an earthquake EWS for the Romanian capital (Wenzel *et al.*, 1999). The design relies on specific seismotectonic properties of the intermediate depth Vrancea earthquakes that determine the hazard for Bucharest. These specifics include the epicentral stationarity of the hazardous earthquakes and their consistent source mechanisms. Together these features allow to design a simple, cheep, and robust system.

Within the last 60 years Romania has experienced 4 strong Vrancea earthquakes (Oncescu and Bonjer, 1997): Nov. 10, 1940 (M_w = 7.7, 160 km deep); March 4, 1977 (M_w = 7.5, 100 km deep); Aug. 30, 1986 (M_w = 7.2, 140 km deep); May 30, 1990 (M_w = 6.9, 80 km deep). The latter event was followed by a M_w = 6.3 aftershock on May 31, 1990. The 1977 event had catastrophic character with 35 high-risk buildings collapsed and 1500 casualties, the majority of them in Bucharest. The epicenters of the instrumentally well-located intermediate depth (80 - 200 km) seismicity are confined to a region of 30×70 km, with an average epicentral distance to Bucharest of about 130 km. This geometric relationship between hypocenters being confined to a small source volume and at a fixed distance to the capital allows the design of an EWS with a warning time of about 25 s for all potential intermediate deep earthquakes. A Romanian EWS would thus be similar to the Mexican case, where the site of strong earthquakes is constrained to the plate boundary at significant distance from Mexico City. For both sites a fairly constant warning time can be made available, although Bucharest can only utilize about one third of the time available to Mexico City.

The expected level of ground motion in Bucharest expressed in horizontal peak ground acceleration strongly depends on magnitude and depth of the event. As a rule of thumb, an increase of 50 km source depth can be balanced by an increase in magnitude by 0.5 units to give the same PGA in the capital. Thus, the prediction of ground motion in Bucharest from source parameters requires a fairly precise determination of both magnitude and focal depth, which in turn requires the operation of an extended seismic network, communication between stations and a central processing facility, and data being adequately processed. With regard to an EWS it becomes a complex and vulnerable system and a lot of time would be lost for data processing before a warning could be issued. If, on the other hand, only the P-wave amplitude of an epicentral station suffices as indicator for a strong earthquake, the system design becomes very simple and thus reliable, and precious time can be saved. The prediction of the level of ground motion the capital will experience must then be established on the basis of scaling relations between P-wave amplitudes at the epicenter and ground motion parameters in Bucharest.

The geophysical basis for the existence of scaling relations is found in the consistent fault plane solutions of all strong and most moderate and weak Vrancea earthquakes. All events have a very similar radiation pattern. Simple relationships empirically derived from strong and moderate motion records are proposed by Wenzel *et al.* (1999). Fig. 1 shows the example of a scaling relation between the maximum epicentral P-wave amplitude and seismic intensity in Bucharest-Magurele.

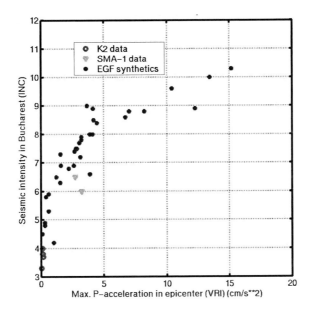

Figure 1: Scaling relationship between maximum P-wave amplitude P_{epi} (1-2sec filtered) at the epicenter and seismic intensity I in Bucharest. The best fit is obtained for $I = 6 \cdot P_{epi} ** 0.2$.

The minimum configuration for an EWS for Bucharest consists of 3 accelerometers operated in the epicentral area and located closely to each other to allow direct cable connection with the processing unit. Essentially only one instrument is required. For back-up purposes and to avoid accidental triggering, several instruments should record continuously and the processing unit should react upon coincident high amplitudes measured by a subset of the instruments (say two out of three). All instruments and the processing unit form one facility at an epicentral location. Thus access and maintenance can be handled easily and no communication between remote stations and computers is required in the decision process. It is possible to use direct UHF radio communication for the alerting message because antennas could be installed between the mountainous epicenter and Bucharest along a 'line-of-sight'. The communication with users of the warning message could occur with pagers or by other means such as mobile phone systems.

3 Rapid Information on Ground Motion – Shake Maps

Shake maps provide rapid information on ground shaking parameters in a specified area within minutes after an earthquake. The routine based generation of so-called ShakeMaps in Southern California (Wald *et al.*, 1999) is one of the most recent developments in real-time seismology. The cornerstone of shake maps is seismological instrumentation and real-time communication. The area of a shake map can be national, regional and urban. The parameters that quantify ground shaking are intensity, horizontal peak ground acceleration

(PGA) and -velocity (PGV), and spectral values of acceleration at specified periods. As site effects play a critical role in the spatial distribution of ground shaking, their understanding and quantification appears to be critical for the design of a shake map, specifically if the level of instrumentation is not very high and interpolation between observations at the station network has to be done in an 'intelligent' way. Typical hypocentral distances of the intermediate depth Vrancea earthquakes that determine the hazard for Bucharest are in the range of 250 km. Compared to such distances the spatial extent of the city is relatively small. From this source-to-site geometry we conclude that the lateral variations of ground motion in the city are mainly caused by site effects. Furthermore the fairly large hypocentral distances are responsible for the fact that the maximum ground motion that has been observed (in 1977) amounts to 0.2 g only. It is thus the potential of site specific amplification within the city and the vulnerability of the building stock that contribute predominantly to the high damage potential.

Strong earthquakes in 1977, 1986 and 1990 were recorded by a network of analogue SMA-1 and SMAC-B instruments. These stations were operated by INCERC, NIEP (National Institute of Earth Physics) and ISPH/GEOTEC (Institute for Hydroenergetic Studies and Design). Fig. 2 shows the distribution of seismic stations in Bucharest by squares and coded with three letters. Installation of a modern network of digital Kinemetrics-K2 instruments by the University of Karlsruhe and NIEP (Bonjer *et al.*, 2000) began in 1997. Fig. 2 also shows the distribution of these stations within the city by triangles and a three-letter code. In March 2000 part of the stations were relocated in order to direct the main focus of the seismic observations to the study area of engineering investigations in the center of Bucharest. In Fig. 2 active stations of the network are indicated by black, closed stations by white symbols.

In the following we suggest an approach to interpolate ground motion parameters in order to generate ground motion maps. Studies of site effects by Bonjer et al. (1999), Bonjer et al. (2002), and Wirth *et al.* (2003) indicated that ground motion in Bucharest do not vary significantly at frequencies below 2 Hz. Thus, low frequency ground motion maps can be realized through simple interpolation of the observations. The benefit of applying more sophisticated procedures, as described in the following, is restricted to higher frequencies. However, taking into account the given building stock, frequencies up to about 10 Hz are of interest to civil engineering.

The basis for the generation of PGA-maps in Bucharest is given by PGA values of an earthquake recorded by the K2-stations. PGA at SMA-1, SMAC-B and closed K2 sites can be estimated by multiplying the observation at INCERC with the mean PGA-ratios determined from previous weak and strong motion data by Wirth *et al.* (2003). To use INCERC as a reference site for the whole town is a natural choice because INCERC is the site with the most records up to now. Thus PGA-ratios with respect to INCERC can be determined with a higher redundancy than with respect to any other site. We treat both horizontal components separately, but consider only the higher of both PGA values in the map. PGA at every grid point is computed as the average of the three nearest observations.

Because intensity is directly correlated to damage, intensity-maps are much more descriptive than PGA-maps. For the determination of seismic intensity from the Fourier amplitude spectrum (FAS) we make use of a method proposed by Chernov (Chernov and Sokolov, 1983, 1988; Sokolov and Chernov, 1998) and further developed by Sokolov (Sokolov, 2002; Sokolov and Wald, 2002). FAS at SMA-1, SMAC-B and closed K2 sites can be estimated by multiplying the observation at INCERC with the mean FAS-ratios determined from previous weak and strong motion data by Wirth et al. (2003). The intensities can be derived from observed and estimated FAS using the resultant horizontal component. The interpolation method is the same as for the PGA-maps. As an example for our procedure we compute PGA- and intensity-maps for the moment magnitude 5.3 earthquake on April 28, 1999 (Bonjer et al., 2000). This event was recorded at six sites in Bucharest, including the reference site INCERC. The spatial interpolations of the recorded and computed values are shown in Fig. 3.

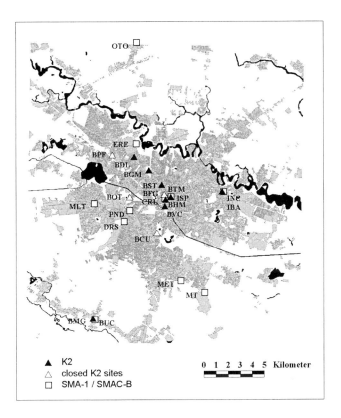

Figure 2: Distribution of strong motion accelerometers in Bucharest. Residential and industrial facilities are shaded grey, lakes and rivers black. SMA-1 and SMAC-B strong motion instruments (dark squares with three-letter code) recorded signals from all three major earthquakes: 1977, 1986, and 1990. They are operated by INCERC, NIEP (National Institute of Earth Physics) and ISPH/GEOTEC (Institute for Hydroenergetic Studies and Design). Installation of a network of digital Kinemetrics-K2 instruments by the University of Karlsruhe and NIEP began in 1997. These instruments are shown as triangles with a three-letter code. Black triangles indicate the present constellation of the network, white triangles closed sites.

For the verification of the described procedures intensity and PGA values are also estimated for sites where the earthquake was recorded. The comparison between estimates and observations show that for about half of the PGA values observation and estimate deviate about a factor of 2. Although the deviation is smaller for the other half, this indicates a high inaccuracy of PGA estimation. PGA is not systematically over- or underestimated. Observation and estimation of intensities differ by not more than 0.2 units. Wirth *et al.*

(2003) showed that the estimation of site effect amplification functions generally contains a large amount of aleatory uncertainty. Thus we emphasize the importance of having a database of direct observations as large as possible.

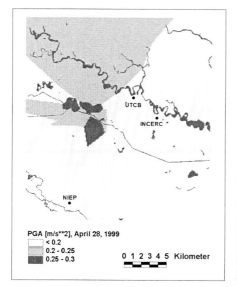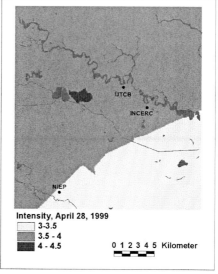

Figure 3: PGA- (left) and intensity-map (right) for the moment magnitude 5.3 earthquake on April 28, 1999. Darker colours indicate higher values of PGA or intensity respectively. The generation procedures are described in the text.

4 Damage Projections

Damage projections for realistic strong earthquakes are evolving as important tool for disaster mitigation. It enables

- to demonstrate the potential damage and loss the society has to cope with
- to assess the necessary amount of resources requested for mitigation
- to develop priorities in disaster mitigation policy on a scientifically sound basis
- to measure the efficiency of mitigation action
- to measure variations of risk with time

The anticipated damage and loss estimation tool for Bucharest consists of four modules that describe hazard, vulnerability, risk to the built environment and loss of lives (Fig. 4). The database for the modules is partly available but needs to be refined before reasonable estimates emerge.

Horizontal peak acceleration for a Vrancea earthquake specified by magnitude and depth can be estimated from attenuation relations (Lungu *et al.*, 1998) together with an appropriate microzonation of Bucharest. An alternative approach is the deterministic computation of the realistic strong ground motion for seismic hazard and microzonation analysis (Radulian *et al.*, 2000, Moldoveanu et al., 2004). Vulnerability curves will provide damage classes for different building types. The European Macroseismic Scale 1998 (Grünthal, 1998) will be used as a guide. On the basis of an inventory of the inner city of Bucharest (buildings, lifelines, bridges, etc.) damage, injuries, and losses can be estimated.

The system will function as a scenario-based tool but also in a probabilistic mode where the probable damage within a given period of time is estimated. Once available it can be utilized as an add-on to the shake map that will be rapidly established after a large earthquake so that disaster relief forces may have rapid damage and loss projections, already some minutes after the event.

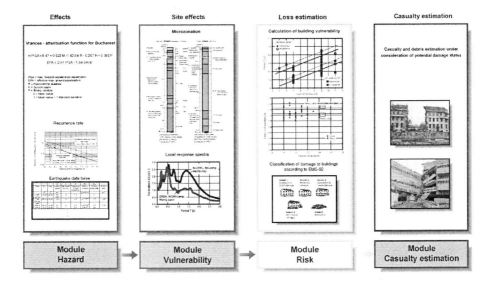

Figure 4: Scheme for modules of the damage and loss estimation tool (after Lungu and Coman, 1994 and Lungu *et al.*, 1998)

References

Bonjer, K.-P., Oncescu, M.-C., Driad, L. and Rizescu, M., 1999. A Note on Empirical Site Responses in Bucharest, Romania, In: Wenzel, F., Lungu, D. and Novak, O. (eds.), Vrancea Earthquakes: Tectonics, Hazard and Risk Mitigation, Kluwer Academic Publishers, p. 149-162

Bonjer, K.-P., Oncescu, L., Rizescu, M., Enescu, D., Driad, L., Radulian, M., Ionescu, M. and Moldoveanu, T., 2000. Source- and Site-Parameters of the April 28, 1999 intermediate depth Vrancea Earthquake: First results from the new K2-network in Romania (Abstract), In: Proceedings on the XXVII. General Assembly of the European Seismological Commision (ESC), Lisbon, p. 53

Bonjer, K.-P., Grecu, B., Rizescu, M., Radulian, M., Sokolov, V., Mandrescu, N., Lungu, D. and Moldoveanu, T., 2002. Assessment of site effects in Downtown Bucharest by recordings of ambient noise, moderate and large intermediate depth earthquakes of Vrancea focal zone (Abstract), In: XXVIII. General Assembly of the European Seismological Commision (ESC), Genova, Book of Abstracts, p. 276

Chernov, Yu.K. and Sokolov, V.Yu., 1983. Some relations between ground motion parameters and felt intensity of the earthquakes. Engineering seismology problems **24**, Moscow, Nauka Publishing House, p. 96-111

Chernov, Yu.K. and Sokolov, V.Yu. 1988. Earthquake felt intensity estimation using the strong ground motion spectra. Engineering seismology problems **29**, Moscow, Nauka Publishing House, p. 62-73, (in Russian)

Grünthal, G. (ed), 1998. European Macroseismic Scale 1998 EMS-98. European Seismological Commission, Subcommission on Engineering Seismology, Working Group Macroseismic Scales, Luxembourg. (4)

Harben, P.E., 1991. Earthquake Alert System Feasibility Study, Lawrence Livermore National Laboratory, Livermore, CA, UCRL-LR-109625

Lungu, D. and Coman, O., 1994. Experience database of Romanian facilities subjected to the last three Vrancea earthquakes. Part I: Probabilistic hazard analysis to the Vrancea earthquakes in Romania". Research Report for the International Atomic Energy Agency, Vienna, Austria, Contract No. 8223/EN. (6)

Lungu, D., Cornea, T. and Nedelcu, C., 1998. Hazard assessment and site-dependent response for Vrancea earthquakes. In: F. Wenzel, D. Lungu & O. Novak (eds.), Vrancea Earthquakes: Tectonics, Hazard and Risk Mitigation, Kluwer Academic Publishers, p. 251-267(7)

Moldoveanu, C.L., Radulian, M., Marmureanu, G. and Panza, G.F. 2004: Microzonation of Bucharest: State-of-the Art. Pure & Applied Geophysics, 161, 1125-1147.

Oncescu, M.C. and Bonjer, K.-P., 1997. A note on the depth recurrence and strain release of large Vrancea earthquakes, Tectonophysics 272, p. 291-302

Radulian, M., Vaccari, F., Mandrescu, N., Panza, G.F. and Moldoveanu, C.L., 2000. Seismic hazard of Romania: A deterministic approach, In: Seismic Hazard of the Circum-Panonnian Region. Panza, G.F., Radulian, M. and Trifu, C.-I. (eds.), Pure and Applied Geophys., 157 1-2, 221-247

Sokolov, V.Yu. and Chernov, Yu.K., 1998. On the Correlation of Seismic Intensity with Fourier Amplitude Spectra, Earthquake Spectra 14 4, p. 679-673

Sokolov, V.Yu., 2002. Seismic Intensity and Fourier Acceleration Spectra: Revised Relationship, Earthquake Spectra 18 1, p. 161-187

Sokolov, V.Yu. and Wald, D.J., 2002. Instrumental intensity distribution for the Hector Mine earthquake: a comparison of two methods. BSSA 92 6, p. 2145-2161

Wald, D.J., Quitoriano, V., Heaton, T.H., Kanamori, H., Scrivner, C.W. and Worden, C.B., 1999. TriNet ShakeMaps: Rapid generation of instrumental ground motion and intensity maps for earthquakes in southern California, Earthquake Spectra, 15, 537-556

Wenzel, F., Oncescu, M.C., Baur, M. Fiedrich, F. and Ionescu, C., 1999. An early warning system for Bucharest, Seismological Research Letters 70 2, p. 161-169

Wirth, W., Wenzel, F., Sokolov, V.Yu. and Bonjer, K.-P., 2003. A uniform approach to seismic site effect analysis in Bucharest, Romania, Soil Dynamics and Earthquake Engineering, 23, 737-758.

EQSIM: A New Damage Estimation Tool for Disaster Preparedness and Response

Frank Fiedrich[1], Johannes Leebmann[2], Michael Markus[1], Christine Schweier[1]
[1] Institute for Technology and Management in Construction, University of Karlsruhe (TH), D- 76128 Karlsruhe, Germany;
[2] Institute for Photogrammetry and Remote Sensing, University of Karlsruhe (TH), D-76128 Karlsruhe, Germany

Abstract

Fast and reliable damage estimation is essential for both disaster preparedness and disaster response. So far most available systems are stand alone tools which are solely applied in just one of these fields. In this paper we present the concept of a new version of EQSIM - a GIS-based damage estimation tool for Bucharest. EQSIM runs in distributed computer networks and is integrated into a disaster management tool (DMT), which is presented in this conference as well. EQSIM has three major components: (1) a calculation component which implements a state of the art damage estimation methodology for building damages and human losses, (2) a GIS-based client program, which can be used for scenario definition and visualization of the results and (3) a central internet database to store the relevant input data and the scenario results. All different components are linked via a software interface for distributed computer systems.

Keywords: Damage Estimation, Disaster Planning, Disaster Response.

1 Introduction

Reliable damage estimations are essential for disaster preparedness and management as well as for the insurance industry. Therefore there is a lot of ongoing research activity in this field (for possible applications see e.g. (Earthquake Spectra, 1997) or (Soil Dynamics, 2001)). Even today the majority of the available models is based on the ATC-13 report (ATC, 1985). Central component of this approach are damage probability matrices, where the damage probabilities are dependent on different building types and earthquake intensities. A disadvantage of this approach is that these matrices require the intensities as an input parameter, although the earthquake intensity is defined on the basis of the resulting damages. More recent approaches try to model the building behavior under earthquake load directly. The most influencing work in this field comes from the Federal Emergency Management Agency (FEMA) which developed in collaboration with the National Institute for Building Sciences (NIBS) a comprehensive loss estimation methodology and implemented it in their loss estimation tool HAZUS. This methodology allows a complete analysis from direct earthquake loss up to the economic consequences (NIBS, 2001). The Collaborative Research Centre 461 "Strong Earthquakes" (Wenzel, 1997) implemented the GIS-based damage estimation tool EQSIM. EQSIM applies a comparable methodology and uses data from Vrancea earthquakes and Bucharest to calculate damage scenarios for a defined test area in the inner city of Bucharest. An initial version of this tool used the ArcView-GIS (Baur et al., 2000) and allows to calculate damage scenarios on building level. Based on this experience and on an analysis of available damage estimation tools the following limitations of up-to-date systems can be defined:

- Limited use during disaster response (e.g. integration of shake map results and observed damages during damage assessment)
- Use of local databases complicates the application during planning and response by different users
- Lack of integration into computer systems during disaster response and into computer based training systems

In this paper the authors present a new version of EQSIM, which uses a different software architecture to overcome these problems. The paper is organized as followed: First the new software architecture is described. This is followed by an explanation of the methodology for the building and casualty estimation. Finally the graphical user interface and the simulation components are described.

2 Software Architecture of EQSIM

The new version of EQSIM uses an approach based on the client-server architecture for distributed computing in computer networks. EQSIM consists of the following components, which will be described throughout this paper:

1. Central database
2. EQSIM server
3. EQSIM clients for planners and operation centre staff, including graphical user interface
4. Interfaces for field personnel such as SAR-teams
5. Decision support components using agent technology
6. Simulation interfaces to use EQSIM in simulation based training
7. Different simulators for the dynamic simulation of disaster response activities

All data, which is necessary to calculate damage scenarios, is stored in a central internet Oracle database, which can be accessed by all other components of EQSIM. The main calculation component is the EQSIM server program, which performs the calculation of the damage scenarios and implements the damage estimation methodology. Calculation requests can either come from client programs where users can define scenarios based on the historical earthquake database respectively defined earthquake parameters (such as location, magnitude and depth) or such requests arise from autonomous software programs which may be used for decision support during disaster response. Because all scenario results are stored in the central database the results of the calculated scenarios can be used by the EQSIM components at any time. During disaster response, field personnel may update the central database with observed damages. These observed damages are input for a groundtruthing-component of the server program to improve the scenario calculations. During a simulated training exercise all components are linked via a distributed simulation, which is based on the distributed High Level Architecture (HLA) – an IEEE standard for distributed simulation systems. In this case the field personnel and the response resources may be simulated by separate HLA-simulators. Figure 1 gives an overview over the possible application fields of EQSIM.

2.1 The Central Database

The central Oracle 9i database stores all relevant information. The database includes tables for different aspects of the damage estimation. The tables fall in the following groups:

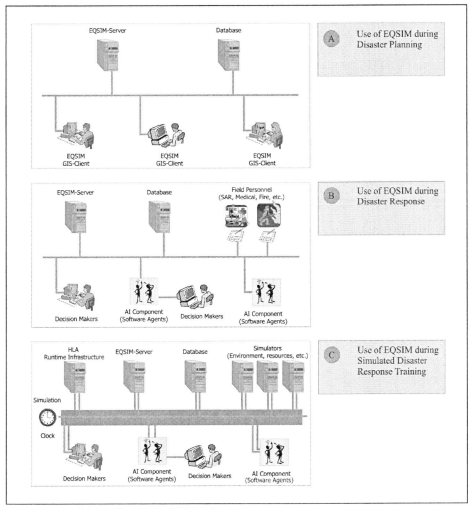

Figure 1: Possible applications of EQSIM

- **Static area information**: This data describes the existing infrastructure of the test area. Static area information includes aspects like soil data based on microzonation or building details (e.g. construction specifications and occupancy data per daytime).
- **Information about building behaviour**: The building damages are calculated with the capacity spectrum method and therefore these tables include capacity and fragility curves for all defined building types.
- **Information for casualty estimation**: These tables define the probabilities for different injury classes dependent on building type and damage class.
- **Earthquake related information**: This includes the historical earthquake database with strong Vrancea earthquakes based on the ROMPLUS catalogue (Oncescu et al., 1999) and tables with available response spectra from recorded time histories.

- *Meta knowledge*: This group of tables includes for example parameters for the calculation of seismic response spectra or parameters for the used attenuation functions.
- *Scenario results*: For scenario analysis all scenario results (e.g. damage and casualty probabilities on building level) are stored in different database tables.

Because the database includes meta knowledge about different aspects of the damage estimation methodology it is flexible to be used with other areas. If for example another attenuation function must be used due to different geophysical conditions, no changes in the source code of the software are required and the scenarios can be calculated easily.

2.2 The Server Program

The EQSIM-Server program implements the damage estimation methodology. This methodology allows to calculate the estimated building damages and casualties. The calculation of building damages is based on seismic response spectra. These spectra can either be obtained from recorded time histories or they may be calculated with earthquake specific parameters. For the second case the peak ground acceleration (PGA) can be calculated with local attenuation functions (Lungu et al., 1999). Local soil conditions and building codes, such as Eurocode 8 (CEN, 2002), are then used to transform PGA to standardized response spectra.

The building behavior for different building classes is described by capacity curves. These curves compare the spectral displacement of a building class with the spectral acceleration. To find the relevant spectral displacement for a specific earthquake scenario the capacity spectrum method is used. In the capacity spectrum method the elastic response spectrum of the earthquake is transformed to different inelastic response spectra. For these inelastic spectra the intersections with the capacity curves are calculated. If the ductility of the building class equals the ductility of the inelastic demand spectrum the respective spectral displacement is determined. For details of the capacity spectrum method see e.g. (ATC, 1995) or (Freeman, 1998). From spectral displacement values the damage probabilities can be calculated by means of fragility curves. These curves depend on the building type as well and contain for possible displacement values the probabilities that a building will be in a certain damage class. If the determined maximum displacement is used the damage probabilities can easily be drawn from the fragility curve (see figure 2).

The methodology for casualty estimation is also based on the HAZUS methodology (NIBS, 2001) and is adapted to Romanian conditions. The casualties can be calculated for three different scenarios (day, night and rush hour) and are broken down into four injury severity classes. The casualty estimation methodology requires as input the damage state probabilities of the different building classes, information about the distribution of building classes in the study region and population distribution data. The population distribution within the test area can be estimated for the three different scenario types with information which is available for each building and is stored in the central database, namely the number of residents, the social function of the building and the building square footage.

To estimate the human losses casualty rates are used. They describe the probability for occupants of buildings to be in any of the four injury severity levels at the time of an earthquake. These probabilities are given per building class for the five possible damage states of buildings. So far EQSIM uses casualty rates published by the FEMA (NIBS, 2001) for comparable structures in the United States, but they will be superseded by values specific to Romania in the near future. The HAZUS methodology was modified in the way, that it allows also the estimation of casualties for single buildings and not only for building classes.

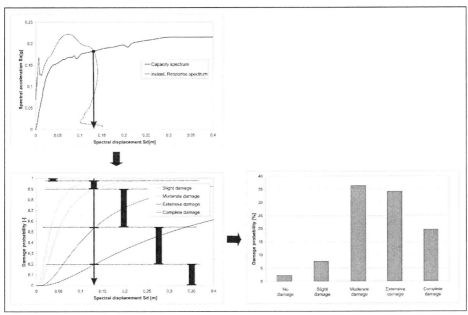

Figure 2: Calculation of building damage probabilities based on the capacity spectrum method

The probabilities that a person residing in a certain building class of a certain damage state in a certain microzone is situated in a certain injury severity class are calculated by means of the casualty rates:

$$\hat{p}_{m,k,j,n}(VK_m / SK_{k,j}^n) = p_{m,k,n} * q_{k,j}^n \forall m,k,j,n \tag{1}$$

with

$\hat{p}_{m,k,j,n}(VK_m / SK_{k,j}^n)$ probability of a person to be situated in the injury class m, if residing in a building of the building class k of the damage state n in the microzone j

$p_{m,k,n}$ probability of a person to be situated in the injury class m, if residing in a building of the building class k of the damage state n (equivalent to casualty rate)

$q_{k,j}^n$ probability of a building of the building class k in the microzone j to be situated in the damage class n (calculated by the damage estimation module of EQSIM)

To calculate the probability of an occupant being injured or killed per building class ($p_{m,k,j}$) the probabilities are summed over the damage states.

$$p_{m,k,j} = \sum_n \hat{p}_{m,k,j,n}(VK_m / SK_{k,j}^n), \forall m,k,j \tag{2}$$

The expected number of occupants injured or killed per building class ($N_{m,k,j}$) is the product of the number of occupants of the building at the time of an earthquake ($N_{k,j}$) and the probability of an occupant being injured or killed ($p_{m,k,j}$). If the number of the occupants of each single building in the study region is known or can be estimated, as in the test area in Bucharest, it's also possible to calculate the casualties at the building level. For this purpose the deterministic damage state of each single building in the test area, computed by the damage estimation module of EQSIM is applied. The casualty rate of each single building derived from the damage state and the building class is multiplied by the number of occupants present at the time of an earthquake. Due to the methodology, the results are more accurate for larger areas, like the test area of Bucharest, and more inaccurate for single buildings.

2.3 The Client Program

The main component of the client program is a Graphical User Interface (GUI) which helps to guide the user through the process of scenario definition and visualization of the results. The features of the GUI can be expressed by use case diagrams. Use case diagrams describe what a system does from the standpoint of an external observer. The emphasis is on what a system does rather than how. In a use case diagram one can differentiate between actors, system boundaries and relations. An actor is who or what initiates the events involved in that task. Actors are simply roles that people or objects play. The connection between actor and use case is a communication association. Figure 3 shows the four different actors which can be recognized in EQSIM: (1) the central database, (2) other data sources, (3) the EQSIM-server program and (4) the user of the client program. The border of the grey box in Figure 1 defines the system boundary of the client program.

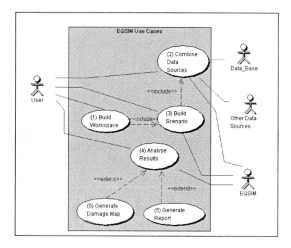

Figure 3: Use cases for the EQSIM Client program

First thing in using the GUI is to build a workspace (1). A workspace can be defined as an environment based on which a user is able to store and compute different scenarios. One has two possibilities, either to work with pre-existing settings and content or to start without any data. It should be possible to save the settings as well as the content to continue the work later on. If one builds a new workspace, one needs to combine various data sources (2) and also needs to choose from other necessary scenario parameters for building a scenario (3). For building a scenario one needs to initiate the calculations by EQSIM. To transfer the data to the calculation tool an communication protocol based on the Extensible Markup Language (XML) is used (for a detailed description of XML see for example (Harald and Means, 2001)). After the calculation of the results, it is necessary to analyze the results (4). This analysis can be done by generating a report (5) of the results, which includes the extent of destruction, area of destruction and destruction to different types of buildings. Next to reports the results can be analyzed visually by generating a damage map (6).

2.4 Simulation Components

EQSIM can also be used in virtual disaster response training by the staff of Emergency Operation Centres (EOC). For this EQSIM is embedded in a distributed simulation of the response activities after strong earthquakes. This simulation uses the IEEE standard High Level Architecture (HLA) for distributed simulation systems as a common framework[1]. In HLA-based simulations each participating simulator must implement a predefined HLA-interface. All EQSIM-components implement this interface as well. This allows different simulation components to send calculation requests to the server during a simulation-based exercise. In the simulation system the disaster environment and the use of resources within the disaster environment can be simulated in real time. Simulated resources include SAR-Teams, ambulances, fire fighting units, recon units and heavy equipment resources for repair work of blocked roads and rescue operations.

To interact with the simulation different elementary actions are defined. These actions can be used during a simulation by humans, such as management-level personnel, to perform either predefined or improvised plans based on the available actions. Additionally a multiagent system can be linked to the simulation environment where the software agents[2] use predefined flexible plans for their reasoning process. Because both the software agents and the computer interfaces for the humans implement the HLA-interface it is possible for them to take advantage of the damage estimation tool EQSIM for the planning of the response activities.

3 Results and Future Work

In this paper we presented a new version of the damage estimation tool for Bucharest. The new client-server architecture permits to use EQSIM simultaneously by different users. It can now be applied for disaster planning, disaster response and for disaster response training. The latter is embedded in a computer simulation of the response activities. The distributed simulators, EQSIM and the user interfaces communicate in case of training basing on a High Level Architecture (HLA) framework. The calculation component of EQSIM can be invoked by different applications using XML standard and it uses a state-of-the-art methodology to estimate building damages and casualties.

The damage estimation tool EQSIM is an essential component of the Disaster Management Tool (DMT) of the Collaborative Research Center *Strong Earthquakes*. The DMT is a software system supporting decision makers, surveillance and intervention teams during disaster response (Markus et al.). Further development of EQSIM will concentrate on improved consideration of Romanian conditions and in the development of a component for cost-benefit analysis for different retrofitting strategies.

---

[1] For an introduction to HLA-based simulation systems see for example (Fujimoto, 2000).
[2] Software agents are computational systems with goals, sensors, and effectors, which decide autonomously which actions to take, and when.

Acknowledgements

The research in this paper is part of different research projects within the Collaborative Research Center (CRC) Strong Earthquakes: A Challenge for Geosciences and Civil Engineering. It is funded by the German Science Foundation (DFG) and the state of Baden-Württemberg.

References

Applied Technology Council (ATC): Earthquake Damage Evaluation Data for California. ATC 13 Report, Applied Technology Council, Redwood City, California, 1985.

Applied Technology Council (ATC): Seismic Evaluation and Retrofit of Concrete Buildings Vol.1. ATC 40 Report, Applied Technology Council, Redwood City, California, 1995.

Baur, M., Bayraktrali, Y, Fiedrich, F., Lungu, D. and Markus, M.: EQSIM - A GIS-Based Damage Estimation Tool for Bucharest. In *Earthquake Hazard and Countermeasures for Existing Fragile Buildings* (Lungu, D. and Saito, T., editors), pp. 245-254, Independent Film, Bucharest, 2001.

CEN - European Committee for Standardization: Design of structures for earthquake resistance, Part 1: General rules, seismic actions and rules for buildings. Eurocode 8, Draft No. 5, European Commitee for Standardization, Brussels, 2002.

Earthquake Spectra: *Loss Estimation Issue*. Earthquake Spectra, 13 (4), 1997.

Freeman, S.A.: The Capacity Spectrum Method as a Tool for Seismic Design. In *Proceedings of the 11th European Conference on Earthquake Engineering*, September 6-11th 1998, Paris, A.A. Balkema, Rotterdam ,1998.

Fujimoto, R.M.: *Parallel and Distributed Simulation Systems*. Wiley Series on Parallel and Distributed Computing 3, John Wiley, New York, 2000.

Harold, E.R. and Means, W.S.: *XML in a Nutshell: A Desktop Quick Reference*. O'Reilly, Cambridge, MA, 2nd Edition, 2001.

Lungu. D., Cornea, T. and Nedelcu, C.: Hazard Assessment and Site-Dependent Response for Vrancea Earthquakes. In *Vrancea Earthquakes* (Wenzel, F., Lungu, D. and Novak, O., editors), pp. 251-267, Kluwer Academic Publishers, Dordrecht, 1999.

Markus, M., Fiedrich, F., Leebmann, J., Schweier, C. and Steinle, E.: Concept For An Integrated Disaster Management Tool. In *Proceedings of the 13th World Conference on Earthquake Engineering*, 1.-6. August, 2004, Vancouver, Canada, forthcoming.

National Institute of Building Sciences (NIBS): *Earthquake Loss Estimation Methodology HAZUS 99 SR2*, Technical Manual, National Institute for Building Sciences, Washington D.C., USA, 2001.

Oncescu, M.C., Marza, V.I., Rizescu, M. and Popa, M.: The Romanian Earthquake Catalogue between 984-1997. In *Vrancea Earthquakes* (Wenzel, F., Lungu, D. and Novak, O., editors), pp. 43-47, Kluwer Academic Publishers, Dordrecht, 1999.

Soil Dynamics and Earthquake Engineering: *Special Issue on Loss Estimation*, 21 (5), 2001.

Wenzel, F.: A Challenge for Geosciences and Civil Engineering – A New Collaborative Research Center in Germany, *Seismological Research Letters*, 68 (3), 438-443, 1997.

Special Contribution

Von Stochastischer Bemessung zum Risikomanagement

Erich J. Plate

em. Prof. Hydrologie und Wasserwirtschaft, ehem. Leiter des ehem. Instituts für Hydrologie und Wasserwirtschaft (IHW,) Universität Karlsruhe (TH)

Zusammenfassung

Die Konzeption des Risikomanagements hat sich am Institut des Autors aus den Arbeiten zur stochastischen Bemessungstheorie entwickelt. Unter dem Einfluss der Internationalen Dekade für Katastrophenvorbeugung (IDNDR) wurden die stochastische Bemessungstheorie unter Einbeziehung einer erweiterten Vulnerabilität zur Risikoanalyse. Der Weg von der stochastischen Bemessung zum Risikomanagement wird einleitend an Hand der Tätigkeiten des Autors verfolgt, und anschliessend fachlich nachvollzogen.

1 Einführung

Zwei Themen aus der Bemessungspraxis von Wasserbauingenieuren haben mich in meiner Karriere immer wieder beschäftigt: die Ermittlung der Versagenswahrscheinlichkeit der nach den Regeln der Technik bemessenen Staudämme (Plate, 1982,1984, Plate & Meon, 1988), und von Deichen (Plate, 1998), und das durch Schadstoffeinleitungen bewirkte Versagen der Selbstreinigungskraft von natürlichen Fließgewässern durch Überschreiten einer kritischen Konzentration (Plate, 1991, Schmitt-Heiderich & Plate, 1995). Das Gemeinsame dieser beiden Probleme ist, dass die Bemessung nach Versagenswahrscheinlichkeit erfolgen sollte - und hierfür wollte ich eine statistische Methodik liefern, aufbauend auf der hinreichend bekannten Zuverlässigkeitstheorie (Ang & Tang, 1984, Plate, 1993, Plate, 2000). Um die Möglichkeit des Einsatzes dieser Methodik auf allen Gebieten der Wasserwirtschaft auszuloten, veranlasste ich die DFG, als damaliger Vorsitzender der Kommission Wasserforschung, ein entsprechendes Werkstattgespräch zu fördern, und ferner organisierten Prof. L.Duckstein und ich im Jahr 1987 ein NATO Advanced Study Institut (Duckstein & Plate, 1987) zu dem Thema. Hieraus entstand ein Versuch, die verschiedenen, auf statistischen Grundlagen beruhenden Bemessungsverfahren zu ordnen, der zur Definition von 4 Stufen der stochastischen Bemessung führte (Plate & Duckstein, 1988, Plate, 1993). Das Risiko als Bemessungsgröße erscheint hierbei in der 4. Bemessungsstufe, der Bemessung nach Kostenoptimierung.

Im Jahre 1988 veröffentlichte die American Society of Civil Engineers in ihrer Hauszeitschrift einen Aufruf an die Wissenschaftler aller Fachrichtungen, sich an der Bewältigung der zunehmenden Katastrophen in der ganzen Welt zu beteiligen und eine Dekade der Vereinten Nationen zum Thema Katastrophenvorbeugung zu unterstützen. Er stammte von Frank Press, dem damaligen Präsident der National Academy of Sciences der USA. Mich hat diese Thematik sehr interessiert, und es gelang mir (unter tätiger Mitwirkung des DFG Sachbereichsleiter Dr. U. de Haar), die DFG zu veranlassen, bereits 1988 eine Arbeitsgruppe zum Thema „Naturkatastrophen" einzurichten.

Die International Dekade zur Katastrophenvorbeugung (International Decade for Natural Disaster Reduction) wurde dann, tatkräftig unterstützt vom damaligen

Aussenminister H.D.Genscher, als Dekade der Vereinten Nationen von 1990 bis 1999 beschlossen, mit einem vom VN Generalsekretär ernannten " Scientific and Technological Council" ausgestattet, dem ich von 1990 bis 1997 angehörte. Das erste Treffen des zuerst von Dr. J. Bruce, Kanada, geleiteten STCs fand auf dem Petersberg in Bonn statt, mit großen Erwartungen, die sich allerdings nur teilweise erfüllten. Bei der großen IDNDR Midterm Konferenz in Yokohama 1994 spielte das STC keine große Rolle, immerhin entstand die Yokohama Declaration unter ihrer Mitwirkung. Ich schlug damals vor, zum Abschluss der Dekade die Hauptthemen der Dekade: „Frühwarnung, Katastrophenpläne und Landesplanung, Risikomanagement" in drei Vorkonferenzen durchzuführen. Der Vorschlag wurde in Deutschland aufgegriffen, und das Deutsche Auswärtige Amt übernahm, in Zusammenarbeit mit derm Geo-Forschungszentrum Potsdam, die Aufgabe, die Frühwarnkonferenz im Jahr 1998 in Potsdam durchzuführen (Zschau & Küppers, 2003) - eine Nachfolgekonferenz fand im Jahre 2003 in Bonn statt.

Vom Auswärtigen Amt wurde für die Betreuung der Dekade ein Deutsches Nationalkomitee gegründet, das nacheinander von Botschafter a.D. van Well, Bundesminister a.D. B.Wischnewski, Bürgermeister a.D. H.Koschnick, Bundesminister a.D. N.Blühm und Bundesministerin a.D. I.Schwaetzer geleitet wurde. Die DFG Arbeitsgruppe wurde zum Wissenschaftlichen Beirat der DFG für das Deutsche Komitee für die IDNDR, aber bereits in den frühen Jahren der Dekade wurde er ganz vom IDNDR Komitee übernommen. Unterstützung bei der Leitung des Beirats erhielt ich zunächst von Dr. W.Kron in meinem Institut in Karlsruhe, später übernahm das GFZ Potsdam die Betreuung des Wissenschaftlichen Beirats, und nach dem Ausscheiden von Dr. Kron wurde Dr. Bruno Merz bis zum Ende der Dekade der Sekretär. Der Wissenschaftliche Beirat sah seine erste Aufgabe darin, eine Bestandsaufnahme der Forschung zum Thema Naturkatastrophen zu machen (Plate et al., 1993). Zehn Jahre später nach dem Ende der Dekade entstand noch einmal eine zusammenfassende Darstellung, in der die Erfahrungen der Dekade und der Stand des Wissens am Ende der Dekade zusammengestellt wurden. (Plate et al., 2001). Die wichtigste Funktion des Wissenschaftlichen Beirats bestand darüber hinaus in der Anregung von Forschungsvorhaben - ich glaube, dass die Einrichtung des Forschungsnetzwerkes Naturkatastrophen aus dieser Arbeit hervorgegangen ist, wie auch die Idee von regelmäßig zu veranstaltenden Foren, mit denen das inzwischen als DKKV weitergeführte Deutsche IDNDR Komitee jedes Jahr vor die Öffentlichkeit tritt.

Schon sehr bald wurde klar, dass Katastrophenforschung nicht als ein rein technisches Problem zu sehen ist, dass vielmehr der Mensch als Verursacher und Leidender im Vordergrund stehen muss. Die Frage der Erfassung der Vulnerabilität wurde immer mehr zum Schlüssel für zukünftige Forschung: selbstverständlich zusätzlich zu den technischen und organisatorischen Fragestellungen, die zur Verbesserung von Vorhersage und Katastrophenvorbeugung beantwortet werden müssen. Daher muss heute der Begriff des Risikos erweitert werden: wir können nicht mehr nur die finanzielle Seite des Risikos betrachten: wir müssen Bewertungskriterien finden für das zumutbare Risiko und Wege finden, wie verhindert wird, dass ein extremes, den Bemessungsfall überschreitendes Ereignis zur Katastrophe führt. Schon der Begriff der Katastrophe muss klarer definiert werden. Katastrophenvorbeugung muss nicht nur Sache einer Nation oder Region bleiben, sondern jede kleine Gemeinde muss Risiko bewusst werden: statt die Augen vor dem möglichen Auftreten eines Extremereignisses zu verschliessen, sollten wir lernen, mit dem Risiko zu leben. In diesem Sinne habe ich sehr gern die Aufgabe aufgegriffen, die Planungsaufgaben für ein Institut der Universität der Vereinten Nationen (UNU mit Sitz in Tokyo) mit dem Titel „Environment and Human Security" in Bonn zu leiten. Im Dezember 2003 wurde dies Institut gegründet, Prof. Janos Bogardi zum ersten Direktor ernannt, der das Thema Vulnerabilität und Vulnerabilitätsbewertung zum Kernproblem des Arbeitsprogramms gemacht hat: die Verhinderung bzw. Bewältigung von Katastrophen durch extreme Ereignisse, die auf eine sich durch menschliche Einflüsse veränderte und verändernde Landschaft und Gesellschaft treffen.

2 Von der Stochastischen Bemessung zum Risiko

In diesem Abschnitt soll umrissen werden, welcher Zusammenhang zwischen unserer üblichen Baupraxis und der allseits geforderten Risikoanalyse, oder noch weitergehend, dem Risikomanagement besteht. Bekanntlich wird ein Bauwerk bemessen, in dem nachgewiesen wird, ob ein Bauwerksentwurf in der Lage ist, die auf es einwirkenden externen Belastungen S infolge eines Extremereignisses u aufzunehmen und schadlos abzuführen. Die Belastbarkeit R des Bauwerks ist die Fähigkeit, die Belastungen durch interne Eigenschaften des Bauwerks aufzunehmen, z.B. durch die inneren Reibungskräfte eines Erddamms oder die Materialeigenschaften von Bauteilen. Die Belastbarkeit wird berechnet auf der Basis von durch Erfahrung oder Experimente vorgegebenen zulässigen internen Größen, z.B. maximal zulässige Spannungen in Bauwerksteilen.

Aus wirtschaftlichen Gründen können die Belastungen nicht beliebig hoch angesetzt werden, es muss immer damit gerechnet werden, dass noch höhere Belastungen auftreten können. Daher muss eine gewisse, geringe Versagenswahrscheinlichkeit in Kauf genommen werden. Diese sollte für verschiedene bauliche Anlagen mit gleicher Bedeutung gleich sein. Es ist daher naheliegend, die allein auf Erfahrungen beruhenden Belastungsgrößen durch solche zu ersetzen, die zu einer vorgegebenen Versagenswahrscheinlichkeit P_V führen, und als Bemessungsregel eine zulässige Versagenswahrscheinlichkeit P_{vzul} zu fordern, d.h. es muss gelten:

$$P_V \leq P_{vzul} \,. \tag{1}$$

Eine Bemessung nach Versagenswahrscheinlichkeit wird stochastische Bemessung genannt. Der Vorteil der Verwendung von Gl.1 als Bemessungskriterien liegt darin, dass hierdurch einerseits Unsicherheiten, andererseits aber auch Lastkombinationen berücksichtigt werden, indem z.B. Belastungskollektive mit gleicher Gesamtüberschreitungswahrscheinlichkeit definiert werden können. Insbesondere bei der Aufstellung von Normen ist es von großem Vorteil, wenn eine einheitliche Bemessungsgröße für alle möglichen baulichen Anlagen festgelegt werden kann, wie es die Gru Si Bau (Grundlagen der Sicherheit von baulichen Anlagen, 1981) fordert.

Für große bauliche Anlagen ist dieses Vorgehen jedoch unbefriedigt, weil das Problem der Entscheidung nur verlagert ist auf die Ebene der Festsetzung der zulässigen Versagenswahrscheinlichkeit. Dieses Problem kann nur überwunden werden, wenn statt der Versagenswahrscheinlichkeit ein objektives Entscheidungskriterium verwendet wird. Es bietet sich hier an, das Nutzen - Kosten Verhältnis zu verwenden, ausgedrückt durch:

$$Z = \frac{1}{T} \sum_{t=1}^{T} \frac{1}{(1+i)^{tn}} \left(B(t) - C_A(t) - C_s(t) - RC(t) \right)\Big|_D \tag{2}$$

(nach Crouch & Wilson, 1982). Hierin ist Z der mittlere jährliche Nettogewinn, mit $B(t)$ als Nutzen, $C_A(t)$ als Kosten der Anlage, $C_S(t)$ bezeichnet andere, monetär zu bewertende Kosten, z.B. soziale Kosten, und RC ist der Erwartungswert der Schäden, die trotz der Anlage zu erwarten sind. T ist die Bemessungslebensdauer der Anlage, t der Zeitindex t = 1,2,....T in Anzahl der Jahre, i der Zinssatz, und D bezeichnet die Entscheidungsvariable, in diesem Fall die verschiedenen Optionen für eine Lösung des Problems, z.B. die Dimensionen eines Bauwerks. D kann aber auch für ganz verschiedene Lösungsmöglichkeiten stehen, die dasselbe Ziel erreichen sollen: andere Arten von Anlagen, oder auch die Lösung, garnichts zu tun. Nach der einfachsten Anwendung dieser Formel muss eine Anlage gewählt werden, für die Z positiv ist. Eine befriedigendere Lösung wird jedoch erhalten, in dem Gl.2 als Zielfunktion eines Entscheidungsprozesses gesehen wird, bei dem es gilt D so zu bestimmen, dass Z ein Maximum wird.

Das Risiko RC erscheint in Gl.2, als Erwartungswert der Schadenskosten, der durch das Versagen der baulichen Anlage entstehen kann. Für die Nutzen - Kosten Analyse muss daher abgeschätzt werden, welche Konsequenzen durch das Versagen der baulichen Anlage entstehen - hier wird in erster Linie an monetäre Schäden gedacht: Schäden an der Anlage selber, aber auch Schäden für die Umwelt und zusätzliche Sachschäden als Versagensfolgen. Es ist üblich, die Risikokosten binär aufzufassen: es entsteht kein Schaden, wenn der kritische Zustand unterschritten, wird, es entsteht ein Schaden $\varphi \cdot K$ wenn der kritische Zustand überschritten wird. Formelmässig ausgedrückt ist dies das Restrisiko, das z.B. die Versicherungs industrie ihren Prämien zugrunde legt:

$$RC = \varphi \cdot K \cdot P_v \qquad (3)$$

unabhängig von der Zeit, und gegeben durch das Produkt aus relativer Vulnerabilität oder Exposition φ, maximal möglichem Schaden K und Überschreitungswahrscheinlichkeit für den kritischen Zustand. Genauer wird RC über die Gl.4 berechnet:

$$RC = \int \varphi(t, u) \cdot K(t) \cdot f_u(t, u) \cdot du$$
$$(4)$$

wobei $f_u(t,u)$ die dem schadenserzeugenden Ereignis u zugeordnete Wahrscheinlichkeitsdichte funktion ist, und sowohl Exposition als auch maximale Kosten hängen von der Größe u ab und können auch zeitlich veränderlich sein. Damit wird erfasst, dass bereits für ein Ereignis u, das unter dem Bemessungsgrenzwert u_{grenz} liegt, ein Schaden auftreten kann, oder aber auch dass ein Übergang von unbeschädigt, $\varphi = 0$ bis zur vollen Zerstörung, $\varphi = 1$ besteht.

3 Risiko im Risikomanagement

Der Schritt vom Berechnungsverfahren für RC als Bemessungsgröße zur Risikoanalyse wird dann gemacht, wenn das Bauwerk an den Bauherrn übergeben wird. RC ist vom Bauherrn aus gesehen ein Restrisiko, das er tragen muss. Er kann sich dagegen versichern, er kann jedoch auch durch andere Maßnahmen sein Risiko vermindern. Wenn die bauliche Anlage von großer Bedeutung ist, wie z.B. eine Talsperre oder eine Deichanlage, eine Brücke oder ein Küstenschutzbauwerk, dann kann man nicht, wie es in der Vergangenheit üblich war, unbekümmert mit dem Restrisiko leben und hoffen, dass nichts passiert, sondern es muss durch geeignete Maßnahmen gewährleistet werden, dass trotz eines möglichen Versagens der Anlage eine Katastrophe verhütet wird. Die Methodik hierfür ist das Risikomanagement.

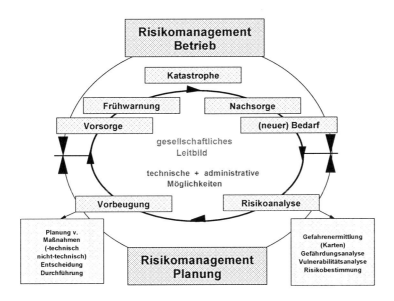

Abb.1: Der Kreislauf von Planung und Betrieb von baulichen Anlagen, durch deren Versagen eine Gefährdung für Menschen entsteht.

Risikomanagement ist die Summe aller Aktivitäten, die bei einer durch Extremereignisse gefährdeten baulichen Anlage ergriffen werden können, um Schäden zu minimieren und insbesondere die Gefährdung von Menschenleben zu reduzieren. Es erfolgt in einer zeitlichen Sequenz, die als Kreis nach Abb.1 dargestellt werden kann.

Der Kreislauf deutet an, dass es sich beim Risikomanagement um eine zyklische Aufgabe handelt, die es z.B. im Hochwasserschutz nach jedem größeren Hochwasser und in jeder Generation neu zu lösen gilt. Es muss unterschieden werden zwischen zwei von einander zu trennenden Bearbeitungsteilen: der Planung und dem Betrieb. Nach jedem Auftreten eines großen Schadensereignisses wird zunächst die Katastrophenhilfe einsetzen, einschliesslich der Phase des Wiederaufbaus und der emotionalen Bewältigung der Schrecken der Katastrophe. Nicht nur am direkt betroffenen Ort, sondern auch in Orten mit ähnlichen Gefährdungen wird die Frage neu gestellt, ob die vorhandenen Schutzmaßnahmen den Ansprüchen an die Sicherheit noch genügen. Eine Analyse der Situation wird zeigen, ob neue Erkenntnisse neue Planungen erfordern. Diese Untersuchung ist die Risikoanalyse (siehe z.B. Plate, 2002).

Der Planungsprozess findet seinen Abschluss mit der Entscheidung für zu ergreifende Maßnahmen und deren Umsetzung. Nach Erreichen des Planungszieles wird die Anlage dem Betreiber übergeben, der durch geeignete Maßnahmen dafür Sorge zu tragen hat, das die Anlage funktionsfähig bleibt, dass z.B. ein Frühwarnsystem so funktioniert, wie es geplant ist, und z.B. durch ständige Wartung des Beobachtungsnetzes wie auch der Schutzanlagen zur Bewältigung eines Extremereignisses gerüstet zu sein. Besonders darauf hingewiesen sei, dass das Risikomanagement eine kontinuierliche Aufgabe ist, die immer wieder überprüft werden muss. Neue Erkenntnisse, neue technische Möglichkeiten, aber auch neue gesellschaftliche Entwicklungen müssen berücksichtigt werden - in diese Kategorie gehört auch die Berücksichtigung der Ergebnisse der Klimaforschung, um eventuell drohenden Auswirkungen von Klimaänderungen begegnen zu können.

Heute wird gefordert, dass der Umgang mit dem Restrisiko eingeplant wird. Dafür steht der obere Halbkreis der Abb.1. Die Planung geht über in die Vorsorge. Das bedeutet,sich für den Fall zu rüsten, wenn das vorhandene Schutzsystem versagt oder nicht ausreicht. Katastrophenpläne gehören ebenso zur Vorsorge wie die gute Wartung vorhandener Schutzanlagen. Hierbei spielt die Warnung eine besonders wichtige Rolle. In dem Kreislauf nach Abb.1 von Katastrophe und Nachsorge zu Vorsorge, bei dem sich Planung und Betrieb abwechseln, steht die Frühwarnung an prominenter Stelle. Die modernen technischen Möglichkeiten einer Vorhersage sollten voll genutzt werden, wobei zu bedenken ist, dass eine Vorhersage der geringere Teil eines Warnsystems ist: die Weitergabe der Vorhersage an verantwortliche Stellen zur Anfertigung einer Warnung ist ein weiterer wesentlicher Teil, aber noch wichtiger ist, dass die Warnung die Bevölkerung erreicht und genügend genau ist, um nicht durch falschen Alarm Vertrauen in zukünftige Warnungen zu zerstören. Und schliesslich muss die Warnung früh genug erfolgen, dass entsprechende Vorbeugemaßnahmen getroffen werden können (Zschau & Küppers, 2003).

4 Schlussbemerkung

Das Thema Risikomanagement ist ein direkter Überbegriff auch für die Aufgabe der stochastischen Bemessung, da das Risiko den Erwartungswert für die Folgen des Versagens der baulichen Anlagen darstellt. Anders jedoch als bei der Bemessung sind die Schadenserwartungswerte beim Risikomanagement nicht nur auf die Anlage bezogen. Ferner muss der Begriff des Versagens weiter gefasst werden und auch nicht nur die Zerstörung des Bauwerkes oder der baulichen Anlage beinhaltet, sondern auch das Versagen, das dadurch entsteht, dass die Anlage ihren Zweck nicht erfüllen kann.

Gemeinsam ist der stochastischen Bemessung mit Nutzen - Kosten Analyse und der Planung von Schutzmaßnahmen die Methodik für die Ermittlung des Risikos. Für die Bestimmung des Risikos brauchen wir die Wahrscheinlichkeitsdichteverteilungen der Zufallsgrößen. Der Unterschied in den beiden Definitionen des Risikos liegt in der Vulnerabilität, die bei der Bemessung durch den Wert der baulichen Anlage bestimmt ist, während sie im Risikomanagement die Eigenschaften der EARs erfasst - erschwert besonder, wenn die Gefährdung nicht nur Sachen betrifft, sondern wenn Menschen betroffen sind anders behandelt werden muss. Gerade in der schwierigen Definition der Vulnerabilität sehen wir die größte Forderung an die Katastrophenforschung.

Heute definiert man eine Katastrophe als ein Ereignis, bei dem die betroffenen Menschen so in ihren Handlungsmöglichkeiten geschädigt sind, dass sie sich mit den eigenen Kräften -seien sie körperlicher, gesellschaftlicher oder wirtschaftlicher Art - nicht in den Zustand vor dem Extremereignis zurückversetzen können: sie sind auf Hilfe von außen angewiesen (Plate et al., 1993, Plate & Merz, 2001).

Dieser allgemeine Katastrophenbegriff trifft gleichermaßen für eine Familie oder eine Gemeinschaft, oder auf ein ganzes Land, sie unterscheiden sich nach der Anzahl der Betroffenen, und nach der Größe des Extremereignisses. Am wirtschaftlichen Schaden allein darf eine Katastrophe nicht gemessen werden. Es müssen auch die nicht-materiellen Schaden und die Folgen, die langfristig entstehen, berücksichtigt werden. Das heisst nichts anderes, als dass die Katastrophe erst dann vollständig überwunden ist, wenn auch die Folgekosten und die Folgen für die Gesellschaft abgeklungen sind. Es sind grundsätzlich drei Faktoren zusammenzusehen, die den Begriff der Vulnerabilität ausmachen: (Blaikie et al. 1994): die Gefährdung (engl.: hazard), ausgedrückt durch die Stärke des auslösenden Ereignisses, die Verletzlichkeit der Menschen gegen das Ereignis (engl.: exposure), und schliesslich die Kraft der Menschen, die Folgen eines Extremereignisses zu überwinden (engl.: coping capacity).

Literatur

Ang,A.H. and W.H.Tang (1984): "Probability concepts in engineering planning and design, Vol.2" J.Wiley, New York

Blaikie, P.T., T.Cannon, I.Davis and B. Wisner, 1994: At risk: Natural Hazards, people's vulnerability, and disasters. Routledge, London

Crouch, E.A.C., R.Wilson 1982: Risk Benefit Analysis, Ballinger Publisher, Boston, Mass. USA

Duckstein, L., Plate, E.J. (eds.), (1987): Engineering Reliability and Risk in Water Resources. Nijhoff 1987. NATO ASI Series, Series E: Applied Sciences No. 124

Plate, E.J., (1982): Bemessungshochwasser und hydrologisches Versagensrisiko für Talsperren und Hochwasserrückhaltebecken. In: Wasserwirtschaft Jg.72,1982,H.3,pp.91-97

Plate, E.J., (1984): Reliability analysis of dam safety. In: Frontiers in Hydrology. (Ven Te Chow-Memorial-Vol.) Eds.: W.H.C.Maxwell, L.R.Beard.Littleton,Col.:Water Res. Publ. 1984. pp. 288 - 304

Plate, E.J., (1991): Probabilistic modelling of water quality in rivers. In: Water Resources Engineering Risk Assessment. Ed.: J. Ganoulis, Berlin: Springer-Verl. 1991, (NATO ASI Series G 29), pp. 137 – 166

Plate, E.J. (1993): "Statistik und angewandte Wahrscheinlichkeitslehre für Bauingenieure" (Statistics and applied probability theory for civil engineers) Ernst und Sohn, Berlin

Plate, E.J., (1996): Stochastische Bemessung bei Wassergüteproblemen. In: Wasser im System Boden - Pflanze - Atmosphäre. Festschrift zum 60. Geburtstag von Prof. Dr. Gerd Peschke. IHI-Schriften H. 2, 1996, S. 20 – 30 (Internationales Hochschulinstitut Zittau)

Plate, E.J. (1998): Stochastic hydraulic modelling - a way to cope with uncertainty. in: K.P.Holz et.al. (ed.)1998: Advances In Hydro - Sciences and –Engineering, Vol. III, Proceedings of the 3rd International Conference on Hydro – Science and – Engineering, Cottbus/Berlin, Germany (CD-ROM)

Plate, E.J. (2000) Stochastic design – has its time come? (keynote lecture) in Z-Y.Wang &S.X.Hu (eds). Proceedings of the 8th International Symposium on Stochastic Hydraulics, July 2000, Beijing. Balkeema , Rotterdam pp.3-14

Plate, E.J. (2002): Flood risk and flood management, J. Hydrology, Vol.267, pp.2-11

Plate, E.J., Duckstein, L., (1988): Reliability-based design concepts in hydraulic engineering. In: Water Resources Bulletin, Vol. 24, 1988, No. 2, pp. 235 – 245

Plate, E.J., Meon, G., (1988): Stochastic aspects of dam safety analysis. In: Proceedings of the Japan Society of Civil Engineers; Hydraulic and Sanitary Engineering, No. 393/II-9, 1988, May, pp. 1 – 8

Plate, E.J., et al., (Hrsgb.) (1993): Naturkatastrophen und Katastrophenvorbeugung. Bericht zur IDNDR. DFG, VCH Verlagsgesellschaft, Weinheim

Plate,E.J. Z-Y. Wang (2001) Flood disaster management, (invited lecture) XXIX Congress, International Association for Hydraulic Engineering and Research, Beijing September 2001. Proceedings, Vol.2 pp.77-87

Plate, E.J. B.Merz (Hrsgb) (2001): Naturkatastrophen - Ursachen, Auswirkungen, Vorsorge. Schweizerbart´sche Verlagsbuchhandlung, Stuttgart

Schmitt – Heiderich, P., E.J.Plate, 1995: River pollution from urban stormwater runoff. In: Statistical and Bayesian Methods in Hydrological Sciences, Paris, UNESCO, Vol.2, Chapter 2 pp.1-17

Zschau, J., A.N.Küppers, 2003: Early Warnng Systems for Natural Disaster Reduction. Springer Verlag

Information and Communication

Integrating Real-Time Information into Disaster Management: The IISIS Dashboard

Louise K. Comfort, Mark Dunn, David Johnson, Robert Skertich, Adam Zagorecki
University of Pittsburgh, Pittsburgh, PA 15260 USA

1 Introduction

Designing an information system to support decision processes in emergency environments presents an extraordinarily complex set of challenges for both technical and organizational managers. Emergency managers need to understand the performance of the response system at several levels of operation simultaneously. While they need to know the detailed requirements for performance at a local site, they also need to recognize the consequences of a failure at any one site for its neighboring sites in the system. Further, managers need to recognize the consequences of cumulative failure at one level of operation that may lead to potential failure at other levels. Without the capacity to understand the interdependencies of a complex technical system, managers cannot anticipate the destructive consequences of the aggregation of apparently minor failures at single sites and make informed decisions to prevent cumulative damage to the performance of the whole system. No response system can be considered reliable if it lacks the tools necessary to aggregate and present information to managers at different levels of responsibility simultaneously to support coordinated decision making under urgent conditions. Innovative information and simulation technologies to support this decision process are essential.

Providing real-time, decision quality information to practicing managers in timely, graphic form that displays the operation of a complex infrastructure system at multiple scalar levels would enable managers at their respective positions within the system to monitor interactions among the components and adjust performance reciprocally to reduce risk. This capacity generates a self-organizing approach to the management of risk that uses local information to achieve a global goal. It also introduces a second dynamic, the system's informed adaptation to changed conditions that serves to interrupt the spread of dysfunction from the disaster event throughout the entire system. It is the interaction between these two dynamics – spreading dysfunction and cascading adaptation – that measures the system's performance – or fragility - under threat. This task can be performed computationally, but requires a sociotechnical approach.

This paper will examine three issues in relation to information management in environments vulnerable to risk. These issues include: 1) development of self-organizing processes among organizations operating at different levels of responsibility and different degrees of exposure to risk; 2) interoperability of both technical and organizational systems that monitor, store, and exchange information regarding the status of communities exposed to risk; and 3) integration of real-time information from technical monitoring systems into an organizational information system to support decision making. These issues, taken together, constitute a sociotechnical approach to managing information for decision makers in environments exposed to risk. The paper will conclude with the design for an executive 'dashboard' that seeks to integrate these three functions into a coherent sociotechnical decision support system for practicing managers.

2 Theoretical Background

The concept of self organization has been well-defined in the literature on complex, adaptive systems. Essentially, it refers to the action taken by an individual or organization in response to incoming information about changing conditions that affect its performance (Kauffman, 1993; Comfort, 1994; Axelrod and Cohen, 1999). In an organizational context, self organization refers to the mutual adaptation among multiple organizations in response to shared information regarding changing conditions that affect all organizations in a given setting or environment. The key factor in precipitating self organization, that is, the ability to initiate action spontaneously, is the recognition of changes in the immediate environment that will affect the continued performance, or survivability, of the actor or actors. The capacity for self organization depends upon the existence of an information infrastructure that can communicate information about changing conditions quickly and accurately, as well as a valid knowledge base from which the actor can infer the consequences of those changes upon the continuing performance of the affected actors or organizations. This capacity for inference, or reasoning from known data to estimated future consequences, is a critical function in self organization. It is a function shared by both computational machines and human managers, and enables a sociotechnical approach to the complex problems of emergency management.

The conditions in which self organization can occur in practice, however, require the "interoperability" of both technical and organizational systems. Interoperability, a term of frequent use in emergency management communications, has been largely used to address the problem of incompatibility among different types of radio systems (National Governors Association, 2002) or different formats for technical decision support systems (General Accounting Office, 2002). The Office of Domestic Preparedness (ODP, 2003, p.2-1) defines interoperability as "the ability of two or more public safety agencies to exchange information, when and where it is needed, even when different communication/information systems are involved." Although ODP extends the concept of interoperability beyond radios to include fixed facilities, mobile platforms, and portable (personal) devices, the focus is still primarily on communications equipment. This focus on communications was broadened in findings from a research workshop sponsored by the U.S. National Science Foundation (Rao, 2003). The interdisciplinary researchers and practicing managers gathered at this Workshop recognized the need for interoperability of the highly diverse, large-scale networks of communications and information exchange used by public, private, and nonprofit organizations to provide societal services. The findings acknowledged the centrality of communications to the capacity of different types of organizations to coordinate their actions to manage risk effectively.

In practice, interoperability in disaster management is a function that includes more than mechanical devices. Crucial to interorganizational coordination is not only the ability to transmit and receive messages, but also the ability to recognize risk (Klein et al. 1993) that may appear in scattered form, or in micro changes in different locales over time. This ability requires a further set of skills, both technical and organizational, that allow the integration of real-time information into a coherent profile of operational performance for a community exposed to risk.

Integration of real-time information into an effective decision support system for practicing managers depends not only on the technical devices to monitor performance of key functions, but also the organizational analysis of the interdependence among these functions as they support the normal operations of a complex, urban community. Designing effective decision support for practicing managers means integrating all three concepts – self organization, interoperability, and integration of real-time information through effective monitoring and reporting systems – into a functioning information system that can be readily understood by managers of diverse organizations at different levels of responsibility, authority, and access to resources. This means designing a sociotechnical system in which the technical components fit the organizational requirements for information and action, and the organizational actors have sufficient technical knowledge and skills to understand the limits and potential points of failure in the technical system. Building such systems requires an interdisciplinary, interjurisdictional approach that can move across the traditional boundaries of operating agencies and engineered systems.

A sociotechnical system links organizations, individual policy makers, groups of clientele, communications networks, and computers into a distinct operational system that uses the flexibility of current information technology to support adaptive behavior by individuals and organizations in a changing environment. The resulting system represents a collection of entities that are capable of adjusting their behavior to one another and to the environment in order to achieve a shared goal (Comfort 1994; Comfort 1999). The technical components extend the knowledge base, memory, and reasoning capacity of the individuals and organizations that participate in the system. The individuals and organizations, in turn, monitor the performance of the technical components to ensure that they are functioning to support interacting organizations and inter-organizational problem solving processes under dynamic conditions

Sociotechnical systems, carefully designed and maintained, create a robustness in information processing and management that enables rapid search for, and dissemination of, information to multiple sources simultaneously. These functions facilitate adaptation in complex organizational networks to evolving conditions or destructive events. This robustness, however, is tempered by the potential fragility of each subsystem – technical and organizational – in which failure in one subsystem may trigger failure in the related subsystem, setting off a cascade of failure throughout the entire system. The propensity for either robustness or fragility is multiplicative. That is, when the sociotechnical system is operating well, it increases the capacity and strength of its components' performance. Conversely, failure in one component, unrecognized, threatens failure to its near neighbors and, untreated, spreads failure throughout the system.

The challenge to researchers working in this field is to design a theoretical architecture for decision support that encompasses the characteristics of both technical and organizational systems, but minimizes the potential fragility of both systems. Borrowing from the analysis of biological feedback mechanisms used in the metabolism process, Csete and Doyle (2004) outline a "bowtie" architecture that captures the essential characteristics of information processes in complex emergency management environments. In this architecture, information "fans in" from multiple diverse sources into a "knot" where a small number of common characteristics among the streams of information are used to break down and analyze the information from multiple streams. This process then redefines the information against the existing knowledge of the environment and produces new reports that then "fan out" to relevant different actors in the emergency management system. The "knot" serves primarily as the information processing center that not only integrates information from diverse sources, but interprets it against current knowledge of a complex community and redirects new packets of information to relevant actors in the system. The process is continuous, as the new information is accepted, acted upon, or rejected, and consequently generates a fresh set of responses that again "fan in" from multiple sources in a feedback-response dynamic to the "knot" for re-analysis and re-interpretation.

This model differs significantly from the hierarchical framework that has dominated organizational structures for information processing in emergency environments. It captures, instead, the multiplicity of sources that generate information simultaneously, and uses a small set of "core variables" to analyze and categorize the content of incoming information streams in terms of its relevance for the different actors at different levels of responsibility in the wider emergency management information system. In substantive ways, the "bowtie" model captures the function of an Emergency Operations Center, where a small number of experienced operations chiefs review incoming streams of information from multiple sources and transform that information into action strategies for the different agencies engaged in response operations. The next challenge is to design the internal information processing functions of the "knot" for emergency managers in a valid, verifiable manner, as well as to establish the linkages for real-time monitoring of critical functions in a community that provide the most effective indicators of performance of the system.

3 The IISIS Dashboard

Currently under development at the University of Pittsburgh, the executive dashboard for crisis management represents a systematic effort to design a set of internal information processes to serve executive managers responsible for mitigating and managing crises. In important ways, this effort seeks to specify the internal functions of the "knot" in the bowtie architecture of emergency information processes. The dashboard will enable practicing managers at different levels of responsibility and in different locations to monitor the status of critical operating conditions and systems on a regional basis. The dashboard will enhance the existing Interactive, Intelligent, Spatial Information System (IISIS) prototype that is being tested at the University of Pittsburgh. The IISIS prototype, designed in accordance with national standards for the National Incident Management System adopted by the U.S. Department of Homeland Security and data representation in visual and graphic display, provides managers with real-time decision support to mitigate, prepare for, respond to, and recover from, extreme events.

4 Stages of Development

Five stages are central to the development of the IISIS dashboard. They are outlined briefly, as follows:

4.1 Characterizing the Context of Operations

The first stage for the development of an executive dashboard is the characterization of the initial operating conditions, functions, and processes of the context in which it will operate. This means identifying the physical conditions of the specific region of study, using standard measures that are applicable across jurisdictions. Second, it means characterizing the technical infrastructure that is critical to the daily operation of the region, using standard measures, and identifying the relevant sources of data for monitoring that infrastructure. Third, it is necessary to identify the relevant sources of data to characterize the organizational policies, procedures and processes that govern the operations of the region. These data will produce a basic knowledgebase for the region, the technical systems that perform essential daily operations, and the organizational policies and processes that manage the technical systems. The data will be stored in an Oracle database that supports ready access and analysis for information functions.

4.2 Creating the "Knot": Identifying the Core Information

Using the data collected for the knowledgebase, the research team, advised by practicing managers, identifies the core information needed to operate each critical system, and the core technical nodes through which this information must flow. This analysis produces the small number of core variables that are essential to maintain operations of the system. This set of variables makes up the analytical content of the "knot" in the bowtie architecture of the system.

4.3 Data Collection and Transmission Functions: "Fanning In"

The next stage is to design a means of transmitting information regarding performance of critical functions into the core IISIS system for analysis, interpretation, representation, and retransmission. These are the separate streams of data that "fan in" to the "knot" for processing. In most instances, this transmission of data will require the creation of a secure bridge from an already operating system to the IISIS master server. Figure 1 shows an example of a secure bridge between a 911 dispatch system and the IISIS master server.

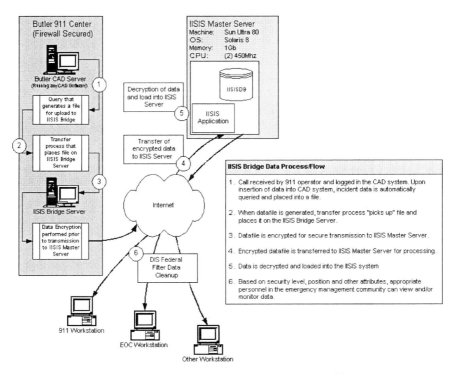

Figure 1: An Example of a Secure Bridge: The IISIS Secure 911 Bridge

Other critical functions included in the design of the fan for transmission of real-time data, in our current research, will be data from traffic monitors of the regional transportation system, current reports on occupancy and capacity of the regional hospital system, and status reports from the National Weather Service. The secure bridge provides a framework for these multiple data sources to transmit data to the IISIS server for induction, analysis, manipulation and display. The bridge will be designed to utilize existing protection measures (e.g. firewall) while increasing transmission security by using encryption and compression. The secure bridge will be a robust and redundant framework that ensures data security with minimal downtime.

Reliable transmission of data from the distributed intelligent agents into the database is essential to ensure timely access for emergency managers. Different transmission technologies can be utilized. Field responders can enter data into personal data appliances (PDAs) that transmit data to the database using wireless transmission. Field responders can use public or private voice communications, wired or wireless, to report data to personnel in the emergency operations center who can manually enter the data. Intelligent agents monitoring environmental conditions can also use wireless technology Public data such as road conditions can be obtained over the public internet. Private data can be securely transmitted over the public internet using virtual local area network technology. Private leased lines or virtual private networks can be utilized to improve reliability. Due to the unpredictable availability of public communication services during a crisis, backup transmission options will be considered for each data source.

Analysis of the operations of each system, validated through review by practicing managers of the systems, will identify the thresholds of risk – or points at which failure occurs – for each system. This information becomes the indicator for raising or lowering the level of risk that is displayed on the dashboard. Continuing this analysis, the research team will identify the organizational and technical interdependencies among the component systems for the region under study. Mapping these interdependencies will allow the identification of primary and alternate routes for transmitting information regarding the operation of any single system to the dashboard, as well as the integration of the different sources of information for the visual display of the status of the region on the dashboard.

The data are integrated into the IISIS system using standard interfaces, such as Oracle and ARC IMS, as well as custom application programming interfaces (API). For example, personnel and logistics databases would have standard interfaces, as most implementations would involve people and equipment. However, new and emerging data sources and types, such as intelligent agents monitoring environmental hazards, could be brought into the system using custom APIs.

4.4 Analysis and Integration of Data on Critical Functions: "Knot Processes"

Data that is transmitted over the secure bridge will be placed on the server for decryption and loading into the database. A standard data interface will accept data, extract/transform/load the data, and prepare the data for intelligent reasoning modeling tools. Custom data sources and types will enter into the system through custom Application Protocol Interfaces (APIs). Once the data has been loaded into the database, it will be immediately available for data mining and display on the dashboard. Risk thresholds, defined in the organizational analysis, will be entered as parameters into the intelligent reasoning components of the prototype. The thresholds will indicate the levels of risk or tolerance for the larger system and will, in turn, trigger changes dynamically and graphically display these escalation events to decision-makers.

4.5 Transmission of Integrated Information to Policy Makers: "Fanning Out"

The rapid and timely transmission of integrated information to relevant decision makers is the key to improving the decision-making process. Due to the sensitive nature of information that is being displayed, security is an important consideration in the development of the dashboard. Pre-defined and authorized users of the dashboard system will receive information at the desktop that corresponds to their organization role or responsibilities. After each user is authenticated into the system, s/he will receive an overall dashboard status report. As people, resources, or circumstances demand, decision-makers will receive messages directed to one or more addresses or devices. The dashboard will be browser-based, using standard HTML technology

The five phases outlined above constitute the primary development stages in a "bowtie architecture" for a sociotechnical decision support system for practicing managers. The actual implementation of the dashboard also focuses on the graphical design of the interface between the computational model and the users.

5 Design of Graphical Interface between the Computational Model and Users

The IISIS dashboard will present decision makers with a comprehensive overview of the critical measures central to decision making during crisis operations. A web server will act as the access mechanism for the dashboard. A browser based interface will provide a real-time overview of critical data mined from the data warehouse. Effective visualization of the extensive amount of data available will enable decision makers to respond quickly during crisis operations. Standard practices to ensure the design of a coherent graphical user interface will be utilized. The display must be intuitive and easy to use. Colors will be used to indicate areas that are experiencing problems. More detailed layers of information for problem areas reveal graphics such as gauges and meters displaying information from the various data sources for that region.

Threshold levels that trigger visual alerts on the dashboard will be pre-set based on initial analysis. For example, green, yellow and red regions will be used to indicate varying levels of risk to the region. A single isolated incident may not be displayed, but an incident occurring at multiple locations and requiring a range of responding resources will trip the thresholds and set graphical indicators in the affected regions. Users will have the ability to customize their threshold values as they gain experience with the system. The demand on the system will increase depending on the severity of the crisis. Use of multi-processor servers in a clustered or distributed architecture will be investigated. The processing of the incoming data will be separated from the processes that send data to the dashboard in order to improve performance.

Figure 2 shows an example of a dashboard designed to be implemented for emergency response organizations at the municipal level. Each municipality will show the status of critical operating systems, with real-time data fed from relevant organizational sources. The interface will be tailored to the needs of the crisis managers accessing the system who will need to make quick decisions when responding to a crisis. The proposed dashboard will be reviewed and analyzed by practicing emergency managers for clarity and ease of understanding. The test of feasibility for the dashboard will be a demonstration of its design and functions to a set of selected emergency personnel, and their objective appraisal of its operation.

6 Contribution of the Dashboard to Disaster Management

The Dashboard will integrate real-time information from multiple sources in a disaster environment into a more advanced metric for assessing the level of risk to the whole community. The need for such a DSS is increasing with growing exposure to risk in metropolitan regions, critical interdependencies among lifeline systems, and greater uncertainty regarding threats from natural, technical, and deliberate disasters.

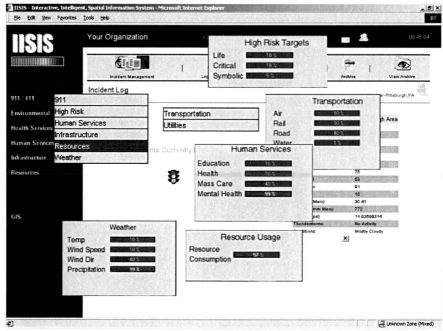

Figure 2: An Example of an Executive Dashboard for Crisis Operations

Acknowledgments:

This research is supported by National Science Foundation grant, CNS ITR#0325353, Secure CITI: A Secure Critical Information Infrastructure for Disaster Management.

References:

Axelrod, Robert and Michael D. Cohen. 1999. Harnessing Complexity: Organizational Implications of a Scientific Frontier. New York: The Free Press.

Comfort, Louise K. 1994. "Self Organization in Complex Systems." *Journal of Public Administration Research and Theory*, Vol. 4, No. 3, July:393-410.

Comfort, Louise K. 1999. *Shared Risk: Complex Systems in Seismic Response.* Amsterdam and New York: Pergamon Press.

Csete, Maria and John Doyle. 2004. "Bowties, Metabolism, and Disease." Pasadena: Unpublished Manuscript, California University of Technology. (doyle@caltech.edu)

General Accounting Office. 2003. *Homeland Security: Effective Intergovernmental Coordination is Key to Success.* Washington, DC: U.S. General Accounting Office. GAO-02-1013T.

Kauffman, Stuart.A. 1993. *The Origins of Order: Self-Organization and Selection in Evolution.* New York: Oxford University Press.

Klein, Gary A. 1993. "A Recognition Primed Decision Making (RPD) Model of Rapid Decision Making. In Gary Klein, Judith Orasanu, Roberta Calderwood, and Caroline E.

Zsambok, eds. *Decision Making in Action: Models and Methods.* Norwood, NJ: Ablex Publishing Corporation. 138-147. National Governors Association, 2002

Office of Domestic Preparedness,. 2002. *Developing Multi-Agency Interoperability Communication Systems: User's Handbook.* Washington, DC: U.S. Department of Homeland Security, p.2-1.

Public Safety Wireless Network Program. 2000. *Fire and EMS Communications Interoperability.* Washington, DC: Department of Justice and Department of Treasury. Information Brief.

Rao, R. 2003. *Cyberinfrastructure Research for Homeland Security.* University of California, San Diego: U.S. National Science Foundation Workshop Report. (February 26-27).

Rhetorical Choices: Words and Images in Flood-Risk Management

Charlotte Kaempf
Institute of Water Resources Planning, Hydraulic and Rural Engineering (IWK), University of Karlsruhe (TH), D- 76128 Karlsruhe, Germany

Abstract

Collaborative risk management depends on effective communication not only among members of authorities, consultancy, and the public, but also among members within any one domain. This paper examines tasks for a holistic approach in flood-risk management. Social interaction and scientific formalism are discussed in context with risk communication modes. A typology of technical visuals relevant for scientific risk communication is defined. Finally, results collected from a survey on the use of technical visuals by experts are discussed.

Keywords: data graphics, diagrams, flooding, hazard, illustrations, risk, technical reports, technical visuals, uncertainty, vulnerability.

1 Introduction

The extreme floods in Europe during Summer 2002 showed again the importance of well-concerted strategies among authorities, managers, stakeholders, and scientists for effective crisis intervention in the case of a flood, but also for proactive flood management to alleviate disastrous effects of future floods (State of Saxony, 2002). Flooding is a natural, regional hazard that potentially causes disaster (i.e., substantial damage). It is crucial to distinguish between given hazards and vulnerabilities, which we can influence by decisions (Weichselgartner, 2002). For example, flood risk management differs from management of hazardous workplaces, since we can control hazardous technologies but not hazardous rainfall events. On the other hand, the implementation of safety measures is controllable. Classification of hazards considers attributes such as cause and impact (see Table 1).

Attributes (poles marking continuum or dichotomy)	Examples for attributes (cause + spatio-temporal scale)
– **cause**: natural versus anthropogenic (given versus controllable) – **extent (spatial scale)**: global versus regional versus local international versus national – **duration (temporal scale) and intensity**: sudden event versus continuous change – **trend**: dynamic versus static – **frequency**: low versus high – **occurrence**: multiple versus singular; regular versus irregular. – **impact of effects**: low versus high	– **natural—regional—event**: water-related hazards—floods, snow storms, hurricanes, tidal waves (tsunamis), typhoons; other—tornadoes, fires, earth quakes, volcanic eruptions, – **natural—regional—process**: water-related hazards—droughts – **anthropogenic—local—event**: accidents in mining industry, terrorist attacks – **anthropogenic—regional—event**: accidents in chemical industry, crude oil transport, or space travel; – **anthropogenic—regional—process**: pandemics such as HIV/AIDS or BSE – **anthropogenic—global—process**: species extinction, chemical pollution of the commons (atmosphere, oceans): ozone depletion

Table 1: Classification of hazards

In environmental risk management, the dissemination of expertise is challenged in all discourse communities (domains), whether they concern the management of natural hazards and their effects (floods, fires, or earthquakes) or anthropogenic hazards such as in electrical engineering (control mechanisms for production and energy plants), or (bio-)chemical engineering (diffuse distribution of genetically modified organisms and toxic substances). Expert opinion on hazards and vulnerabilities is made accessible to peers and the public through documentation in reports and manuals as well as oral presentations at conferences and (in-)formal meetings. Communication about hazards and vulnerabilities of individual instances of disaster events has been discussed in the scholarly literature of communication sciences through detailed case studies. These publications focus on workplaces of hazardous technologies, emphasizing intuitive risk perception, acceptance of safety measures, ethical decision-making, and (in-)efficiency of information pathways in organizations (e.g., mining industry, see Sauer, 2003) or space travel, see Winsor, 1988).

In this article, I will discuss objectives in flood-risk management—focusing on event-oriented and preventive tasks (section 2). Thereafter, I review how risk-management experts communicate "risk" and related concepts within and among domains (section 3). A typology of technical visuals (section 4) will serve to define which kind of visual information will be suited to convey the meaning of "risk" in interdisciplinary expert groups. An ongoing research project analyzing expert reports is based on this typology (section 5). I conclude the paper with an outlook on the future path of this research project (section 6).

2 Objectives and Tasks in Flood-Risk Management

Flood risk has become a research focus and main management agenda worldwide since the UN initiated in 1991 the International Decade for Natural Disaster Reduction (IDNDR; see, for example, Plate, 2002). Kofi Annan, UN Secretary General, emphasized in his address for World Water Day 2004 that prevention should ensure that future hazards do not turn into unmanageable disasters (UN, 2004).Under the paradigm ("*Leitbild*") of sustainable development, organizations engaged in flood-risk management pursue a holistic model of event-oriented and preventive task clusters (see Figure 1).

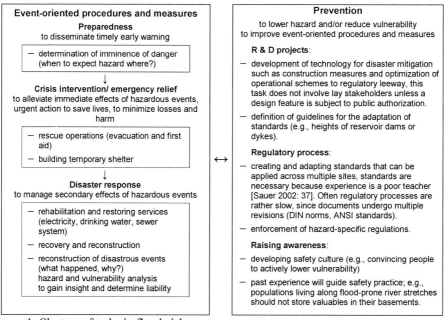

Figure 1: Clusters of tasks in flood-risk management.

All these tasks rely on statistical analysis to assess correlation and covariance of factors, or probabilities of hazardous extreme values such as HQ_{100}, which indicates a run-off peak that is expected to recur every 100 years. Statistical accounts for average loss/injury for various options are represented as arguments for particular policies, standards, explicit rules, or design drafts. Flood management experts (regional planner, hydrologists, hydraulic engineers) will analyze events to assess risk potentials for future events ("what-if" scenarios). They will identify deficiencies in logistics of emergency aid and complex anthropogenic cause-effect relationships, which may have led to loss, injury, and harm of lives and goods. For example, immediately after floodwaters recede, they will use gauge data to calculate the peak water level and duration of high-flood situation. Experts will fill in missing data, specify uncertain data, and collect local calibration data through on-site inspections. Possibly they will capture information contained in narratives of those people whose houses have been flooded. This also implies that experts need to transform descriptive, less precise data records in a second step into reports.

High-impact disaster events mark turning points in policy making. Politicians whom the affected public appeal to meet with officials of agencies, representatives of communities, non-governmental organizations (NGOs), re-insurances, operating companies, and subject matter experts (e.g., regional planners, hydrologists, hydraulics engineers) to better prepare local populations for a future run-off event of similar magnitude (e.g., Kaempf et al., 2000). Supranational institutions favor holistic models and critique approaches that are exclusively event-oriented (UNISDR, 2002: iv), which imply an orientation toward trans-disciplinary approaches, based on cross-disciplinary methods. The collaborative nature of risk management for both analytical and regulatory tasks shows the importance of communication among experts of various disciplines. In the next section, I will explore how partners convey main concepts related to risk management through words and images as well as how the choice of medium influences risk communication.

3 Communication of Risk Within and Among Societal Domains

The concept of "risk" is an inherent feature of our complex socio-technical environments. Nevertheless, there is no consensus on the definition of its meaning. Different people will define the concept "risk" in different ways (see Table 2). Risk-related language is domain-specific, such as regulatory language, academic jargon, or engineering symbols. "Risk" and related concepts (concomitant or opposite) are represented through domain-specific factual knowledge (publications), expertise (tacit or cultural knowledge), and individual embodied sensory experience (pit sense; sure instinct; speech and gesture; Sauer, 2002: 20, 219ff).

Risk	Concomitant concepts
− scientific and technical subject matter experts view "risk" as a function of statistical hazard (probability of occurrence) and assumed vulnerability (proneness to suffer loss of lives and goods) (Kron, 2002); − a measure of the expected loss, injury, and harm due to a hazard event of a particular magnitude occurring in a given area over a specific time period; − the conditional probability and magnitude of loss, injury, and harm attendant on exposure to a perturbation or stress; − **concomitant concepts**: hazard, vulnerability, uncertainty, threat, danger, exposure; and − **opposite concepts**: chance, safety, security.	− **hazard**: probability of an event (threat of an accident to occur); − **vulnerability**: the extent to which a community, structure, service or geographic area is likely to be affected by a specific hazard; − **uncertainty** (after Pollack, 2003) (a) *unfamiliarity*: new data, observation that does not match past experience; (b) *variability* (← *inherent randomness*): semantic of terminology; structure (organization) and dynamics (processes) in socio-technological systems; (c) *imprecision*: measured data; and (d) *incompleteness* (← *availability, existence*): gaps in data sets, criteria of rhetorical analysis.

Table 2: Risk-related lexical fields ("*Wortfelder*")

The concept of "probability of event frequency" is easiest to communicate, since it matches our understanding of chances in a lottery game. Sequential "cause-effect relationships" is another concept that deserves attention in communicating risk occurring in complex natural and socio-technical systems: for example, rain fall will cause flooding, which will cause failure of construction measures such as reservoirs and dams. In general primary effects are easy to imagine, but secondary effects are harder to grasp. Nevertheless, understanding secondary effects is important for our decisions about anthropogenic slow change (Pollack, 2003). Global hazards may pose regional and local risks; e.g., global warming may cause patterns of severe precipitation events to increase in Central Europe, which will in turn lead to increased local flood disasters.

The prevalent type of discourse is instrumental, which is often critiqued for its unidirectional orientation from experts to lay people and its emphasis toward technological progress. For example, hydrologic flood-relevant information is represented through mathematical formalism of statistics for extreme values for run-off in flood-prone areas (e.g., Plate, 2002, Helms et al., 2002). Another option of representing hydrologic information is based on geographical information systems (GISs) that are linked to a problem-specific database of interdisciplinary factual knowledge, which is maintained by a content management system (CMS). GIS-maps may present extent, frequency, and severity of flooding hazards and vulnerabilities. Experts use such GIS-based decision support systems (DSSs) to present risk classifications of flood-prone areas and scenarios of potential risk to political decision makers (Oberle et al., 2000; Fuerst, 2004). Such representations challenge equality with respect to fair public participation (Kaempf, 2001).

The optimal model for risk communication has been the subject of numerous scholarly publications: Jeff Grabill and Michele Simmons suggest dialectic discourse, a model emphasizing interactive communication among partners including the public (Grabill and Simmons, 1998). Donald Zimmermann discussed Ajzen's Theory of Reasoned Action, Witte's Extended Parallel Processing Model (EPPM), and the dialectic discourse model (Zimmermann, 2001). Beverly Sauer proposes a "reflexive" approach drawing on local knowledge and experience of lay audience (Sauer, 2002: 118). In practice, public participation is established through informal instruments such as EPA's Community-based Environmental Protection (EPA, 2004) or procedural authorization instruments such as "consensus meetings" in environmental risk assessment legislation, where the local population may participate as both effector and affectant. Stakeholder-oriented risk communication explores how people perceive the possibility of loss or injury and respond to and cope with the consequences of disaster. Peter Slovic argued in his landmark paper that those who manage health and safety, technical experts and decision makers, need to understand how lay people think about and respond to risk (Slovic, 1987).

Sauer analyzed risk communication specific to the sector of mining and identified six steps to stabilize risk-related data. She conceptualizes a cyclic model for the management of risk-related documentation, which considers all societal domains involved (Sauer, 2002: 72–3). She showed how information related to accidents in the mining industry moves among societal domains. For example, local knowledge of workers embedded in narratives is collected and transferred to the domain of science and engineering to reconstruct mining disasters. Because data are constantly transcribed, abstracted, summarized, synthesized, and (re-)combined to meet new needs, Sauer addresses texts within the documentation cycle as living documents (Sauer, 2002: 39). Kim Campbell stressed that risk documentation evoke confidence and neither anger nor ridicule. She showed that stakeholders and experts perceive risk differently. The public does not perceive severity of risk on a continuum between "low" and "high" but as extreme values at opposite poles (Campbell, 2001).

Since risk communication challenges the dichotomy between lay and expert knowledge, the public needs support to act as an equal partner in negotiations. Sauer found that risk specialists recognize the need of lay audiences for background information to understand numerical estimates of risk (Sauer, 2002: 12–13). On the other hand, she also found that with respect to interpreting statistical data experts are likely to share many of the biases that affect lay judgments about risk. Biases include sampling errors, availability errors (assessing the frequency of a class, or the probability of an event, by the ease with which instances or occurrences can be brought to mind), and anchoring errors (making estimates on the basis of adjusting values of an initial variable).

The status of research on risk communication within and among societal domains such as consultancy, academe, authorities, or lay people implies to investigate which types of technical visuals experts would use for documentations on anthropogenic risk in flood management. In the next section I will describe a typology of visuals used for such a study.

4 Typology of Technical Visuals (Tables and Figures)

For the systematic analysis of static technical visuals in expert risk communication, I classified visuals according to their purpose of representation. A review of the pertinent literature (e.g., Tufte, 2001a; Tufte, 2001b; Brasseur, 2003; Sachs-Hombach, 2003) implied the following types, clearly emphasizing the display of statistical attribute values:

Tables: to visualize logical structure of content in continuous text as

- *one-dimensional:* (ordered) lists of words or phrases (see Tables 1 and 2 in this article), outlines, or tables of content.
- *two-dimensional:* matrices or grids (e.g., spreadsheets). Tables are superior to graphs in showing exact numerical values and reporting on small data sets of 20 numbers or less.

Two-dimensional images or pictures: to present the likeness of physical objects as

- *photo(graph)s:* created via light in visible or infra-red range or electron microscopy.
- *(engineering) drawings, or illustrations*: to represent physical objects through lines in various views such as orthographic projection, section, and detail.

Diagrams, or schemes: to clarify the topic of concepts, constructions, and relations in a simple and structured way. In technical reports the most typical are

- *flow charts* that visualize interlinked processes, procedures, through boxes and arcs, which may form chains or webs (see for example Figure 1 in this article); and
- *Venn diagrams* (circles) that visualize relations between and among members of sets.

Data graphics: to support reasoning about quantitative information through

- *maps* (spatial data) classify real-world objects such as basic environmental parameters (e.g., weather map, topographic map); GIS maps display, for example, extent, frequency, and impact of flooding hazards; section views such as a city map are addressed as *plans*; the user does not take a perspective viewpoint as in drawings;
- *graphs*: display of (x, f(x)) data collection in a Cartesian coordinate system. (a) *time series* to understand the underlying theory of the data or to make forecasts (multiples: small graphical repeated to enhance comparison); and (b) *(bivariate) scatter plot* or *s. graph* serves the comparison of two sets of related quantitative data.
- *charts* represent relative magnitudes or frequencies of cases that fall into each of several specified categories, which are intervals of some variable: *(a) histograms,*

(b) *bar charts*, which include the Gantt chart that shows time-related project progress, and (c) *pie charts* (percentages); and

- *box plots* display the five-number summary in descriptive statistics, which consists of the smallest observation, lower quartile, median, upper quartile and largest observation.

Confections: to render the abstract more concrete, to add emotional support for decision making, pictograms, or infographics, are inserted into graphs (Tufte, 2001b: 121-51; Dragga and Voss, 2001). This type of visuals shows a close relationship between words and visuals.

5 Use of Technical Visuals: Preliminary Study

To assess the use of technical visuals in intra-domain risk communication I analyzed expert reports written in German: (a) proceedings papers (Horlacher and Martin, 2004) written by engineers engaged in reservoir management to prove successful application of expertise for problem solving; members of this community are diverse with respect to their field of practice (academe, consultancy, and authorities); and (b) progress reports (Plapp et al., 2003) written by doctoral students engaged in natural-disaster research projects to document research status; members of this community are diverse with respect to their disciplinary background (natural sciences, engineering, economics, and computer sciences).

A preliminary quantitative and qualitative content analysis on 15 reports for each group revealed that: the majority of visuals represent statistical data; practitioners often use diagrams indicating their ability for overview due to long-term expertise; doctoral students often present large data sets in tables instead of graphs; furthermore, they often use section photographs to depict lab equipment or physical models for experiments; and doctoral students used twice as many visuals as practitioners (0,8 v/p versus 0,4 v/p respectively; v/p: ratio of average number of visuals [v] over average page length [p]). In view of the latter result it is striking that some scientific writing on risks dispenses with technical visuals altogether (e.g., Pollack, 2003).

6 Conclusion and Outlook

My preliminary study showed that different groups of experts use visuals in different ways. That corresponds to the observation of polysemy for "risk," "hazard," and "uncertainty." Uncertainty of data may lead to disagreement and skepticism. Research on how competing representations of reality affect the discourse of science will support a better understanding of "uncertainty." Rhetoric that addresses questions of uncertainty in deliberation, judgment, and evaluation will provide an appropriate frame for the analysis of oral and written genres of risk communication. Furthermore, as art of invention, rhetoric is suitable to strengthen arguments in risk-related prose and images. Therefore, for probing the concepts hazard and vulnerability, both intrinsically uncertain, rhetoric is superior to dialectic—the method of probing truth of an argument based on axioms of formal logic.

The review on risk communication modes implies that partners in multidisciplinary projects will have to consider other disciplines as audiences otherwise these projects will not merit the label "inter-" or "transdisciplinary". Aside from further analyzing intra-domain use of technical visuals, future research may address the following topics:

- effectiveness of different genres of technical visuals to convey "risk"; for example, do graphs give more room for interpretation than tables or illustrations? Or, is the use of a specific genre determined by available technology?

- experts' writing for various audiences; for example, a comparison of intra-domain discourse versus writing for popular scientific journals such as *Natural History*, *Smithsonian*, or *Scientific American* in the US and *Spektrum der Wissenschaften* or *Naturwissenschaften* in Germany.

Aside from words and the five types of static visuals, discussed in this article, experts often supplement presentations with digital film clips in real time or fast motion. Viewing an amateur video of a flash flood in a village after dam failure makes the audience understand "hazard" and "vulnerability" more directly than a series of photographs.

Finally, any research that focuses on risk communication is challenged to point out ways for a cosmopolitan understanding of transnational risks as Ulrich Beck postulated twenty years ago at the wake of the first reports on Earth's warming and other anthropogenic global changes (Beck, 1986).

Acknowledgment

Thanks are due to Ken Baake, Department of English at Texas Tech University, Lubbock, TX, U.S.A., and Juergen Ihringer, Institute of Water Resources Management at University of Karlsruhe (TH), Germany, for valuable discussion.

References

(all Web references: last access 6/29/04)

Beck, Ulrich. *Risk Society: Towards a New Modernity*. London: Sage, 1992 (original German 1986)

Brasseur, Lee. *Visualizing Technical Information—A Cultural Critique*. Amityville, NY: Baywood, 2003.

Campbell, Kim Sydow. "The Effects of Information Design on Perception of Environmental Risk." *Proceedings of the 45th Annual Conference* (date, place). Arlington, VA: Soc Tech Comm, 4p, 2001.
Available at: http://www.stc.org/confproceed/1998/PDFs/00064.PDF

Dragga, Sam and Dan Voss. "Cruel Pies: The Inhumanity of Technical Illustrations." *Technical Communication* 48 (3): 265–78, 2001

Environmental Protection Agency (EPA). *Community-based Environmental Protection*. Homepage. 2004. Available at: http://www.epa.gov/ecocommunity/

Fuerst, Josef. *GIS in Hydrologie und Wasserwirtschaft*. Heidelberg: Wichmann, 2004.

Grabill, Jeff and Michele Simmons. "Toward a Critical Rhetoric of Risk Communication: Producing Citizens and the Role of Technical Communicators." *Technical Communication Quarterly* 7.4: 415–41, 1998.

Helms, Martin, Bruno Buechele, Ute Merkel, and Juergen Ihringer. "Statistical analysis of the flood situation and assessment of the impact of diking measures along the Elbe (Labe) river." *Journal of Hydrology* 267: 94–114, 2002.

Horlacher, Hans-B. and Helmut Martin (eds.). *Risiken bei der Bemessung und Bewirtschaftung von Fliessgewässern und Stauanlagen—Risks in Design and Management of Rivers and Reservoirs.* Dresdner Wasserbauliche Mitteilungen 27. Dresden, Germany: Inst. für Wasserbau und Technische Hydromechanik, 2004.

Kaempf Ch, W Buck, M Helms, B Buechele, O Evdakov, P Oberle, St Theobald, J Ihringer and F Nestmann. "Environmental Management Methodology to Support Sustainable

River Basin Development - Features of a Goal-Oriented Decision-Support-System (DSS)." - In: *Decision Support Systems for River Basin Management* (Int. Workshop, BfG Koblenz; April 6, 2000, Koblenz), BfG Veranstaltungen **4**, 88–109, 2000.

Kaempf, Charlotte. "Decision Support Systems (DSSs) for Environmental Management: Web-Based Communication Modules to Enhance Public Participation." *Proc. 48th Ann. Conf.* (May 2001, Chicago). Arlington, VA: Soc Tech Comm, 338-43, 2001. Available at: http://www.stc.org/ConfProceed/2001/PDFs/STC48-000145.pdf

Kron, Wolfgang. "Flood risk = hazard x exposure x vulnerability." (keynote lecture). In: Wu et al. (eds), *Flood Defence 2002*. New York: Science Press, 82–97, 2002. Available at: http://www.cws.net.cn/cwsnet/meeting-fanghong/v10108.pdf

Oberle P., S. Theobald, O. Evdakov, and F. Nestmann. "GIS-supported flood modelling by the example of the river Neckar." *Kassel Reports Hydraulic Engin* 2000 (9): 145–55, and Potsdam Institute for Climate Impact Research (ed.), *Proc. Eur. Conf. "Advances in Flood Research"*, 652–64, 2000.

Plapp, Tina, Christian Hauck, and Makky Jaya (eds.). *Ergebnisse aus dem interfakultativen Graduiertenkolleg "Naturkatastrophen"—Zusammenstellung ausgewählter Veröffentlichungen und Forschungsberichte (1998–2002)*. Karlsruhe, Germany: Inst. für Technologie und Management im Baubetrieb, Universität Karlsruhe (TH), 2003.

Plate, Erich. "Flood Risk and Flood Management." *Journal of Hydrology* 267: 2–11, 2002.

Pollack, Henry N. *Uncertain Science…Uncertain World*. Cambridge, 2003.

Sachs-Hombach, Klaus. *Das Bild als kommunikatives Medium—Elemente einer allgemeinen Bildwissenschaft*. Cologne, Germany: Halem, 2003.

Sauer, Beverly A. *The Rhetoric of Risk: Technical Documentation in Hazardous Environments*. Mahwah, NJ: Lawrence Erlbaum, 2003.

Slovic, Paul. "Perception of Risk." *Science* 236, 280–85, 1987.

Tufte, Edward R. *The Visual Display of Quantitative Information*, 2nd ed. Cheshire, Ct: Graphic Press, 2001a.

Tufte, Edward R. *Visual Explanations—Images and Quantities, Evidence and Narrative*. Cheshire, Ct: Graphic Press, 2001b.

United Nations (UN). The Secretary General (Kofi Annan). "Water and Disaster: Be Informed and Be Prepared" (Message on World Water Day 2004). 2004. Available at http://www.waterday2004.org/docs/UN_sgmessage.doc

United Nations International Strategy for Disaster Reduction (UNISDR; José Antonio Ocampo and David Johnson, eds.) *Guidelines for Reducing Flood Losses*. 2002. Available at: www.unisdr.org

Weichselgartner, Juergen. "Flutkatastrophe: Störfall der Natur oder der Vernunft." *GAIA* 12.4: 245–48, 2003.

Winsor, Dorothy. "Communication Failures Contributing to the Challenger Accident: An Example for Technical Communicators," *IEEE Transactions on Professional Communication*, 31.3: 101–7, 1988.

Zimmermann, Donald E. "Risk Communication—Lessons from Communication Science." *Proc. 48th Ann. Conf.* (May 2001, Chicago). Arlington, VA: Soc Tech Comm, 2001. Available at: http://www.stc.org/confproceed/2001/PDFs/STC48-000064.PDF

Hochwassermeldedienst am Rhein zur Hochwasservorsorge
Flood Warning Service at the Rhine-river for Precautionary Action of Flood

Dr.-Ing. Dieter Prellberg,
Ltd. Baudirektor, Landesamt für Wasserwirtschaft Rheinland-Pfalz, Mainz
State office for water resources Rhineland Palatinate, Germany

Kurzfassung

Der Hochwassermelddienst warnt vor Wassergefahren. Im Rahmen der Verhaltensvorsorge ist er ein wirkungsvolles Instrument zur Begrenzung von Hochwasserschäden.

Am Rhein werden von der Schweiz, über Deutschland bis in die Niederlande Hochwassermeldungen von vier Zentralen verbreitet. Sie haben ihre Zusammenarbeit vertraglich vereinbart und stimmen die Vorhersagen an den Schnittstelen ab.

Die Hochwasserzentralen verbreiten im Internet, über das Fernsehen (Videotext) und im Rundfunk für jedermann stündlich aktuelle Wasserstände und mehrmals am Tag Vorhersagen der weiteren Entwicklung (24 bis 72 Stunden) für alle Pegel am Rhein. Eingangsdaten für die Modelle zur Hochwasservorhersage sind Wasserstände und gefallene Niederschläge der eigenen Messnetze und vorhergesagte Niederschläge der Wetterdienste. Künftig sollen auch radargemessene Niederschläge genutzt werden.

Abstract

The flood warning service warns about flood-risks. Within the scope of the behavioral precaution it is an effective instrument to limit flood damages.

At the Rhine-river four flood warning centres – in Switzerland, Germany (2) and the Netherlands – disseminate flood information. The co-operation agrees upon contracts and the forecasts s will be harmonized at the boundaries.

The flood centres disseminate hourly latest water-levels and several times per day flood forecasts for the next 24 to 72 hours at all gauging sites of the Rhine-river. These warnings and forecasts are accessible for everyone by internet, videotext and broadcast. Input for the flood forecast models are water levels, measured precipitation and quantitative precipitation forecasts provided by the weather services. In near future radar measured precipitation will be used.

1 Allgemeines

Der Hochwassermeldedienst warnt vor Wassergefahren. Im Rahmen der Verhaltensvorsorge ist er Teil einer weitergehenden Hochwasservorsorge, die neben der natürlichen Hochwasserrückhaltung und dem technischen Hochwasserschutz ein wirkungsvolles Vorsorgeinstrument zur Begrenzung von Hochwasserschäden ist (LAWA, 1995).

In der Bundesrepublik Deutschland fällt die Regelung des Hochwassermeldedienstes in die Zuständigkeit der Länder. Rechtliche Grundlagen dazu enthalten die

Landeswassergesetze (Wassergesetz, 2004) oder sie werden aus den landesrechtlichen Regelungen über den Katastrophenschutz abgeleitet. Zur inhaltlichen Konkretisierung wurden in den meisten Bundesländern Hochwassermeldeverordnungen erlassen (Hochwassermeldeverordnung, 1986).

Der Hochwassermeldedienst umfasst in der Regel das Beobachten der Niederschläge, Wasserstände und Abflüsse in den Einzugsgebieten, für die ein Meldedienst besteht. Diese Beobachtungen werden zu Hochwassermeldungen ausgewertet und in der Form von aktuellen Wasserständen und Hochwasservorhersagen nach festgelegten Meldeplänen weitergegeben. Die Weitergabe dieser Meldungen erfolgt entweder auf der Verwaltungsebene an die für den Katastrophenschutz zuständigen Kreis- und Gemeindemeldestellen - diese warnen wiederum nach örtlichen Warnplänen (Rahmen-Alarm- und Einsatzplan, 1995) die Bevölkerung ihres Gebietes in geeigneter Weise. Oder die Hochwassermeldungen werden auf möglichst vielen Wegen direkt an die Betroffenen vor Ort verbreitet. Hierzu werden Rundfunkmeldungen, Videotext im Fernsehen und Internet genutzt. Das Hochwassergeschehen läuft in den einzelnen Einzugsgebieten oder auch in Teilabschnitten der großen Flüsse unterschiedlich ab. Damit regionale Besonderheiten bei der Erstellung von Hochwasservorhersagen berücksichtigt werden können und die interessierten lokalen Behörden sowie die Öffentlichkeit rasch und umfassend informiert werden können, ist eine dezentrale Organisation der Hochwassermeldezentren erforderlich und hat sich bisher bewährt. Dieses wurde auch von einer von der Internationalen Kommission zum Schutz des Rheins (IKSR) eingesetzten Expertengruppe bestätigt (Bestandsaufnahme, 1997).

2 Hochwasservorhersagen

Im Einzugsgebiet des Rhein (185 000 km²) sind eine Vielzahl von Vorhersage- und/oder Meldezentralen eingerichtet. Auf den ersten Blick könnte das den Schluss zu lassen, dass ein koordinierter Melde- und Vorhersagedienst damit nicht möglich ist. Tatsächlich sind diese Zentralen aber alle untereinander verbunden und vernetzt und es bestehen Vereinbarungen über die Weitergabe von Informationen.

Die Erstellung von Hochwasservorhersagen setzt die Kenntnis über die Entwicklung im Einzugsgebiet oberhalb des vorherzusagenden Bereiches voraus. Damit diese Kenntnisse aktuell verfügbar sind, werden die gemessenen Wasserstände und Niederschläge automatisch registrierender Stationen regelmäßig in die Hochwassermeldezentren fernübertragen (Bild 3). Zusätzlich werden – sofern vorhanden – die Hochwasservorhersagen obenliegender Zentralen ausgewertet. Sind entlang einer Flussstrecke mehrere Zentralen für die Erstellung der Hochwasservorhersagen zuständig, werden nur untereinander abgestimmte Werte an die Medien und die Öffentlichkeit weitergegeben.

Zur operationellen Wasserstands- und Abflussvorhersage werden empirische Verfahren oder Modelle eingesetzt. Empirische Vorhersageverfahren sind Pegelbezugslinienverfahren, zeitgerechte Abflusssummierung sowie die Kombination beider Verfahren. Als Modelle kommen in Abhängigkeit von der Größe des Einzugsgebietes Niederschlag-Abfluss-Modelle und Wellenablaufmodelle zum Einsatz. Beim Niederschlag-Abfluss-Modell wird der fallende Niederschlag, als hochwasserauslösend und innerhalb des Vorhersagezeitraumes den Hochwasserablauf bestimmend, für die Vorhersage berücksichtigt. Mit Wellenablaufmodellen wird die Verformung der Hochwasserwelle entlang einer Gewässerstrecke vorhergesagt, ohne Berücksichtigung des innerhalb des Vorhersagezeitraumes fallenden Niederschlages. Eine weitere Modellgruppe ist die der statistischen Filterverfahren, bei denen die Modellparameter aus historischen Hochwassern berechnet werden.

Der Zeitrahmen für relativ genaue Vorhersagen im Hochwasserfall liegt je nach Größe des Einzugsgebietes und der Fließzeit im vorherzusagenden Bereich beim heutigen Stand der angewendeten Vorhersageverfahren in Größenordnungen von 6 Stunden (z.B. Nahegebiet) bis 72 Stunden (z.B. Deltarhein in den Niederlanden). Dieser Zeitraum kann erweitert werden, wenn beim Einsatz von Niederschlag-Abfluss-Modellen statt der gemessenen Niederschläge quantitative Niederschlagsvorhersagen der meteorologischen Dienste verwendet werden. Entsprechend der Güte dieser Vorhersagen ist bei den damit erreichten verlängerten Vorwarnzeiten die Treffsicherheit der Hochwasservorhersage jedoch reduziert. Die so ermittelten Wasserstands- oder Abflusswerte sind nur noch als Frühwarnsystem einzustufen, das eine Abschätzung möglicher Entwicklungsszenarien in entsprechender Schwankungsbreite liefert. Auf der Basis von Wellenablaufmodellen erstellte Vorhersagen können erweitert werden, wenn geeignete hydrologische Vorhersagemodelle für die seitlichen Zuflüsse in das Ablaufmodell eingebunden werden.

3 Vorhersagekette am Rhein

Am Rhein erstellen vier Zentralen Hochwasservorhersagen:

- die Schweizer Landeshydrologie in Bern betreibt ein Modell bis Rheinfelden – künftig bis Basel
- die Hochwasservorhersagezentrale Baden Württemberg (HVZ) in Karlsruhe rechnet von Rheinfelden bis Worms
- das Hochwassermeldezentrum Rhein (HMZ Rhein) in Mainz rechnet von Karlsruhe bis Emmerich – das HMZ Rhein ist eine gemeinsame Einrichtung des Landes Rheinland-Pfalz (Landesamt für Wasserwirtschaft) und der Wasser- und Schifffahrtsverwaltung des Bundes (WSD Südwest) und ist im Hochwasserfall mit Personal beider Dienststellen besetzt
- die niederländische Zentrale RIZA in Lelystad rechnet von Andernach bis zur Rheinmündung

Ergänzt wird die Kette demnächst durch eine eigenständige Vorhersage für den Bodensee, in die neben der HVZ und der Landeshydrologie dann auch das österreichische Bundesland Vorarlberg eingebunden ist. Die Zentralen stehen unter einander in Verbindung und stimmen sich hinsichtlich der Vorhersagen an den Schnittstellen ab. Die Hochwassermeldungen von Rheinfelden bis Emmerich werden im Internet und Videotext durch das HMZ Rhein verbreitet, so dass sich nach außen die Darstellung für die gesamte deutsche Rheinstrecke als homogen darstellt.

Jede Zentrale übernimmt jeweils die Vorhersage des Oberliegers als Input für die Erstellung der Vorhersagen in seinem Bereich. Eigenständige Vorhersagen für Zuflüsse werden dabei ebenfalls integriert; z.B. übernimmt das HMZ Rhein Vorhersagen für die Mosel vom HMZ Mosel.

Die parallele Berechnung im Bereich Karlsruhe bis Worms durch HVZ und HMZ Rhein beruht darauf, dass international vereinbart das Vorhersagemodell der HVZ Grundlage für die Steuerung der Rückhaltungen am Oberrhein ist. Deren Wirkung ist auf den Pegel Worms ausgelegt. Die beiden Zentralen haben sich aber darauf verständigt, dass die Vorhersagen für diesen Bereich durch das HMZ Rhein nach außen verbreitet werden, ggf. nach vorheriger Abstimmung bei unterschiedlichen Berechnungen.

Als gemeinsame Plattform bieten die Meldezentren am Rhein eine Übersicht auf der Internetseite IKSR (www.iksr.de). Über einen Link gelangt man von dort auf die Homepage des jeweils zuständigen Meldezentrums.

4 Maßnahmen zur Verbesserung der Hochwasservorhersagen

Es muss deutlich darauf hingewiesen, dass der Nutzen von Vorhersagen entscheidend von deren Güte und Verlässlichkeit abhängt. Eine lange Vorhersagezeit mit zwangsläufig geringerer Verlässlichkeit kann einen Vertrauensverlust der Betroffenen in die Vorhersage bis hin zum möglichen Ignorieren von Warnungen bei extremen Hochwasserereignissen bewirken und wäre damit dem Ziel der Schadensminderung kontraproduktiv. Für den effektiven Einsatz von Hochwasservorhersagen ist eine Abwägung zwischen Vorhersagezeit und Vorhersagegüte anhand des Bedarfs vor Ort erforderlich.

Als Teil des Aktionsplans Hochwasser der IKSR sollten auch Wege zur Optimierung und Verknüpfung der Hochwassermeldedienste sowie zur Verbesserung der längerfristigen Hochwasservorhersage aufgezeigt werden.

Die Schwerpunkte der Maßnahmen waren bisher:

- Ausbau und Betrieb eines Kommunikationsnetzes zum Datenaustausch,
- Abschluss der erforderlichen Vereinbarungen zum Daten und Vorhersageaustausch,
- Weiterentwicklung und Einsatz der notwendigen hydrologischen Vorhersagemodelle,
- Verbesserung der hydrometeorologischen Daten zur Hochwasservorhersage.

Zu Letzterem wurde mit dem Deutschen Wetterdienst vereinbart, dass die Hochwassermeldezentren täglich die auf der Grundlage hochauflösender Rechenmodelle (Lokal-Modell) in Stundenschritten über einen Zeitraum von 48 Stunden erstellten quantitativen Niederschlagsvorhersagen erhalten. Die Niederschlagsvorhersagen enthalten den Gebietsniederschlag in flüssiger und fester Form in einem 7 km -Raster für das gesamte Rheineinzugsgebiet. Damit wird es möglich, Tendenzen denkbarer Hochwasserentwicklungen auch über mehrere Tage anzugeben. Die damit möglichen längeren Vorwarnzeiten liefern - entsprechend der Unsicherheit der Niederschlagsvorhersage – zunächst nur eine Abschätzung und ersetzen nicht die quantitative Hochwasservorhersage mit ihrer hohen Treffsicherheit. Zur Zeit arbeitet der Deutsche Wetterdienst an der Verbesserung des Lokal-Modell, dessen Maschenweite auf 2,8 km gesenkt werden soll. Außerdem sollen dabei über ein Bodenmodell der Einfluss des Wassergehaltes im Boden auf die Niederschlagstätigkeit sowie das horizontale Verdriften des Niederschlages berücksichtigt werden. Mit einem Aktionsprogramm 2003 hat der Deutsche Wetterdienst die Verbesserung der Wetter- und damit auch der Niederschlagsvorhersage zur Schwerpunktaufgabe erklärt.

Zur weiteren Erhöhung der Treffsicherheit der Vorhersagen ist die umfassende Kenntnis des tatsächlich gefallenen Niederschlages erforderlich. Damit diese Daten operationell verfügbar sind und in Vorhersagemodelle eingebunden werden können, rüsten die Länder ihr Niederschlagsmessnetz auf automatisch registrierende Messgeräte mittels hochauflösendem elektronischen Wägesystem und integrierter Datenübertragung (Ombrometermessnetz) um. Diese Maßnahme erfolgt in Abstimmung mit dem Deutschen Wetterdienst, der ergänzend sein Messnetz zur Zeit ebenfalls entsprechend umstellt (Messnetz 2000). Es wird dann flächendeckend-repräsentativ ein gemeinsames Ombrometermessnetz zur Verfügung stehen, auf das die Hochwassermeldezentren und der Deutsche Wetterdienst jederzeit Zugriff haben. Ein entsprechendes Messnetz besteht bereits in Baden-Württemberg und Rheinland-Pfalz. In Bayern, im Saarland, in Hessen und in Nordrhein-Westfalen wurde mit dem Aufbau begonnen.

Dieses Ombrometermessnetz ist auch Voraussetzung dafür, dass künftig über Radarmessungen quantitative Niederschlagsdaten in Echtzeit gewonnen werden können.

Hierzu hat der Deutsche Wetterdienst ein Radarverbundnetz errichtet, in das auch angrenzende Stationen der Nachbarstaaten einbezogen werden. Das Radar liefert in einem Radius von 125 km Daten, aus denen durch Aneichung die Rohdaten der Radarmessung in quantitative Niederschlagsdaten mit einer räumlichen Auflösung von ca. 1 km * 1 km umgewandelt werden. Diese Aneichung ist erst mit Hilfe der automatischen Niederschlagsmessgeräte möglich.

Die erforderliche Entwicklungsarbeit wird von der Länderarbeitsgemeinschaft Wasser (LAWA) in dem Projekt RADOLAN gefördert und soll im Jahr 2004 abgeschlossen werden. Derzeit liefert das qualitative Radarbild bereits wertvolle Informationen über die räumliche und zeitliche Variabilität des Gebietsniederschlages (örtlich differenzierte Niederschlagshöhen, Zugrichtung).

Davon ausgehend, dass in absehbarer Zeit quantitative radargemessene Niederschlagsdaten verfügbar sind, soll in einem weiteren Projekt RADVOR-OP mit finanzieller Unterstützung der LAWA eine radargestützte, zeitnahe Niederschlagsvorhersage für den operationellen Einsatz (Niederschlag-Nowcasting-System) entwickelt werden. Die zur Zeit täglich erstellten Vorhersagen (basierend auf 00, 12 und 18 Uhr (UTC)-Beobachtungen) des numerischen Wettervorhersagemodelles, die ca. fünf Stunden nach dem Beobachtungstermin zur Verfügung stehen sind als Input-Daten für das hydrologische Vorhersagemodell der Wasserstände bzw. Abflüsse je nach Vorhersagezeitpunkt zwischen sechs und 17 Stunden alt.

Sofern also innerhalb dieses Zeitraumes Abweichungen zwischen der tatsächlichen Wetterentwicklung und den entsprechenden Berechnungen des Wettermodelles auftreten, können diese nicht berücksichtigt werden und bei der Hochwasservorhersage wird bisher ggf. auf einer zum Vorhersagezeitpunkt unzutreffenden Niederschlagsvorhersage aufgesetzt. Daher sollen die aktuellen radargemessenen Niederschlagsfelder in die Datenassimilation des Wettervorhersagemodelles beim DWD integriert werden, um die Niederschlagsvorhersagen bis zu einem 15-stündigen Vorhersagezeitraum weiter zu verbessern. Damit wird es auch möglich, das Wettermodell im Hochwasserfall häufiger (on demand) zu rechnen. Darüber hinaus sollen ergänzend hierzu die kurzfristige Entwicklung und Zugbahnen radargemessener Niederschlagsfelder in einem Vorhersagezeitraum bis max. vier Stunden über sogenannte Radar-Tracking-Verfahren abgeschätzt werden.

5 Zusammenfassung

Der Hochwassermeldedienst warnt vor Wassergefahren. Im Rahmen der Verhaltensvorsorge ist er Teil einer weitergehenden Hochwasservorsorge. Im Rahmen des Meldedienstes verbreitete Hochwasservorhersagen sind neben der natürlichen Hochwasserrückhaltung und dem technischen Hochwasserschutz ein wirkungsvolles Vorsorgeinstrument zur Begrenzung von Hochwasserschäden.

In der Bundesrepublik Deutschland fällt die Regelung des Hochwassermeldedienstes in die Zuständigkeit der Länder. Rechtliche Grundlagen dazu enthalten die Landeswassergesetze oder sie werden aus den landesrechtlichen Regelungen über den Katastrophenschutz abgeleitet. Zur inhaltlichen Konkretisierung wurden in den meisten Bundesländern Hochwassermeldeverordnungen erlassen.

Der Zeitrahmen für relativ genaue Vorhersagen im Hochwasserfall liegt je nach Größe des Einzugsgebietes und der Fließzeit im vorherzusagenden Bereich beim heutigen Stand der angewendeten Vorhersageverfahren in Größenordnungen von 6 Stunden (z.B. Nahegebiet) bis 72 Stunden (z.B. Deltarhein in den Niederlanden). Dieser Zeitraum kann erweitert werden, wenn beim Einsatz von Niederschlag-Abfluss-Modellen statt der gemessenen Niederschläge quantitative Niederschlagsvorhersagen der meteorologischen Dienste verwendet werden. Entsprechend der Güte dieser Vorhersagen ist bei den damit erreichten verlängerten Vorwarnzeiten die Treffsicherheit der Hochwasservorhersage jedoch reduziert. Die so ermittelten Wasserstands- oder Abflusswerte sind nur noch als Frühwarnsystem einzustufen, das eine Abschätzung möglicher Entwicklungsszenarien in entsprechender Schwankungsbreite liefert. Auf der Basis von Wellenablaufmodellen erstellte Vorhersagen können erweitert werden, wenn geeignete hydrologische Vorhersagemodelle für die seitlichen Zuflüsse in das Ablaufmodell eingebunden werden.

Der Nutzen von Vorhersagen hängt entscheidend von deren Güte und Verlässlichkeit ab. Eine lange Vorhersagezeit mit zwangsläufig geringerer Verlässlichkeit kann einen Vertrauensverlust der Betroffenen in die Vorhersage bis hin zum möglichen Ignorieren von Warnungen bei extremen Hochwasserereignissen bewirken und wäre damit dem Ziel der Schadensminderung kontraproduktiv. Für den effektiven Einsatz von Hochwasservorhersagen ist eine Abwägung zwischen Vorhersagezeit und Vorhersagegüte anhand des Bedarfs vor Ort erforderlich.

Schwerpunkte der Maßnahmen zur Verbesserung der Hochwasservorhersage sind:

- Errichtung redundanter Messsysteme und Datenübertragung
- Ausbau und Betrieb eines Kommunikationsnetzes zum Datenaustausch,
- Verbesserung der hydrometeorologischen Datenbasis zur Hochwasservorhersage und der Niederschlagsvorhersage,
- Weiterentwicklung der notwendigen hydrologischen Vorhersagemodelle.

Literatur

Bestandsaufnahme der Meldesysteme und Vorschläge zur Verbesserung der Hochwasservorhersage im Rheingebiet; Bericht Nr. II – 12 der KHR, 1997. Internationale Kommission für die Hydrologie des Rheingebietes, Lelystad (NL) Internationale Kommission zum Schutze des Rheins, Koblenz

Hochwassermeldeverordnung vom 26. Februar 1986, Gesetz- und Verordnungblatt für das Land Rheinland-Pfalz, Nr. 7/1986, Seite 69.

LAWA-Länderarbeitsgemeinschaft Wasser Leitlinien für einen zukunftsweisenden Hochwasserschutz, Umweltministerium Baden-Württemberg, Stuttgart 1995

Rahmen-Alarm- und Einsatzplan "Hochwasser" Ministerium des Innern und für Sport, Rheinland-Pfalz, Mainz 1995

Wassergesetz für das Land Rheinland-Pfalz (Landeswassergesetz - LWG-), in der Fassung der Neubekanntmachung vom 22. Januar 2004 (GVBl. S. 53)

Social Structure, Trust and Public Debate on Risk

Elke M. Geenen
Institute for Sociology, University of Kiel, D-24118 Kiel, Olshausenstr. 40, Germany

Abstract

This paper is concerned with the importance of social distribution of knowledge in context of public debate on risk. The phenomenological attempt of Alfred Schütz' sociology of knowledge gives the instrument to differentiate on the side of the public in risk communication processes, while up to now in scientific, political and administrative discourses the citizens are widely constructed as a mass or a unified total subject. Additionally it has to be seen that part of the social structure is an unequal distribution of cultural capital (Pierre Bourdieu). Using Niklas Luhmanns differentiation between trust and confidence it will be examined on which state of knowledge trust or confidence are predominating in dependence of cultural capital.

Keywords: risk communication, sociology of knowledge, relevance, trust, confidence, expert, cultural capital, social inequality.

1 Introduction

In sociological discourses on risk cultural capital (Pierre Bourdieu) generally is no topic up to now. This means: An emphasis on the question what risk means to individuals with different equipment in cultural capital is missing. In this respect the people in public risk communication processes are thought as a unit - as a mass or a crowd. In just the same way public is conceptualised by politicians in risk discourses. Still we are not arrived at, by Norbert Elias (1987) so called, "The society of individuals".

2 Social Distribution of Knowledge

2.1 Quality of Knowledge and Structures of Relevance

To make a difference in this respect would already be easy, if we have a look on Alfred Schütz' phenomenological concept of the social distribution of knowledge. Not only the quantity of available knowledge varies individually but the quality of knowledge which shows different degrees of clarity, exactness and familiarity (Schütz, 1971, p. 16). Schütz distinguishes between simple knowledge and knowledge which shows familiarity. In every day situations it is enough to know that things are functioning. Action may be habitually, automatically and only in part conscious, so that people have typical solutions for typical problems (Schütz, 1972, p. 65). Background assumptions (Hintergrundannahmen) about the constancy of the world make it possible to move every day in a world, which seems to be granted and is largely taken unquestioned and confidentially, unless countermand is seen. Schütz distinguishes between 4 zones of knowledge and three types of ways to handle information.

The individual knowledge is divided into zones of different and reduced relevance which are pervading each other (for a rough model see Figure 1). 1. We have know-how in the zone of primary relevance (within our reach), 2. In a second zone we have instruments ready to use, 3. We take relative irrelevant zones unquestioned as granted (unless countermand), 4. There are zones which seem to be absolutely irrelevant to us, because their change does not influence our problems, as we believe. As passive recipients of events out of our control relevance is imposed on us. We may transform them into events of essential relevance as a result of our chosen interests. Imposed relevance is not completely transparent for the recipient (Schütz, 1972, pp. 94 f.). Growing anonymity in modern societies reduces the zone of common essential relevance while the zone of political, economical and social imposed relevance increases.

Schütz distinguishes between three ideal types how to handle knowledge: "the man in the street", the "expert" and the "well informed citizen", all of them are in dependence of the topic of knowledge at the same time in each individual (Schütz, 1972, pp. 87 f.).

The *man in the street* has a vague, for practical purposes adequately precise knowledge about trustworthy recipes which he follows as if they were ritual procedures. He lives so to speak naively in his own essential relevances and considers imposed relevances only as conditions of his action without understanding their origin and their structure. His opinion therefore is ruled more by feeling than by information.

The knowledge of the *expert* is concentrated on a limited subject. His opinions base on secured allegations. By his special field he is directed towards a special system of imposed relevance, which he accepts as important for his thinking and acting.

The *well informed citizen* tries to gain reasonable and well-founded opinions and believes himself to be qualified to decide whether someone else is an expert or not. About the genesis of relevance actually or potentially imposed to him he has to collect as much of knowledge as possible (Schütz 1972, p. 97). In his effort to restrict the zone of irrelevance, he has to take into consideration that things which are of no relevance to him at the moment, tomorrow may be imposed to him as primary relevant, and that even the sphere of absolute irrelevance can reveal as threatening.

Schütz does not consider, that the individual participation in the three ideal types is not equally distributed. This means, we have to consider an inter-individual difference of cultural capital in the social distribution of knowledge. The chance to be a well informed citizen, who has the ability to decide reasonable and well-founded whether someone is an expert and or not, or to develop criteria which judgement out of several of experts should be preferred, crucially depends on the availability of a methodical knowledge about how to examine given information.

People whose cultural capital is not sufficient are outside of the zone of their essential relevance mainly dependent on confidence in sense of Luhmann.

Is the corresponding cultural capital absent, knowledge inevitably stays vague and imposed relevance can only be tolerated or suffered as conditions for ones own actions, without understanding. This means that in risk communication a passive position and may be a feeling of helplessness is the result.

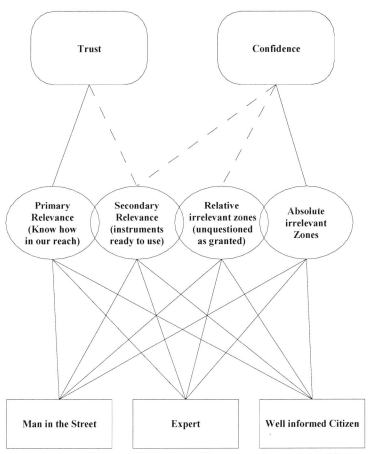

Figure 1: Zones of Relevance, Types how to Handle Knowledge, and Confidence vs. Trust.

2.2 Confidence and Trust

Niklas Luhmann discriminates between confidence (Zuversicht) and trust (Vertrauen). To have *confidence* means to believe that ones own expectations are not disappointed. To have no confidence means to have to live in a world of permanent uncertainty. *Trust* affords a situation at risk. Trust means to have the chance to choose rationally between alternatives (cf. Luhmann, 2001, pp. 147 f.). The feeling of missing influence on decisions may turn trust into confidence. On the other hand, the chance of individual influence may change confidence into trust. If even confidence is missing, a feeling of alienation (Entfremdung), non satisfaction or anomie follows (Luhmann, 2001, p. 156).

In Luhmanns concept trust is tightly bounded to the term of risk. But for the individual ability to differ between danger and risk a specific knowledge is necessary. For the man in the street Aids or BSE appear as dangers, about which he is not able to develop a well founded opinion, while they are risks for the physician, the analytical chemist or the well informed citizen, who are able to consider carefully in decision situations, select rationally and based on criteria out of public or specific information pools.

In scientific papers, debates on risk communication and in thinking about trustworthy concepts, it is not seen yet, that a concept for "the well informed citizen" has to fulfil other criteria than for "the man in the street".

Further: From disaster situations result, that at least partly, involved persons or victims turn from "men in the street" to "well informed citizens". In disaster subcultures they possibly may become "experts", if we consider that in the aftermath of a striking event risk consciousness increases considerably. An indicator for this is the search for information of stricken or involved persons.

The well informed citizen is sensible, if offers are uncovered. This means in here, that stricken persons who have changed in the concerned topics from men in the street to well informed citizens examine the public debate, respectively statements of politicians and acts of the concerned administration how risks are described and whether pronounced intentions to act are followed by deeds.

Only a small amount of people continues for a long time to stay in the state of well informed citizen, if we look on a specific risk. For most people the specific risk looses the state of primary relevance. But some may be characterised as continually sensitised to detect uncovered offers. For example, they learn to distinguish whether political promises to enforce mitigation are uncovered or whether political and administrative action is directed towards the realisation of mitigation. In the political system the communication code is voting (by elections). This communication code is a kind of noise in public discourse on risk because the attempt to get electoral votes is interfered by the underlying political concept of public as an undifferentiated mass.

If we regard cultural capital in context of risk discourse as the ability to select specific information about risks and events and to be able to decide if trust in political and administrational coping strategies is justified, an asymmetric structure of expectations will become obvious. While the political system widely proceeds from the assumption of the citizen as the man in the street, the citizens cultural capital differs widely and the individuals are in the spectrum between the expert, the well informed citizen and the man in the street. So, for the well informed citizen public announcements about risks seem to be vague or, if proofed in information seeking process, as even wrong so that, as a kind of message results, the political and administrational system appears as at least partly incompetent to estimate risks or real risks appear as veiled. So, in debate on risk on side of political actors often a nearly passive and receptive and not reflective public is wrongly assumed. An example is the myth assuming panic of the "mass", following strong events or in a disaster situation or if warnings are officially released.

In consequence of inadequate debates on risk trust and even confidence as essential symbolic capitals of the political and administrative system are spent.

3 Conclusion

In scientific as well as in public debate on risk three factors generally are not recognised:

First, that societies are societies of individuals in the sense of Norbert Elias and that they are not to be modelled as an ideal unified total subject.

Secondly, the distribution of knowledge in society is unequal because the cultural capital of individuals differs considerable and the individual amount and quality of cultural capital influences the ability to select information out of different information pools and to participate in debates on risk.

Thirdly, the intra-individual knowledge is not homogenous. So we are men or women in the street on one subject and experts or well informed citizens on other topics.

To say it polemically: in context of risk and in risk debate the structures of the earth are conceptualised more differentiated than the public which mostly still seems to be a homogenous crowd or mass. Politicians, administrators and scientists somehow behave as if public would be a glass box. In fact they are confronted with a black box. Instead of looking at them as strangers they are pseudo-stylised (Geenen, 2002). One of the results is, that forecasting or predicting the public reactions in general or those of individual citizens or of social groups in situations at risk are widely failing. One example is (besides the missing ability of natural scientists to predict earthquakes), that the estimations of public behaviour in case of an earthquake prediction or warning are much more situated in the realm of myth than based on well founded theoretical and empirical assumptions and results (Geenen, 1995). So the political and administrational anxieties are projected into an imaginary mass. One result of this wrong going societal perception and risk communication is distrust and even loss of confidence on the side of the citizens, who are in their search for information confronted with an opaque mixture of truth, symbolic policy, missing or superficial knowledge and withhold information and delayed released information.

Georg Simmel points out that trust and confidence are one of the most important synthetic powers in societies (Simmel, 1992, p. 393). If trust and confidence are lost by processes pointed out in here, the synthetic ability and cohesion of society can be reduced considerably. In the sociology of knowledge Alfred Schütz has shown us one way to overcome this dilemma.

References

Bourdieu, Pierre: Entwurf einer Theorie der Praxis auf der ethnologischen Grundlage der kabylischen Gesellschaft. Übersetzt von Cordula Pialoux und Bernd Schwibs, Suhrkamp, Frankfurt a. M. 1979.

Elias, Norbert: Die Gesellschaft der Individuen, hgg. von M. Schröter, Frankfurt a. M. 1987.

Geenen, Elke M.: Soziologie der Prognose von Erdbeben. Katastrophensoziologisches Technology Assessment am Beispiel der Türkei, Duncker & Humblot, Berlin 1995.

Geenen, Elke M.: Soziologie des Fremden. Ein gesellschaftstheoretischer Entwurf, Leske + Budrich, Opladen 2002.

Luhmann, Niklas: Vertrauen. Ein Mechanismus der Reduktion sozialer Komplexität, 2. erw. Aufl, Ferdinand Enke Verlag, Stuttgart 1973.

Luhmann, Niklas: Vertrautheit, Zuversicht, Vertrauen: Probleme und Alternativen. In: Martin Hartmann und Claus Offe (Hg.), Vertrauen. Die Grundlage des sozialen Zusammenhalts, Campus Verlag, Frankfurt, New York 2001, p. 143-160.

Schütz, Alfred: Gesammelte Aufsätze I. Das Problem der sozialen Wirklichkeit. Mit einer Einführung von Aron Gurwitsch und einem Vorwort von H. L. van Breda, Martinus Nijhoff, Den Haag 1971.

Schütz, Alfred: Gesammelte Aufsätze II. Studien zur soziologischen Theorie. Herausgegeben von Arvid Brodersen. Übertragung aus dem Amerikanischen von Alexander von Baeyer, Martinus Nijhoff, Den Haag 1972.

Simmel, Georg: Das Geheimnis und die geheime Gesellschaft, in: Soziologie. Untersuchungen über die Formen der Vergesellschaftung, hgg. von Otthein Rammstedt, Suhrkamp, Frankfurt a. M. 1992, S. 383-455.

Geospatial Data Acquisition by Advanced Sensors in Disaster Environments

H.-P. Bähr, A. Hering Coelho, J. Leebmann, E. Steinle, D. Tóvári
Institute of Photogrammetry and Remote Sensing (IPF), University of Karlsruhe, 76128 Englerstr. 7, Karlsruhe

Abstract

New possibilities for the support of common disaster management tasks have been arisen by the development of new sensor technologies. On the one hand, the development of real time kinematic GPS and on the other hand the improvement and miniaturisation of orientation sensors made techniques like Augmented Reality (AR) and airborne laser-scanning (ALS) possible.

This paper shows three applications of these new techniques in the field of disaster management:

(1) Using augmented reality for decision support when appearing floods or earthquakes, and using airborne laserscanning for (2) improvement of digital terrain models for flood simulations and (3) automatic detection and interpretation of damaged structures.

In the following, the three different applications are explained in more detail. (1) An augmented reality system (ARS) solves problems of flood preparedness and earthquake disaster response. In the context of disaster management it is useful to represent different invisible disaster-relevant information and to overlay it with the image of reality.

(2) Digital Terrain Models (DTM) play a very important role in hydrological numerical models. A very accurate DTM is needed when analysing the run-off in an area. Besides topography, terrain objects also have influence on the water flow. The main goal of the task is to improve the quality of the DTM generated from ALS data.

(3) Laser-scanning data can be used to detect changes in urban areas after strong earthquakes. Detection and classification of these changes are used to recognise building damages as an important information input for a disaster management system based on GIS techniques.

Keywords: laserscanning, Augmented Reality, classification, visualization, earthquake, flood.

1 Introduction

In case of natural disasters usually the lack of information hinder the rescue teams from a fast and efficient aid. Information about the situation in the affected area or the expected impact of the disaster are necessary for handling the situation. Mostly, the rescuers don't have any knowledge about the location, which can seriously slow down the speed of assistance. The lack of training has a similar effect. For these reasons, reliable and fast data acquisition methods are needed. A fast and accurate perception facilitates the decision making. New applications are developed in this field at the IPF based on the data acquired by some advanced sensors. These sensors are the parts of the augmented reality and airborne laserscanner systems.

The aim of an augmented reality system (ARS) is to superimpose a real-world scenery with a virtual extended version of itself in real time, while airborne laserscanning (ALS) is a common tool of high accurate and dense topographic mapping . The following work aims to encourage the use of ARS and ALS in the field of disaster management. The investigations show promising results in recognition of earthquake caused damages and in flood predictions and simulation.

2 Technical Aspects

First of all, a short overview about both techniques will be given.

2.1 Airborne Laserscanning (ALS)

The laserscanner is mounted on an aircraft, and emits laser. pulses directed to the terrain. The emitted pulse is reflected back by the ground or an object on the surface (building, vegetation). From the travelling time the distance of the instrument and the measured point can be calculated. The position of the instrument is provided by GPS system and the direction of the instrument is provided by an inertial navigation system. From these measurements 3D position of the measured points can be calculated. The measurement density is up to 5 point/m². In Figure 1 some principles of this technique are visualized. A basic element of the measurement system is the equipment for using *differential* navigation in the *Global Positioning System (dGPS)*. This is a precisely positioning methodology based on the satellite navigation system GPS. The accuracy can be increased up to some cm using the differential mode.

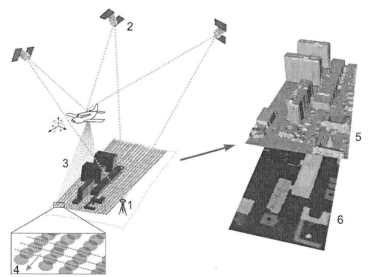

Figure 1: Principles of airborne laser scanning; 1 & 2 - dGPS equipment, 3 – distance measurement rays, 4 – scan pattern on ground, 5 & 6 - 3D and 2D visualizations of digital elevation models (DEMs)

Additionally, navigation sensors for registering the aircrafts movements, so-called *Inertial Navigation Systems*, are installed onboard.

A possible laser scanning technique is the so-called *pulsed* laser scanning. This means that a short pulse of an electromagnetic wave, here in the near infrared, is emitted from the system. The time that has passed until backscattered signals parts can be detected by the system is registered and used to ascertain the covered distances. This information, together with the knowledge about the emitting characteristic of the system and the sensor position, enables to compute co-ordinates in 3D of the measured points.

Inside the footprint several objects with different heights might be hit. Therefore, a laser pulse is generally not uniformly backscattered, most often various reflections reach the sensor system. It can be selected which backscattered part should be regarded as the measurement signal. Mostly, either the signal part first reaching the sensor or the one arriving last is used and the measurement is called done in *first pulse* respective *last pulse mode* (see e.g. [Steinle and Vögtle, 2000]). Last pulse data is suitable for *digital terrain model (DTM)* generation, while first pulse data can be used e.g. for vegetation detection. The reflected energy (intensity) can be measured as well, and it depends on the reflectors surface, size and shape.

In post-processing *digital elevation models (DEM)* are produced based on the 3D point co-ordinates. They are produced by interpolating the heights of the knot points of an overlaid regular grid, typically 1m spaced, from the neighboured measured points.

This technique can be applied in poor weather conditions and at night as well. Through this unique feature, laserscanning might be the most useful airborne mapping technology in disaster prevention and management.

2.2 Hardware Components of the Augmented Reality System

Augmented Reality (AR) systems superimposes virtual objects to the reality and in that way extend the reality. This can be used e.g. to visualize or acquire with spatial data in natural environments.

In principle, a mobile ARS consists of devices for measuring position and orientation, computing the virtual scene, displaying the combined result and, eventually, a digital video camera to capture the images of reality.

For position tracking, differential GPS is used. The orientation is measured by the inertial measurement unit (IMU).

The user of the ARS can choose from two ways of displaying the superposition of virtual and real images. One can use a portable computer with a video camera or a retinal display (see figure 2.).

The framework for holding all the components is a backpack rack. A pole for the GPS-antenna, a tripod and an aluminium suitcase are mounted on the backpack. A tripod that is mounted on the backpack carries the camera and an IMU measuring the orientation of the camera. A second IMU is attached to the retinal display.

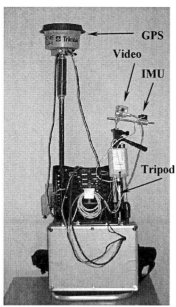

Figure 2: Left: Detailed view of the ARS equipment. Handheld version
(middle), see-through-version (right) of the system

3 Applications in the Context of Disaster Environments

3.1 Recognition of Earthquake Caused Damages

A prerequisite to recognise earthquake caused damages at buildings is to detect changes in general. The recognition process consists of classifying the changes into different groups and to interpret each group, i.e. to connect a semantic. The definition of these groups can be done already before the recognition process, but generally it limits them to already known phenomenon. Depending on the task, this needn't be a disadvantage, e.g. if a damage detection after earthquakes should serve as basis for rescue measures, which is the case for the presented project, it is even necessary. The division into the different classes is here application- and not occurrence-driven, i.e. the partitioning is carried out according to the needs of the rescue organizations and not according to what can be found in the data in the special case.

In order to detect changes in the buildings geometry, a reference of the initial state must exist. This means that a data base of house models should be set up in regions were change detection is planned, e.g. earthquake-endangered areas. When the methodology need to be applied, building models of the present state must be created, e.g. by carrying out a laser scanning flight and using automatic building modelling procedures based on the resulting digital elevation model (DEM).

After having obtained two models in 3D (pre- and post-earthquake), they are overlaid and change measures are computed. The used measures are dependent on the further application. To detect building damages, an important change indicator is the difference in volume of both models. Other change measures can be defined e.g. based on the alteration in orientation, position, size and continuity of single planes and modifications in relations between the planes, e.g. neighbourhood.

However, it is crucial to differentiate changes from alterations in the data due to the measurement features (compare [Steinle & Bähr, 2002]). Using laser scanning derived DEM, changes in position of geometric primitives (points, edges and planes) lower than the grid spacing, mostly 1m, and in the height of a single pixel below 15cm, what is the measurement accuracy, cannot be detected reliably.

3.2 Augmented Reality System for Earthquake Disaster Response

The ARS is part of a disaster management tool (DMT) developed by the Collaborative Research Center 461 (CRC461) at the University of Karlsruhe (TH). The ARS is developed as a piece of specialized equipment for supporting the rescuers that try to find people trapped in the rubble of collapsed buildings. The disaster management tool is thought to support search and rescue teams in case of disaster as well to train them that they are prepared for rescue tasks. ARS component provides a very detailed view. It represents different invisible disaster relevant information and overlays it with the reality on the same scale.

In the preparing phase before an event the rescuers could be trained with simulated damage situations displayed by an ARS. Another important measure to become prepared for a possible event is to establish consciousness in the population for the risks they are confronted with. Simulated damage superimposed with reality could be a tool to show these risks. By doing this, it could improve the readiness to spend money for preparedness measures e.g. to strengthen the buildings and in that way to reduce the number of victims.

As an example of an application a simulated collapse based on a detailed 3D-model of a building was used. A digital surface model of the campus of Karlsruhe University measured by a laser-scanner was also available. In figure 3 and figure 4 parts of these data sets can be seen. Then the overlay was generated using the video camera based ARS. The camera was directed towards the corner at which the damage was simulated. The superposition of the camera image of the corner and the laser-scanner data is displayed in figure 5.

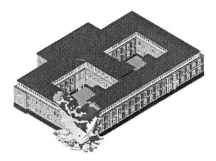 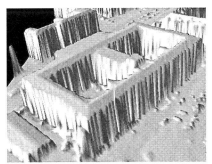

Figure 3: Simulation of building damage   Figure 4: Laser data of the building studied
(Schweier et al.,2003).   (DSM).

Figure 5: Superposition of video and laser-scanning model (transparently shown)

3.3 Digital Terrain Models for Flood Simulation

Flood water simulations are based on digital terrain models. The most efficient data acquisition method for DTMs is nowadays the ALS technique. The gathered data perform a model of the surface, which includes the buildings and vegetation as well. These objects should be removed by a filtering process. The filtering process and the subsequent DTM generation using airborne laserscanning data can be significantly improved by classification of non-terrain objects. This high resolution terrain models and the determination of vegetation areas (position, size, density and height of trees etc.) has to be carried out to model hydrologic processes, runoff models to simulate floods. For these purposes it is necessary to classify all 3D objects on the surface of the earth, i.e. mainly buildings and trees/bushes, in some cases also terrain objects like rough rocks which may be additionally included in the detected objects.

All features are derived exclusively from laserscanning data itself without additional information like spectral images or GIS data. This is caused by specific restrictions in context of disaster management, where data acquisition has to be carried out also during night time and poor weather conditions.

A 3D object classification method has been developed that detects and classifies objects into 3 main categories: vegetation, buildings and terrain. In this phase the method is based on raster data. Using laserscanning data, the pixel-wise classification can't be enough reliable. Therefore, first of all a segmentation of 3D objects has been realized. For each segment, object-specific features are extracted and used for the above mentioned classification process, like height texture, border gradients, first/last pulse height differences, shape parameters or laser intensities. The Fuzzy-logic method is used to obtain a reliable building, vegetation and terrain classification based on these features.

By using a-priori knowledge about the characteristics of 3D objects in laserscanning data for definition and extraction of object-relevant features, suitable results can be achieved using fuzzy logic. The classification may significantly improve the quality of the DTM, by adding the terrain objects.

Vegetation can be segmented e.g. by the watershed algorithm. Since the segments are usually parts of a whole canopy, these need to be merged. Taking segment shape, relative location and height of local maximums into consideration, the segments that possibly belong to the same tree can be joined.

3.4 AR for Disaster Relief: High Water

Flood simulations basically rely on calculations based on measurements of rain and topography. In literature (Casper, Ihringer, 1998) several different solutions for flood simulation are presented. Airborne laserscanning provides a solution for this purpose.

Results of high water simulations can be used as a basis for planning disaster response measures. Visualisation of results is needed in order to help the local forces interpreting the simulations. Traditional visualisation methods like maps or 3D simulations are not suitable to show all details, which can be very important in disaster situation. 3D visualisation with AR can be useful to expand the comprehension of the situation (fig. 6). This example shows, which gates and windows are endangered by the increasing water-level.

In the city of Cologne authorities installed a mobile flood protection wall, to protect the old town against flooding from the river Rhine. In order to simulate the effect of such mobile walls, the user can draw a polygon in a 2D view of the flood zone which represents the virtual position of a mobile wall (fig. 7 left). In the 3D view of the area (fig. 7 right) the user can see the water surface around the protected area.

Figure 6: Virtual water surface nearby a real building, result of occlusion processing

Figure 7: Left: 2D view of polygon representing an area protected with mobile walls. Right: the effect of the protected area in the 3D view of the flood zone

4 Conclusions

The advanced sensors provide high quality data acquisition before and after natural disasters. The knowledge of both scenes is necessary to understand the situation.

Airborne laserscanning technology supplies highly detailed elevation models. The augmented reality is an appropriate technology, in case real-time additional information is needed. The enhanced vision of real world can assist a better understanding of the processes in disaster situations. This can help to prepare for natural disasters like flood. Very fast data acquisition allows efficient decision making and simulations. After strong earthquake events, the visualisation of the original scenes can help the rescue teams to find the survivors.

The advanced sensors data can also be used together in disaster management applications: this true for flood prediction, where the surface model is based on ALS data and the simulation is carried out by the AR system.

References

Casper, M. and Ihringer, J., 1998. GIS-gestützte Regionalisierung von Abflüssen in Gewässern Baden-Württembergs unter Einsatz von Arc/Info und Arc-View. http://www-ihw.bau-verm.uni-karlsruhe.de/members/casper/region.htm, 2003.

Hirschberger S., Markus M., Gehbauer F., Grundlagen neuer Technologien und Verfahrenstechniken für Rettungs-, Bergungs- und Wiederaufbaumaßnahmen; Berichtsband des Sonderforschungsbereiches 461 für die Jahre 1999-2001, Karlsruhe, 2001, pp. 639-686 .

Leebmann J., A stochastic analysis of the calibration problem for augmented reality systems with see-through head-mounted displays. ISPRS Journal of Photogrammetry and Remote Sensing, Special Issues on Challenges in Geospatial Analysis and Visualization, 2003, pp. 400-408.

Markus, M., Hirschberger, S. and Schweier, L., 2002. Building Damage Classification for Damage and Loss Assessment. Proceedings of the *International Conference on Earthquake Loss Estimation and Risk Reduction, Bucharest, Romania*, 2002

Steinle, E. and Bähr, H.-P., 2002. Detectability of urban changes from airborne laserscanning data. *International Archives of Photogrammetry and Remote Sensing (IAPRS)*, Proceedings of the International Symposium on Resource and Environmental Monitoring, Hyderabad, India, December 3-6, 2002

Steinle, E. and Vögtle, T., 2000. Effects of different laserscanning modes on the results of building recognition and reconstruction. *IAPRS*, Amsterdam, The Netherlands, Vol. XXXII, Part B3, p. 858-865.

C. Schweier, M. Markus, and E. Steinle, Simulation of earthquake caused building damages for the development of fast reconnaissance techniques, Natural Hazards and Earth System Sciences, accepted September 2003.

Tóvári, D., Vögtle, T., 2004. Classification methods for 3D objects in laserscanning data. *ISPRS Conference*, Istanbul Turkey, 12-23 July 2004.

Voegtle, T., Steinle, E., 2003. On the quality of obeject classification and automated building modelling based on laserscanning data. The International Archives of Photogrammetry, Remote Sensing and Spatial Information Sciences, Dresden, Germany Vol. XXXIV, Part 3/W13, 8-10 October 2003, ISSN 1682-1750

Wursthorn S., Hering Coelho A., Staub, G., 2004: Applications for mixed reality. *ISPRS Conference*, Istanbul Turkey, 12-23 July 2004.

Integration and Dissemination of Data Inside the Center for Disaster Management and Risk Reduction Technology (CEDIM)

Petra Köhler, Matthias Müller, Manuela Sanders, Joachim Wächter
GeoForschungsZentrum Potsdam (GFZ), Telegrafenberg, D- 14473 Potsdam, Germany

Abstract

This document describes the development of an information infrastructure as foundation for the formulation of a "Risk Map Germany" inside the Center for Disaster Management and Risk Reduction Technology (CEDIM). Specific modules and components are defined and implemented to enable common data and information management and scientific networking.

Keywords: data and information management, geographic information systems, information infrastructure, risk mapping.

1 Introduction

The virtual Center for Disaster Management and Risk Reduction Technology (CEDIM) aims at the analysis of risks due to different natural and man-made hazards and at the development of new methodologies and tools for risk reduction. CEDIM was founded in December 2002 by GeoForschungsZentrum Potsdam and University of Karlsruhe (TH). It combines various disciplines reaching from geology and geophysics to engineering, actuarial sciences and economics as well as geomatics.

During the first project a digital „Risk Map Germany" is going to be generated basing on integrated assessment of risk. Germany and Baden-Württemberg in particular is taken as an example to develop procedures for the identification, quantification and evaluation of risks for human beings, societies and infrastructures and for the visualization of the results in specific but comparable modules merging to one final risk map. Following hazards are examined: earthquakes, floods, storms, space weather and man-made hazards. The related working groups are supported by working group "Data management and GIS" responsible for the establishment of common data and information management and for "risk mapping".

2 Data and Information Management in CEDIM

2.1 Collecting and Organizing Data

Basing on reference and on thematic data spatial interrelations can be detected, so the provision of a suitable and common data base forms a basic requirement for substantiated risk assessment. Various spatial data are acquired, centrally prepared and managed.

Following data sources are actually used in the context "Risk Map Germany:

- Reference data of the Surveying and Mapping Agency of Baden-Württemberg (ATKIS, DEM)
- Digital Elevation Models generated from GTOPO30 and SRTM
- CORINE: Land cover (Federal Statistical Office Germany)
- Socio-economical data (Statistical Offices etc.)

- Demographics, buying power, characteristics of buildings etc. (infas GEOdaten)

The resulting heterogeneous data stock must be harmonized in different ways to achieve a uniform supply for the particular working groups and, thus, the foundation for the results' comparability. Differences according to data formats, geometric mismatches, reference systems and spatial and timely resolution are dissolved to achieve a common data base. For example, all the data is projected according to the recommendations of the Arbeitsgemeinschaft der Vermessungsverwaltungen der Länder der Bundesrepublik Deutschland (AdV) and the initiative of INSPIRE, the Infrastructure for Spatial Information in Europe, which is aiming at establishing uniform guidelines to collection, management and handling of spatial data whole over Europe.

Further measures of data integration and harmonization are

- conversions of different file formats to an accepted format,
- transformations from the raw data (e.g. x, y, z data) to readable data (e.g. GRID),
- adjustment of different geometries,
- elimination of erroneous geometry types resulted from the ongoing workflow in the working process,
- completion and combination of reference data and attribute data like statistics and calculations.

This time CEDIM data is organised and stored in a file based system. But in the view of the cumulative growing amount of data it has to be the aim for the future to realize data management by a special geo-database.

2.2 Data Dissemination

The integrated data base is made available by a central and web-based data distribution service (figure 1) and provides selected visualization and GIS functionalities.

The Arc Internet Map Server (ArcIMS, ESRI) implementation allows project members to inform themselves about the characteristics of existing data and its applicability for their specific questions and tasks. Suited data can be downloaded and further processed in their own geographic information systems (GIS).

Furthermore, the service includes a metadata component. Metadata – „data about data" – comprise information on the owner of the data, its content, quality, restrictions in usability etc. and in CEDIM follow the world-wide approved standard „ISO 19115: Geographic Information – Metadata". These standardized metadata are the foundation for an open and exchangeable documentation of data and maps and the technological implementation of the so-called catalog service (Nebert, 2001) which in future will allow a comprehensive overview of data and finally complex data retrieval by the specification of user-defined search criteria. The final architecture of the data distribution service is shown in figure 2 and illustrates the specific components and relations of data and metadata management (data tier), procedures and services (business logic tier) and user access (client tier).

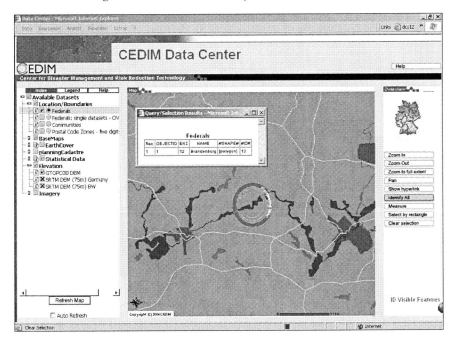

Figure 1: The data distribution service in CEDIM

The data distribution service is linked to the web-portal of CEDIM serving on the one hand as representation outwards providing general information to the public and on the other hand as internal information and communication platform providing tools for document management, exchange of information and discussions etc.

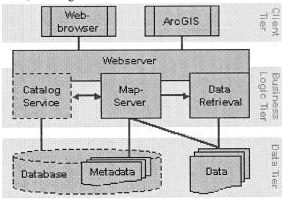

Figure 2: Architecture of the CEDIM data distribution service

2.3 Result: Information Infrastructure

The described components merge in an information infrastructure (FEMA, 2001, Köhler & Wächter, 2004) consisting of

- a technological infrastructure basing on approved architectures and standards
- an integrated data base

- applications and information systems for the solution of complex tasks and questions
- the completion of requirements due to the users' working processes

Following this holistic approach basic data supply, data processing, mapping of information flows and last but not least the transfer of scientific results to the public can be fostered and scientific networking warranted.

3 Risk Mapping

As the allocation of risks due to different hazards is not consistent, the results of risk assessment will be documented by a "Risk Map Germany". In this context "Risk mapping" means the graphical illustration of spatial facts and relations due to a defined area.

Initially existing approaches are identified and compared by working group "Data management and GIS". Suitable methodologies are tested and enhanced and new methodologies for the visualization of risks are developed and realized together with the other sub-projects. Cartographic approaches and GIS techniques like spatial surface analysis, the usage of geo-statistical procedures for spatial allocation and the development of scenarios are used to achieve most acceptable results. Figure 3 demonstrates an example of the first result of the earthquake working group in a choropleth Map.

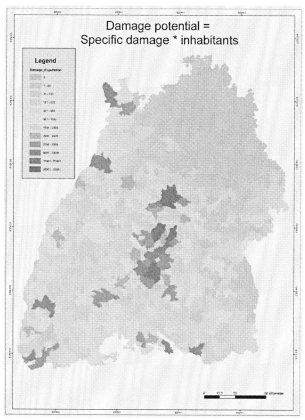

Figure 3: Example map: Damage potential per inhabitants in Baden-Württemberg by earthquake events

The comparability of hazard and risk visualization with its specific characteristics plays an important role to finally allow the transfer of the particular results in one common risk map.

References

Federal Emergency Management Agency: *Information Technology Architecture, Version 2.0 - The road to e-FEMA (Volume 1)*. Available electronically at http://www.fema.gov/pdf/library/it_vol1.pdf, 2001.

Köhler, Petra and Wächter, Joachim: Towards an open information infrastructure for disaster research and management: Data management and information systems inside DFNK. *Natural Hazards, Special Issue* (accepted)

Nebert, Doug, editor: *Developing Spatial Data Infrastructures: The SDI Cookbook*. Available electronically at http://www.gsdi.org/pubs/cookbook/cookbook0515.pdf, 2001.

Risk Communication via Environmental Education. Case Study Galtür, Austria

Angela Michiko Hama[1,2]**, Michael Seitz**[1,2]**, Johann Stötter**[2]

[1] *alpS – Centre for Natural Hazard Management, A-6020 Innsbruck, Austria;*
[2] *Institute of Geography, University of Innsbruck, A-6020 Innsbruck, Austria*

Abstract

Risk communication is most relevant for a sustainable development of mountain regions. Environmental education is an effective way of communicating risk associated with natural hazards, as it interacts with the general public in a target-group specific manner. This paper introduces an environmental education concept, which was developed for the community of Galtür. Insight into the approach and its programmes on alpine natural hazards will be given and its aptitude for risk education discussed.

Keywords: environmental education, risk communication, natural hazards, Galtür, Alps, sustainable development.

1 Introduction

Creating awareness, understanding and responsibility for hazard and risk related issues is fundamental for a sustainable future of the Alps, especially as mountain regions experience global change and its effects at first hand. Due to a significantly increasing human impact and climate change processes, the Alps are at higher risk of being affected by natural hazards than many lowland areas (Becker and Bugmann, 2001; Stötter et al., 2002a).

The community of Galtür is located 1600m above sea level in the Paznaun valley, Tyrol. Having undergone significant socio-economic and ecological changes since the mid-twentieth century, Galtür has turned from farming village to tourist resort, representing the typical transformation of most high Alpine settlements in the Eastern Alps. Situated in a hazard-prone area, Galtür has often been struck by avalanches. Yet in February 1999, an avalanche event of catastrophic dimension took place, which caused 31 casualties and severe direct and indirect financial losses (Heumader, 2000; Stötter et al., 2002b). This experience triggered a change in the perception of alpine natural hazards in the village and lead to the establishment of an institution called 'Alpinarium'. Focusing on the interaction between man, society and the alpine environment and emphasising natural hazards, this institution will function as an exhibition and environmental education centre in order to contribute to a sustainable development of the Alps.

In the following, the scientific concept being developed for environmental education activities at the Alpinarium will be presented. After explaining the link between environmental education and risk communication, the structure of the concept will be introduced and an example of its implementation with respect to natural hazards and risk given. Finally, the need for environmental education-based risk communication as an essential part of innovative risk management will be discussed.

2 Environmental Education as a Means of Risk Communication

Agenda 21 devotes a chapter to sustainable mountain development and calls for enhancing the knowledge-base on these fragile ecosystems. Additionally, the Agenda recognises the importance of education for sustainable development. Thus, the Agenda appeals strongly for disseminating information and creating a sense of stewardship for the environment in general as well as mountain regions in particular (UN, 1992). The 2002 World Summit on Sustainable Development reconfirmed these postulations, and, as a consequence, the 'UN Decade of Education for Sustainable Development 2005-2014' was adopted. By doing so, the role of environmental education is strengthened on a supranational level. Moreover, the goals of modern environmental education are furthered: Education *about* the environment, *from/in* the environment and *for* the environment (Palmer, 1998). This means that in addition to disseminating knowledge, awareness needs to be raised via personal experiences and people empowered by developing action- and decision-making competences.

The UN/ISDR (2004) states that efficient disaster risk reduction can only be achieved with regard to sustainable development and addresses risk communication as one of its future challenges in the context of 'knowledge management'. Studies on risk awareness and risk communication show that serious misconceptions of risk exist and that there is a pressing need for more public-centred communicative approaches (e.g., Lave and Lave, 1991; Gutteling, 1996; Walker et al., 1999; Morgan et al., 2002). Moreover, research places more emphasis on perception and awareness studies rather than on applied risk communication strategies.

Due to its integrative and interdisciplinary nature, environmental education is a promising tool for communicating risk and it contributes to the concept of resilience as part of proactive adaptation. Effective environmental education follows the principles of experiential learning by incorporating cognitive, affective and psychomotoric dimensions and takes up the constructivist approach of so-called 'situated learning' (Reinmann-Rothmeier and Mandl, 2001). Thus, it closely observes general psychological models of learning as well as models of risk information and seeking (e.g., Griffin et al., 1999). Moreover, successful environmental education aims to gain a significant outreach and developes therefore target-group oriented programmes in order to cater for the varying needs of the general public. So far, disaster risk communication has hardly ever utilised environmental education as a format and has mainly targeted stakeholders in the narrower sense rather than the general public (as the *actual* stakeholder in society).

The environmental education concept for Galtür supports the idea of risk communication and concentrates on this issue of highest concern for the population of the Alps. When dealing with risk in combination with environmental education in the Alps, risk related to natural hazards is considered. Following definitions from the technical and natural sciences, risk is consequently defined as the probability of occurrence of a specific process and the height of the related damage potential. By applying a set of selection criteria (target group orientation, current issues and affairs, spatial context, feasibility of implementation, interdisciplinarity, experiential approach), three core themes were chosen for the programmes of the concept. These are the topics of 'headwater', 'socio-economic change' and 'natural hazards'. All three themes comprise the aspect of risk, yet the theme 'natural hazards' addresses it most explicitly. The research methodology of the environmental education concept for Galtür will further illustrate how the goal of risk communication as a part of sustainable development can be reached starting from a predominantly local and regional level.

3 The Concept and its Application

The holistic approach takes both man and the environment into account and incorporates geographical, sociological as well as psychological, pedagogic and didactic methodologies. In addition, new methods are being devised. The concept is split into three integral parts, which encompass analysing the educational potential of the Alpine environment, recording the educational interest in the Alpine environment and developing implementation strategies for environmental education programmes. In the first study, the informative value and meaningfulness of a landscape in respect of educational issues is assessed by a newly developed GIS-based multistep-multicriteria procedure (see e.g., Poschmann et al, 1998; Bastian und Steinhardt, 2002). The educational interest of society in Alpine landscapes is extracted by applying methods of empirical social research, such as standardised questionnaires, semi-standardised interviews as well as expert interviews (Kromrey, 1994). The third part of the concept combines and compares the results of the previous two studies, considers current environmental issues and logistic matters and adapts adequate teaching and learning methods for the programmes. Community participation and capacity building is attained via training local residents for eventually running the centre. These three studies answer the central research problem, which is: "What *can* society, what does society *want to*, and what *should* society learn from an Alpine environment?" (cf. Hama et al., forthcoming).

Each of the core themes disposes of several sub-themes, which are structured in a modular system. Every module consists of a basic module and several additional modules, with the basic module providing a general overview of the respective topic and the additional modules highlighting specific aspects. Depending on the target-group and their accompanied needs, interests and predilections, the components of the module can be used accordingly. Observing the selection criteria introduced above and including the results of the studies lead to the contents of the modules, which are, naturally, interrelated. In the following, the module 'avalanches' is exemplified with respect to risk communication, as avalanches are the most prominent hazard in the Galtür region:

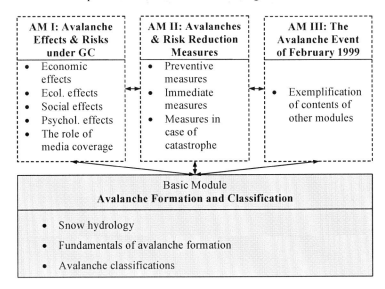

Figure 1: The 'avalanche' module.

The module on avalanches represents a sub-theme of the core theme 'natural hazards' and engages in communicating risk associated with avalanches. The basic module deals with the fundamentals of snow science and avalanche formation, which is a prerequisite for understanding the complex triggering mechanisms of avalanches in theory and in the field. Three additional modules complete the avalanche topic: Number I tackles the subject of the effects and risks of avalanches under changing climatic and socio-economic conditions. Number II gives insight into various risk reduction measures, both on the institutional/political and individual level. In Number III, the contents of the other modules are exemplified by applying them to the avalanche event of 1999, paying respect to the local history. For putting these diverse contents into practice, a pool of teaching and learning methods is available. This pool ranges from snow measurements, role plays on decision-making strategies, discussions, excursions, interviews with experts, lectures and field exercises to workshops on, e.g., spatial planning strategies.

The choice of these contents and methods is backed up by first results of the studies in progress. The assessment of the educational potential of the environment of Galtür shows that the region offers many possibilities for environmental education programmes focusing on the topic of avalanches. In the winter season, many spots in the permanent settlement area as well as in the nearby ski resort can be easily accessed, while avalanche-prone zones need to be avoided. In summer, both valley bottom and slopes function as potential locations for activities, which render, for example, avalanche fences in the starting zones within reach.

Figure 2: View of Galtür and surroundings. Source: Seitz: private photograph.

The semi-standardised interviews conducted with 200 tourists in the summer of 2003 and winter 2003/04 reveal that 88 percent of the interviewees are interested in learning more about environmental issues such as natural hazards. This high percentage calls for fostering this interest and supports the focus of the environmental education concept for the Alpinarium. However, a significant majority underestimates the risks associated with natural hazards: Even though nearly all of the persons know about the avalanche event of 1999, only a minority of them changed their behaviour relating to avalanches. Furthermore, only approximately ten percent enumerate the topics of global change and natural hazards when asked about the types of information attainable in mountain regions. These results clearly demonstrate that improving the public knowledge about and coping capacities for natural hazards and disaster risk is to be considered a top priority issue (Seitz, personal communication).

Regarding the implementation strategies, avalanches and their risk are a manageable topic that offers multiple options for environmental education programmes in Alpine regions. With the community commemorating the fifth anniversary of the 1999 catastrophe by featuring an exhibition on this specific avalanche event, programmes are being developed for accompanying this exhibition. First test-runs with local children had an extremely high acceptance rate and displayed the adolescents' eagerness to deal with disaster risks.

In the following, an example of a programme component of the 'avalanche' module is given. The event forms a part of the 'Additional Module II' (see Figure 1) on avalanches and risk reduction measures and concentrates on familiarising individuals with prevention and immediate measures relevant for their winter sports activities. Courses aiming at increasing the avalanche awareness of sportsmen are attributed a high value, because the number of people killed by self-induced avalanches exceeds the casualties caused by extreme events by far (Lawinenwarndienst Tirol, 2000):

Avalanches: One-Site Decision-Making & Immediate Measures	
Target group	Skiers and snowboarders
Goals	Awareness of avalanche risk
	Knowledge-based decision-making
	Proficiency in avalanche search procedure
Contents	Decision-making theories and application
	Avalanche safety equipment and its handling
	Field exercises including avalanche drill
Terrain	Ski resort Galtür, Alpinarium and surroundings
Logistics	Season: Winter
	Accessibility: by foot, ski and snowboard
	Group size: 4-10
	Group leader: mountain and ski guide

Table 1: Example of a programme with focus on avalanche risk.

The environmental education concept developed for Galtür can be easily transferred to other mountain regions and will be standardised. Obviously, all three studies need to be slightly modified to meet the requirements of the specific region, yet the overall approach and its underlying didactic and pedagogic principles will remain unchanged, as these possess a general validity. Standardising the concept makes it possible for other communities to adopt their own sustainable environmental education programmes and to contribute to efficient and effective risk communication in the field of natural hazards. Westendorf in the Tyrol was selected as a second study area in order to verify the results obtained in Galtür.

4 Outlook

Environmental education-based risk communication needs to be regarded as an essential part of innovative risk management, as it disseminates state-of-the-art knowledge on risk and natural hazards research. Above all, it enables people to act on their behalf, take action for the environment and contribute to a sustainable development.

However, most risk management concepts do not incorporate risk communication sufficiently enough. In the context of the Alps, Kienholz et al. (2004) developed an integral risk management concept for mountain hazards. The concept can be visualised in a spiral consisting of the steps of risk assessment, risk prevention, event management and regeneration. Communication with decision-makers and experts is also included in all four steps, but education of the general public is not allowed for. From the perspective of environmental education, the participation of both active (stakeholders in a narrower sense) and passive (general public) members of society is to be strongly desired. Therefore, communicating risk has to take place on several levels, which are, among others, the general public, experts and decision-makers. The environmental education concept for Galtür takes this into account by developing programmes tailored to the specific requirements of the respective target-group. Unfortunately, structured risk communication approaches have not yet been implemented as an umbrella concept in risk management strategies.

Bottom-up risk communication activities like in Galtür are very valuable on a local and regional scale and put the idea of a Local Agenda into practice. Besides, they bridge the constantly widening gap between research, policy-making and the broad public. Nevertheless, more top-down initiatives are needed in order to advocate risk communication internationally and enhance the importance of public outreach. The newly founded European Disaster Science Initiative (EuDiSI) sets an example with the topic of 'dissemination, communication and education' playing a major role in its agenda. Most successful risk communication can only be made when bottom-up and top-down initiatives cooperate and feed into each other's approaches by generating multipliers on both sides.

Acknowledgements

The authors thank the Alpinarium Galtür Dokumentation GmbH for the financial support of this research project.

References

Bastian, Olaf and Steinhardt, Uta, editors: *Development and Perspectives of Landscape Ecology.* Kluwer Academic Publishers, Dordrecht, Boston and London, 2002.

Becker, Alfred and Bugmann, Harald, editors: *Global Change and Mountain Regions: The Mountain Research Initiative*, IGBP Report 49. IGBP, 2001.

Griffin, Robert J., Dunwoody, Sharon and Neuwirth, Kurt: Proposed Model of the Relationship of Risk Information Seeking and Processing to the Development of Preventive Behaviours. *Environmental Research Section A*, 80: 230-245, 1999.

Gutteling, Jan M. and Wiegman, Oene: *Exploring Risk Communication.* Kluwer Academic Publishers, Dordrecht, Boston and London, 1996.

Hama, Angela M., Seitz, Michael, Sansone, Anja and Stötter, Johann: An Environmental Education Concept for Galtür, Austria. *Journal of Geography in Higher Education*, forthcoming.

Heumader, Jörg: Die Katastrophenlawinen von Galtür und Valzur am 23. und 24. 2. 1999 im Paznauntal/Tirol. In *Proceedings of the International Symposium Interpraevent,* pp. 397- 409, Villach, 2000.

Kienholz, Hans, Krummenacher, Bernd, Kipfer, Andy and Perret, Simone: Aspects of integral risk management in practice – Considerations with respect to mountain hazards in Switzerland. *Österreichische Wasser- und Abfallwirtschaft,* 56 (3-4): 43-50, 2004.

Kromrey, Helmut: *Empirische Sozialforschung,* UTB, Opladen, 1994.

Lave, Tamara R. and Lave, Lester B.: Public perception of the risk of floods: Implications for communication. *Risk Analysis,* 11 (2): 255-267, 1991.

Lawinenwarndienst Tirol, editor: *Schnee und Lawinen 1999-2000.* Land Tirol, Innsbruck.

Morgan; Granger M., Fischhoff, Baruch, Bostrom, Anne and Atman, Cynthia J.: *Risk Communication: A Mental Models Approach.* Cambridge University Press, Cambridge, 2002.

Palmer, Joy A.: *Education in the 21st Century: Theory, Practice, Progress and Promise.* Routledge, London and New York, 1998.

Poschmann, Christian, Riebenstahl Christoph and Schmidt-Kallert, Einhard: *Umweltplanung und –bewertung.* Klett-Perthes, Stutgart, 1998.

Reinmann-Rottmeier, Gabi and Mandl, Heinz: Unterrichten und Lernumgebungen gestalten. In Krapp, Andreas and Weidenmann, Bernd: *Pädagogische Psychologie,* 4th edn, pp. 601-646. Beltz, Weinheim, 2001.

Stötter, Johann, Meissl, Gertraud, Ploner, Alexander and Sönser, Thomas: Developments in natural hazard management in Alpine countries facing global environmental change. In Steininger, Karl W. and Weck-Hannemann, Hannelore, editors: *Global Environmental Change in Alpine Regions,* pp. 113-130. Edward Elgar, Cheltenham, UK and Northampton, MA, 2002a.

Stötter, Johann, Meissl, Gertraud, Keiler Margreth and Rinderer Michael: Galtür. Eine Gemeinde im Zeichen des Lawinenereignisses von 1999. In Steinicke, Ernst: *Geographischer Exkursionsführer Europaregion Tirol, Südtirol, Trentino: Spezialexkursionen im Bundesland Tirol,* Innsbrucker Geographische Schriften 33/2, pp. 167-184. Institute of Geography, University of Innsbruck, Innsbruck, 2002b.

United Nations: *Agenda 21.* Available electronically at http://www.un.org/esa/sustdev/documents/agenda21/index.htm, 1992.

UN /ISDR: *Living with Risk: A global review of disaster reduction initiatives.* Available electronically at http://www.unisdr.org/eng/about_isdr/bd-lwr-2004-eng.htm, 2004.

Walker, Gordon, Simmons, Peter, Irwin, Alan and Wynne, Brian: Risk communication, public participation and the Seveso II directive, *Journal of Hazardous Materials,* 65: 179-190, 1999.

Disaster Management

Concept for an Integrated Disaster Management Tool for Disaster Mitigation and Response

M. Markus[1], F. Fiedrich[1], J. Leebmann[2], C. Schweier[1], E. Steinle[2]

[1] *Institute for Technology and Management in Construction, University of Karlsruhe, D- 76128 Karlsruhe, Germany;*
{fiedrich}{markus}{schweier}@tmb.uni-karlsruhe.de
[2] *Institute for Photogrammetry and Remote Sensing, University of Karlsruhe, D- 76128 Karlsruhe, Germany; {leebmann}{steinle}@ipf.uni-karlsruhe.de*

Abstract

The Disaster Management Tool (DMT) was developed based on EQSIM, a damage estimation tool, to support efficient and integrated disaster management. Main tasks are fast and reliable damage and casualty estimation, damage detection based on airborne laserscanning data and support of disaster management personnel with communication and information tools. A decision support system helps to coordinate the allocation of rescue personnel and machinery. Onsite rescue operations will be supported by an expert system analyzing damage information combined with data about the buildings construction and occupancy. Further development of the DMT will comprise optimization of reinforcement of buildings in a given area with a delimited budget. The DMT is developed for pre-event training and mitigation tasks and post-event disaster management. The development status and further goals will be presented.

Keywords: disaster management, response, mitigation.

1 Introduction

When urban areas are stricken by earthquake disasters and experience substantial destruction, in general the operable disaster response teams are overstrained. An efficient and integrated disaster management could support their activities and help to limit human losses. The Disaster Management Tool (DMT) is a multidisciplinary approach of the German Collaborative Research Centre 461: "*Strong Earthquakes, a Challenge for Geosciences and Civil Engineering*", to perform this task.

The Disaster Management Tool is a software system with up-to-date hardware supporting decision makers, surveillance and intervention teams during disaster response. The response actors can access basic data about building stock, residents and resources as well as dynamic data like seismic measurements, damage estimations and damage observations. It assists decision makers as well as rescue team leaders with decision support and intelligent communication tools. The DMT bases on real data from Bucharest, Romania.

At the present stage of development, improvement of disaster response is the main objective of the tool. The DMT can also be used for risk assessment using the damage estimation tool with expected seismic input as well as for the task of preparedness using the damage estimations for disaster response training and to pre-assess the needed resources. In this paper the different components of DMT, its stage of development and further goals are described in brief.

2   Concept of the Disaster Management Tool

The Disaster Management Tool has three main functional parts. The first "simulation part" comprises components for fast damage and casualty estimation, simulation of future progression of the disaster like fire propagation and consequences of decisions. The damage and casualty estimation based on seismic data is performed by the component EQSIM, which is in the most advanced stage of development within DMT (see Baur et al., 2001).

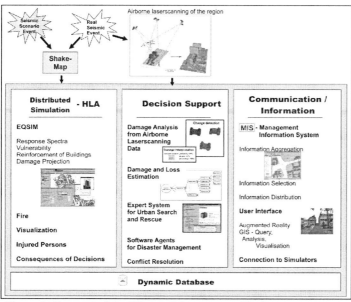

The second part encloses elements for decision support. Main components are a system for damage analysis based on airborne laserscanning, damage and casualty estimation based on building stock and residential data as well as the results of the damage analysis. An expert and information system supports rescue activities at collapsed buildings. A decision support tool for emergency operation centre members

Figure 1: Concept of the Disaster Management Tool

helps to assign the response resources in order to maximize the efficiency of response activities.

To integrate the operations on the different executive levels, the third functional part of the DMT provides means for communication and information. The management information system conducts the aggregation, selection and distribution of information relevant for the specific actors of disaster response. A graphical user interface helps to visualize the mostly geographical related information. And a special augmented reality user interface helps analyzing the situation at rescue site to accelerate response activities.

The dynamic database is the central element of DMT. The different components use the common Oracle database to access static information like building stock data. They also use it to store and exchange dynamic data like observation results from different sources or locations of rescue resources. Backup databases are filed on the local computers depending on the components in use. Figure 1 shows the concept of the DMT.

3   Software Architecture

DMT uses an approach based on distributed computing over computer networks. It consists of components (figure2) that will be described in detail throughout this paper.

The EQSIM server program performs the calculation of the damage scenarios and implements the damage estimation methodology. Calculations are either requested of client programs where users can define scenarios based on the historical earthquake database, respectively defined earthquake parameters (such as location, magnitude and depth) or such requests arise from software agents, which may be used for decision support during disaster response. Field personnel may update the central database with observed damages that are input for a groundtruthing component of the server program to improve the scenario calculations.

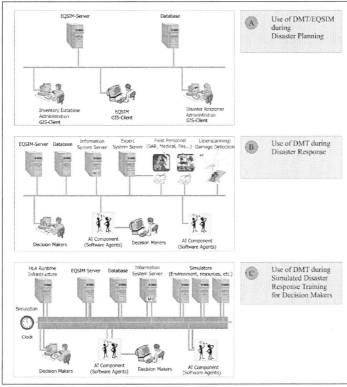

Figure 2: Possible applications of DMT and EQSIM

The Management Information System (MIS) server controls the communication and information flow between users and with the database. It distributes new information to users with relevant responsibilities. The Expert System Servers are located at or near to operation sites. The response teams use mobile computers or PDAs connected via wireless LAN to the servers to obtain case relevant data and advice. Input data from field personnel is sent by these devices to central database. Additional direct access to the database and use of MIS is possible for the field personnel.

During a simulated training exercise, all components are linked via a distributed simulation, which is based on the High Level Architecture (HLA). In this case, the field personnel and the response resources may be simulated by separate HLA-simulators. Figure 2 gives an overview of the possible application fields of the DMT.

4   The Simulation Component

The DMT can also be used in virtual disaster response training by the staff of Emergency Operation Centers (EOC). This simulation uses the High Level Architecture (HLA) as a common framework.

HLA is an Institute of Electrical and Electronics Engineers (IEEE) standard (IEEE, 2000) for distributed simulation systems. A distributed simulation using HLA is called a *federation*, and each single simulator in such a federation is referred to as a *federate*. A federate may be a simulation, live component (e.g., physical device or human operator) or data viewer. The federates interact via a central component which is called the *Run Time Infrastructure* (RTI). To enable this communication, each federate must implement a predefined HLA-interface (compare figure 3).

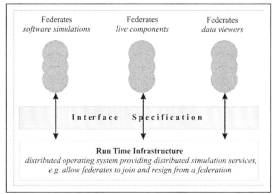

Figure 3:   Main components of an HLA-based
            federation (Fujimoto, 2000, p 212)

A multiagent system can be linked to the simulation environment where the software agents (software agents are computational systems with goals, sensors, and effectors, which decide autonomously which actions to take, and when) use predefined flexible plans for their reasoning process to represent actors in the simulated environment. The disaster environment and the use of resources can be simulated in real time for a simulation-based exercise. Simulated resources include SAR-Teams, ambulances, fire fighting units, recon units and heavy equipment resources for repair work of blocked roads and rescue operations (Fiedrich and Gehbauer, 2004).

5 Test Area in Bucharest

Figure 4. Test area in Bucharest

The Disaster Management Tool is designed for urban areas. Development and testing are carried out in Bucharest as a case study. An area of the city was defined for that data in a high-resolution were acquired, including detailed information for each single building. The test area is part of the downtown district, where most of the damages occurred during the last destructive earthquake in 1977 (Figure 4). The data collected for this test area are compiled in the central database, which is linked to the building layer and street layer of the DMT-GIS. The test area contains 1305 buildings with all kinds of occupancy. Supplementary, 763 small garages or stores, consisting of one storey, were identified in this area. After visual examination, no occupancy was assigned to these buildings.

As data sources served a list of expertises of highly vulnerable buildings from Romanian experts, building data including data of the buildings collapsed in Bucharest on March 4, 1977, acquired by the Technical University of Civil Engineering in Bucharest, a city map in 1:500 scale, results of the census in 1990, and the results of an inspection of the buildings in the test area. The building stock of the test area, which was classified according to HAZUS standard (National Institute of Building Sciences, 1997), consists both of residential buildings and commercial buildings. The northern area is stamped by tall concrete frame buildings of more than 7 storeys. But also many low-rise historical buildings are included in the test area. Other construction types occurring in the test area are *wood* and *steel frame* buildings, *precast concrete*, *reinforced* and *unreinforced masonry* buildings. Altogether, 20 combinations of construction type and height range can be used to describe 97% of the test area building stock.

6 Concept of Rapid Damage Detection

One of the most important factors to plan an efficient use of SAR resources is the knowledge about the location, the extent and the characteristic of totally or partially collapsed buildings. Fast acquisition of such data is essential for rapid assessment of the needed rescue personnel and equipment. Therefore a method based on airborne laserscanning is researched within the CRC (Steinle & Vögtle, 2001).

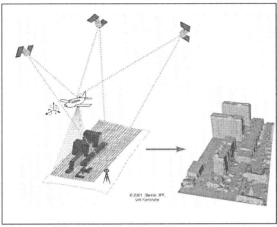

Figure 5: Principle of laserscanning

The airborne laserscanning technology (Ackermann, 1999) allows producing height data sets, e.g. digital surface models (DSM) (see figure 5). Such models are a highly suitable base for the application of automatic procedures for the extraction and geometric modeling of buildings as 3D vector models (Steinle and Vögtle, 2001).

The damage analysis is carried out by comparing building models extracted from post-earthquake laserscanning data with such stored as reference models for the endangered area in the DMT. Differences found between the pre- and post-earthquake geometric models of the affected buildings are quantified in terms of change measures like volume differences, plane orientation change, height change or size alteration. These changes must be further analyzed and interpreted.

As urban environments are undergoing changes all the time, e.g. by construction activities (Steinle & Bähr, 2002), and not only due to catastrophic events like earthquakes, it is necessary to rate the differences in buildings geometry. The changes identified as not being caused by normal urban modifications are further classified in damage types using a so-called damage catalogue. Principally, this is an archive storing observed damages at buildings of former earthquakes, but clustered and characterized according to the needs of the SAR organizations. Each class is described in the catalogue with features that can be extracted from the buildings vector models. Therefore, automatic procedures can use this catalogue to identify the respective damage type at buildings. The most likely damage type will be assigned to the observed building and the ratio of concordance to the damage type stored in the catalogue will be given additionally.

At present, the automatic damage classification process is developed. The tests of this component are planned with laserscanning data of a military exercise area. Further adjustment with the damage catalogue will be necessary.

7 Expert and Information System for Rescue Operation Support and Training

The rescue of trapped victims from collapsed buildings requires a substantial technical, personnel and organizational effort (Coburn and Spence, 2002). For training and operation support, an expert and information system was created and tested with the THW (Technisches Hilfswerk, German civil protection organization) (Markus et al., 2000). It consists of the three components *on-line manual*, *expert system* and *computation component*. After integration into the DMT, communication with control personnel and data-exchange with the central database will be possible

Figure 6: Screenshot of expert system data input

The on-line manual contains information about building types, construction components, rescue equipment and methodology for training and to easily access necessary knowledge during rescue activities. Into the expert system that is linked to it, information relating to a certain case (see figure 6) can be entered. Building stock data from the central database can be taken over. The diagnosis includes advice concerning the processed case and links to specific entries within the information system. Case-relevant checklists are printed and appropriate tools and methods are given. The expert system bases on the D3 Server System application (Puppe et al., 1997), which is provided by the Department of Computer Science VI, University of Würzburg.

With the computing component, calculations of debris masses and masses of building materials such as steel profiles or concrete floors can be performed easily and reliable under operational conditions. This information can be used for strut dimensioning, crane selection and transportation calculations, using again this component.

The expert and information system is a server application that can be used at the rescue site through mobile computers and PDAs (small handheld computer) connected via wireless LAN. It was tested by professional users at model cases. Currently, the connection to DMT and its database is programmed. The functionality will be enhanced.

8 User Interfaces

The standard user interface for DMT users is a graphical user interface (GUI) with GIS functionality. A further user interface bases on augmented reality (AR) technology. An AR system overlays a virtual image and the real image of the scenery in real time. The AR system of the DMT can be used in two variations: (1) with semitransparent displays (see Figure 7), or (2) with a camera to perform the superposition. With GPS devices (used in differential mode) and inertial sensors the position and orientation of the user's head are measured. This information is used to superpose relevant data like residential use of the building or number of storeys using the ARS. But also geometric information can be directly overlaid. At collapsed buildings, the three dimensional view of the undamaged building can be superposed with the actual image of the building (see figure 8). The user can now directly compare the "real" view of the collapsed building with the view of the building before collapse in his spectacle. This helps discovering areas with possible voids in the

Figure 7: Test equipment of the augmented reality user interface

collapse structure or areas where trapped victims are possibly located following their initial locating. This technology, which is only possible with inventory data, can help the rescuers to search selectively and thus to accelerate the rescue process.

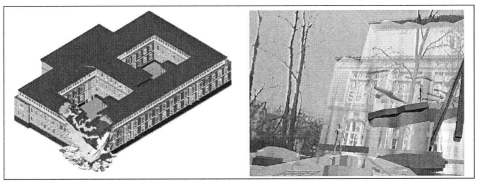

Figure 8: Modified fictive model of a collapsed building at University campus. Right: Superposition of the real view with DSM and with the damage model

First tests of the augmented reality user interface were performed at the university campus. Figure 8 shows the result of the test using a real not collapsed building from DSM and a model of the same building in a partially collapsed state.

9 Conclusions and Future Work

In this paper, the Disaster Management Tool was described as a promising tool for disaster response in an urban environment. The client-server architecture permits to use DMT simultaneously by different users. It can be applied for disaster planning, disaster response and for disaster response training.

The central database allows the users to avail steadily actualized data of the changing disaster environment. It is necessary for data exchange between the DMT components like Management Information System and the Expert and Information System. For simulation of response activities, the distributed simulators, EQSIM and the user interfaces communicate in case of training based on a High Level Architecture (HLA) framework.

The concept of the rapid damage detection and interpretation component, which is using airborne laserscanning technology, was presented and the expert and information system supporting field personnel was described shortly. Then the augmented reality user interface and its planned application within DMT were introduced.

The first test of the whole DMT tool with local disaster management and response units in Bucharest is scheduled for October 2004. The exercise will be held on control center level and on onsite operation search and rescue level. At moment, the different components are adapted to DMT before the tests will be carried out under real conditions.

Acknowledgements

The research in this paper is part of different research projects within the Collaborative Research Center (CRC) *Strong Earthquakes: A Challenge for Geosciences and Civil Engineering*. It is funded by the German Science Foundation (DFG) and the state of Baden-Württemberg.

References

Ackermann, F.: Airborne laser scanning – present status and future expectations. ISPRS (International Society for Photogrammetry and Remote Sensing) Journal of Photogrammetry and Remote Sensing, 54 (1999), pp. 64 - 67, 1999.

Baur, M.; Bayraktrali, Y; Fiedrich, F.; Lungu, D. and Markus, M.: EQSIM - A GIS-Based Damage Estimation Tool for Bucharest. In: Lungu, D. and Saito, T (Eds.): Earthquake Hazard and Countermeasures for Existing Fragile Buildings, Independent Film, Bucharest, 245-254, 2001.

Coburn, A. and Spence, R.: Earthquake Protection, Second Edition, Chichester, England, John Wiley & Sons Ltd., 2002.

Fiedrich, F. and Gehbauer, F.: EQ-RESQUE: An HLA-Based Distributed Simulation System For Disaster Response Activities after Strong Earthquakes. In Proceedings of the Eleventh Annual Conference of The International Emergency Management Society, 18.-21. May, 2004, Melbourne, Australia, 2004.

Fujimoto, R.M.: Parallel and Distributed Simulation Systems. Wiley Series on Parallel and Distributed Computing 3. John Wiley, New York, 2000.

IEEE: Standard for Modeling and Simulation (M&S) High Level Architecture (HLA) - Framework and Rules, IEEE Standard No. 1516-2000; Federate Interface Specification, IEEE Standard No. 1516.1-2000; Object Model Template (OMT) Specification, IEEE Standard No. 1516.2-2000, 2000.

Markus, M., Fiedrich, F., Gehbauer, F. and Hirschberger, S.: Strong Earthquakes, Rapid Damage Assessment and Rescue Planning, in: Kowalski, K.M. and Trevits, M.A. (eds.): Proc. of the 7th Annual Conference of the International Emergency Management Society: "Contingency, Emergency, Crisis, and Disaster Management: Emergency Management in the Third Millennium", Orlando, Florida, May 1999, pp. 369-378, 2000.

National Institute of Building Sciences: Earthquake Loss Estimation Methodology HAZUS. Technical Manual, Vol. I-III., 1997.

Puppe, F.,Bamberger, S., Iglezakis, I., Klügel, F., Kohlert, S., Reinhardt, B., Unglert, T., Wolber, M.: Wissensbasierte Diagnose- und Informationssysteme mit dem Shell-Baukasten D3 –Handbuch, Lehrstuhl für künstliche Inteligenz und angewandte Informatik, Universität Würzburg, 1997.

Steinle, E. and Vögtle, T.: Automated extraction and reconstruction of buildings in laserscanning data for disaster management, in: Proceedings of the Workshop Monte Verita, Switzerland, 10-15 June 2001, Baltsavias, E. P., Gruen, A. and Van Gool, L. (eds.), Automatic Extraction of Man-Made Objects from Aerial and Space Images (III), Balkema (Swets & Zeitlinger), Lisse, The Netherlands, p. 309 – 318, 2001.

Steinle, E. and Bähr, H.-P.: Detectability of urban changes from airborne laserscanning data, in: Navalgund, R. R., Nayak, S. R., Sudarshana, R., Nagaraja, R., Ravindran, S. (eds.): Resource and environmental monitoring, ISPRS commission VII symposium, Hyderabad, India, December 2002.

Economic Efficiency and Applicability of Strengthening Measures on Buildings for Seismic Retrofit: An Action Guide.

Maria Bostenaru Dan

Institute for Technology and Management in Construction, University of Karlsruhe (TH), D- 76128 Karlsruhe, Germany

Abstract

Historical built substance is the basis for all further planning and construction measures in retrofitting earthquake endangered buildings. Actors from the spheres of passive publicity, experts and active affected people are involved in the application of the measures. The focus lays on planning management in the field of experts, with a detailed view on the decision space between scopes, means, function and costs. The interdependencies between the constructive, functional and aesthetical characteristics of a building and a chosen retrofit strategy were researched. The "retrofit elements" concept, developed in order to sustain decisions regarding the economic efficiency and the applicability of the strategy, was employed. It proved suitable to serve the organisational and operational structure in model projects within a strategy to implement retrofit measures in urban areas. Here through the points of view of all participating actors can be involved on a common denominator in the decision process. It is shown that the former conceptual work builds an adequate basis for strategic planning of retrofit measures implementation.

Keywords: seismic retrofit, historical buildings, benefit-costs analysis, decision.

1 Introduction

The reduction of seismic risk through retrofit of existing buildings serves catastrophe prevention. Planning interventions on historical buildings differs from planning of new buildings through an important condition: the existing building is the base for all planning and construction efforts. Some of the most important aspects to be considered are:

- Existing constructions are part of an existing built context;
- More actors are implied in planning an intervention on a building in use;
- An assessment is needed as preliminary study before designing;
- The data available for building assessment is different from that in building design;
- The structure might be pre-damaged through previous earthquakes;
- Construction technology and management has to be tailored in order to cope with particular situations in a historical building.

Providing for safety of existing buildings is thus a complex endeavour. Sandi in Bălan (1982) identified three main categories to be encompassed when dealing with this problem: characterisation, informational and decisional issues, on a three pages summary which unfortunately did not find follow-ups in today's studies. Kevin Lynch presented in his book "The Image of the City" (Lynch, 1960) a theory according to which those who enter a town, perceive it by means of landmarks, paths, districts, nodes and edges. Lynch's (1960) theory was an outgoing point for the "retrofit elements" concept used in this paper.

The author's research regarding economic efficiency and applicability of strengthening measures on buildings for seismic retrofit (Bostenaru, forthcoming_b) provided for basics and a solution approach. This solution approach shows how the task, to plan a strategic retrofit intervention, can be worked up in two stages. The first stage serves for the recognition of problems and opportunities, by means of which a diagnosis is seized and a mission is formulated (Fig. 1). Aim of this paper is to show how the concepts developed in that research can be used for acting within strategic planning of retrofit measures.

The "ingredients" of a strategic plan for seismic risk reduction are action plans, objectives, operative modalities, human resources, time and costs, aiming analysis, evaluation, priority setting and communication. In Table 1 the operations in urban retrofit measures are set into the context of management and decision elements.

2 Identifying Problems and Opportunities: Case Study "Bucharest"

Research was conducted concerning mainly the building stock in Bucharest, Romania, but correlation with possible results in other locations was taken into account, namely Greece. The motivation was the necessity to develop specific retrofit measures for historic buildings in the capital of Romania, characterised through a maze of buildings with different structure, age, state and scale.

A complete and structured review of the building stock on a particular location, using a system which allows for comparisons, was aimed for.

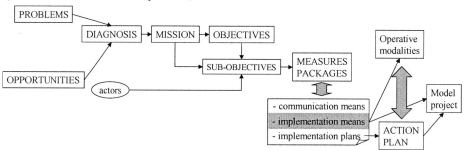

Figure 1: Interdependencies in urban management operations

Management functions	*Operations*	Management modules	Decision elements	Decision phases
Problem analysis	*Data to information*	Problems & Opportunities	Problem definition Evaluation criteria Constraints	Intelligence
Alternatives	*Relationship problems-opportunities*	Diagnosis	Decision matrix	Design
Prediction	*Pathology*			
Evaluation	*Evaluation*		Actor's preferences	
Decision	*Priority setting Goal stipulation*	Mission	Recommendation	Choice
	Main tasks Partial tasks	Objectives & Sub-objectives		
	Measures packages	Measures packages		
	Strategy	Action plan		
Implementation	*Project management*	Model project		
Control	*Feedback for 2nd phase*			

Table 1: Phasing of urban management operations; decision phases after Malczewski (1999)

The building stock in Bucharest was analysed both typologically and area based. Common building types were surveyed. For this purpose first methods used in surveying an urban area were employed, as these allow identifying the morphological types in the given building stock. Then general characteristics supported by observation of a number of buildings, made by the author or by other specialists (in this case based on literature review) was performed. The work also describes the architectural and engineering characteristics as well as the structural damage of differently aged buildings from Bucharest, in comprehensive and interdependent charts.

Earthquake resilient features were identified as problems and seismic deficiencies as opportunities. After evaluating all historical building types it was drawn out that the most vulnerable are residential multi-storey reinforced concrete frame buildings from interwar time. An urban project in existing context begins with a site analysis identifying vulnerabilities as problems and capacities as opportunities. These build the basis to formulate a diagnosis about seismic risk. The step of problem analysis belongs thus in the so called "intelligence" (Malczewski, 1999) phase of the strategic decision (Table 1).

3 Diagnosis Method: "Retrofit Elements" Concept

Interdependencies between functional, aesthetic and constructive characteristics of buildings and the chosen retrofit strategy were investigated. The concept of "retrofit elements" has been developed to support communication among actors making decisions regarding the applicability and economic efficiency of the strategy.

Transforming data into information is an operation involving three activities: inventory, evaluation, classification (Bălan, 1982). Inventory and classification are closely related to decisional issues. Evaluation needs an independent technical approach.

In order to perform the diagnosis a conceptual method was developed by employing a project management instrument for a research topic, namely the structure plan. Like in the theory of Kevin Lynch a building also presents nodes (points), these are its articulations. Its edges (lines) are beams or columns, and its surfaces are built by floors or walls. Recognition characteristics, construction works with duration and needed resources, implied costs for strengthening and reparation as well as earthquake resilient features, seismic deficiencies and earthquake damage patterns can be assigned to members in form of point elements (frame nodes), line elements (columns and beams) and surface elements (floors and walls).

A retrofit element consists of all works which have to be done in order to strengthen, repair, rebuild or even build a structural member. These are spatial elements which are characteristic for the survey, present typical earthquake damages and decisive for a better seismic behaviour in case of retrofitting. With their help the structural performance was assessed stress-strain based, then repair and retrofit costs calculated. The general validity of the prediction has been studied extensively, comparing the computed damage types with real ones and proved to be close to reality. For this purpose a matrix of damages suffered by typical buildings, developed by the author, has been used. The pathology described is supported by sketches. Diagnosis means bringing problems and opportunities in relationship: opportunites which are supporting each other, reciprocally aggravating problems. Earthquake resilient features and seismic deficiencies were put in connection with earthquake damage patterns. Based on this pathology study a diagnosis can be set. Based on the diagnosis the mission can be formulated. The mission includes **goals** without priorities. In this work the mission was "applicability and economic efficiency of seismic strengthening measures on existing buildings".

4 Retrofit Strategy: Operative Means

A retrofit system is implemented through an adequate technical as well as an adequate management strategy. Mitigation interventions result within the strategic planning in formulating a mission in points without priorities. Each point is a sub-objective for a group of actors (Table 2) meant to be reached through measures packages (Table 3). A measure package includes consensus and implementation means (Table 4).

All objectives together define the **applicability**, as this depends on the way the required multi-criteria decision is met (Bostenaru Dan, forthcoming_a). For the reason that the mission points are formulated without priorities, a particular decision method, called balancing method (Strassert, 1995), was considered to be suitable. In decision matrices set up within this method, called data tables, there are no operations between the rows. Rows correspond each to one actor, while the columns correspond each to one alternative.

Actor	Objective	Sub-objectives
Architect	retention of the character of the buildings	respect for historical, building, element, material issues
Engineer	attenuation of the impact of future earthquakes	reduction of vulnerability enhancement of structural performance adequate retrofit design good technical retrofit strategy
Investor	**economic efficiency** of the measures applied	good management retrofit strategy availability of resources optimal values according to indicators
User	good management strategy of the retrofit measures	good conditions during the execution high acceptability of the measure optimal use housing quality

Table 2: Objectives and sub-objectives

Nr.	Measures package	Method	Interest groups
1	improving understanding of the impact of earthquakes	- public presentation - reaction to feedback on findings	Multidisciplinary
2	development of an algorithm for optimisation of retrofit measures	parametrical study	-
3	development of a decentralised decision model	modularisation of a collaborative decision model	Interdisciplinary
4	insights into applicability of retrofit methods	documentation training	Interdisciplinary architecture engineering
5	development of a framework for integral planning	retrofit design model project for the integral planning	Interdisciplinary architecture engineering economics
6	solving contradictions between the objectives of single actors	develop a basis system to administrate modules on different levels	Multidisciplinary
7	highlighting the comprehendsibility of the measures analysed	physical implementation flow along with an education flow for population which has to support the measures	architecture urbanism sociology
8	support changes by political and economic environment:	document existing programmes impact assessment	-

Table 3: Measures packages and collaborative issues

Package	Implementation plan	Implementation means	Consensus means
1	dialogue with other actors	- conferences - publication	familiarisation with earthquake engineering aspects
2	Research on technology	FEM	modelling ability
3	Decision-making on different levels (actors; actor's criteria)	pair wise comparison tools	encompassing actors from different backgrounds involved in the implementation strategy
4	assessment of systems	literature and internet lectures and exercises.	understanding of retrofit systems
5	independent research	exercise	acquiring didactic skills
6	multiobjective systems on urban/building scale	computer tools (spread sheets, GIS)	a database for urban/ building level navigation
7	feedback from the programs	dedicated exhibitions training on pilot projects	presentation to the public
8	conference session	investigation of programme availability	contacts in earthquake prone countries

Table 4: Instruments and their elements

Measures packages for different actors were set up. The solution approach to manage the operations through the method of retrofit elements, to be used in all phases of the retrofit process, can be tested on model projects, which, however, is an area for further research. The measure package chosen for deeper investigation was that of economic efficiency of the measures applied. Five building models of inter-bellum type were designed and modelled together with suitable retrofit measures for fibre based finite elements (FE) simulation. Outlooking to the general methodology of the author to assess post-damage repair versus preventive retrofit costs, damage in following cases was considered: unretrofitted building; retrofit of undamaged and previously damaged building. Within the structural study the seismic performance of reinforced concrete frame buildings subjected to cyclic bending has been assessed. The innovative part lies in the stress-strain based approach applied to models of buildings. Such an analysis allows not only the description of failure modes and determination of limit states eventually reached by the building, but also specific determination of the number and position of structural members suffering different types of damage. In a performance based retrofit approach so-called "costs curves" were computed and the retrofit costs optimised for different "design earthquakes" (Bachmann, 1995).

5 Action Plan: Transition to Building Scale

A measure package in the urban strategy consists of communication means, implementation means (Table 4), and implementation plans on different levels and time horizons (Table 5), taking form in action plans (Table 6).

Time horizon Level	Long term	Short term
Urban	dialogue with other actors	vulnerability assessment at urban scale customisation of the decision system
Building	assessment of innovative systems	applicability assessment
	independent research	model project
Element	research on the technology and management of building retrofit	design and parametrically vary „retrofit elements"

Table 5: Consensus means in an action plan

Step	Goal	Method	Instrument	Measure
1	technical reports on implementation programmes	documentation	investigation training.	4; 8
2	a data table of use for the decision method in the next step	parametrical study	FEM	2
3	support the choices at step 4 and step 1	highlighting comprehensibility	database	7
4	algorithm based on case studies (step 2) for experiments (step 6)	modularisation of the decision model	pair wise comparison	3
5	report about available systems for this purpose	a basis system to administrate modules	computer tools	6
6	trial of educational feasibility (step 3)	project example	exercise	5
7	dissemination of results	presentation	publications	1

Table 6: Retrofit strategy management

Management phase	Technique	Goal	Methods
Planning: problem analysis	Analysis	evaluation criteria	preference structure
		data	*retrofit elements*
define constraints	Decision	Decision space (site, time choice)	screening
interest groups	Participation	Building public ideas	marketing
Alternative planning	Creativity	Different retrofit systems	
Planning: prediction	Prediction	Concept paper (scope, content, organisation, financing)	*retrofit elements*
Information	Decision	set indicators	measurement levels
actors' preferences	Decision	Criterion weighting	Balancing method
Planning/ evaluation	Management	Assessment	*retrofit elements*
Decision	Decision	Ranking alternatives (priority setting)	Analytic hierarchy process
Definition	Design	implementation strategy	*retrofit elements*
Application	Organisation	implement "retrofit elements"	*retrofit elements*
Control	Feedback	Second management circuit	education

Table 7: Phasing in a model project

Implementation means result in model and pilot projects for operations, which are also concretisation forms for action plans. Table 7 exemplifies the management phasing for such a model project: a retrofit strategy for a building. The method based on "retrofit elements" is used on various phases of the project, as common denominator for the actors considered. Collaborative issues in the model project are visualised in Table 8. Although the method of "retrofit elements" is common, the instruments employed within it differ for the actors.

Consensus means are important for the so-called second phase of strategic implementation, when the application of the measure has to spread from pilot or demonstrative projects to „routine" wide ones. As the result of the implementation has to be evaluated from time to time, the planning aim is not a product but an action and learning process for all implied actors (Table 7). Thus also time and human resources cannot be assigned to each objective like the operative modalities were. Within the strategic planning the plan is pictographic vision of the development and orientation for the interaction between all the planning levels and all the participating actors. It is an organisational matter of planning possibilities. The focus lays in the problem itself, in its multifaceted manifestations. Problem based orientation, differentiation of the actors as well as identification and activation of target groups lead in this case to the education mentioned, which accentuates personal implication and action motivation.

Goal	Instrument for			
	architect	**engineer**	**user**	**investor**
evaluation criteria	- analytical study - survey for opinions			
Data	field survey - check list - photography	qualitative assessment	"Raumbuch" (space book)	retrofit provisions
decision space	- resources - regulations			
building public ideas	- advertising instruments - for **passive publicity**			
different retrofit systems	competition			
concept paper	makeability study	vulnerability assessment	operational structure	costs plan
set indicators	- scales - units of measure			
criterion weighting	- data table - advantages-disadvantages table			
Assessment	impact assessment	FE simulation	operational plan	Benefit-Costs Analysis
ranking alternatives (priority setting)	- structure plan - comparative judgement - operational plan			
implementation strategy	conversion design	technical strategy	forum	financing plan
implement "retrofit elements"	write-off	management strategy	workshop/seminar	liquidity
2. management circuit	training for active **affected people**			

Table 8: Issues concerning the organisational structure of the actors in the model project

6 Discussion

Several aspects of strategic management of retrofit operations were detailed. The way these elements of the strategic plan are interrelated is shown in Fig. 2. Outgoing point was a solution approach regarding the applicability and economic efficiency of strengthening measures for seismic retrofit on existing buildings, which provided for new methods and instruments suitable in the first phases of strategic planning. In this contribution that limitation is overcome, and all stages of the developed strategy are exemplified. Further, different "benefit" aspects of retrofit measures (duration, alteration of historic substance, relocation of inhabitants) are taken into consideration.

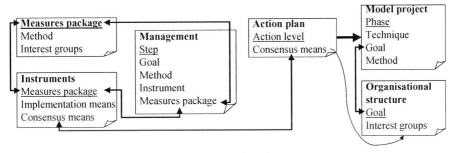

Figure 2: Interdependencies among the action guide elements

An integrated decision support system comprising building survey, structural aspects and calculation of costs, uses the same "retrofit elements" as planning basis. This way the points of view of all participating actors can be brought on a common denominator into the balancing and decision process. The strategy problem addressed is the communication between the actors involved in the implementation of a retrofit measure, regarding its economic efficiency and applicability. Issues concerning the experts (design professionals), like engineers and architects, as well as issues concerning members of the building community, like the investor, were addressed detailedly through specific modelling, while issues concerning affected citizens, like owners or tenants, are merely mentioned in the overall approach. Characteristic for strategic planning is the orientation towards the process and not the final product. A second circuit is explicitly foreseen. But the processual dimension finds even more expression in the operative management approach involving experts, in the simultaneity of implementation means on different levels of detail: urban visions, building projects and element simulations (see also Bostenaru, forthcoming_a).

7 Conclusions

The solution approach developed within the dissertation enjoys methodological applicability. Through the conceptual model of "retrofit elements" managerial instruments like the project structure plan and the organisational structure are brought into incipient phases of the strategic planning. It is an integral planning approach.

Acknowledgements

The research work presented in this paper has been supported by the German Research Foundation (DFG) in frame of the programme Graduiertenkolleg 450 "Naturkatastrophen" with a 36 months fellowship and research infrastructure along the whole research period. A six months stay abroad at the European School for Advanced Studies in Reduction of Seismic Risk (ROSE School), financed by the European Commission in frame of the "Marie Curie" mobility programme as a training site host fellowship, was approved.

References

Bachmann, Hugo: *Erdbebensicherung von Bauwerken*. Birkhäuser Verlag, Basel, 1995.

Bălan, Ştefan, Cristescu, Valeriu and Cornea, Ion, editors: *Cutremurul de pământ din România de la 4 martie 1977* = "The Earthquake in Romania on the 4th of March 1977" (in Romanian). Editura Academiei Republicii Socialiste România. Bucharest, 1982.

Bostenaru Dan, Maria: Multi-criteria Decision Model for Retrofitting Existing Buildings. In: *Natural Hazards and Earth System Sciences*, forthcoming_a.

Bostenaru Dan, Maria: Wirtschaftlichkeit und Umsetzbarkeit von Gebäudeverstärkungsmaßnahmen zur Erdbebenertüchtigung. Grundlagen und Lösungsansatz unter besonderer Berücksichtigung der Situation in Bukarest, Rumänien. PhD thesis, Universität Karlsruhe (TH), Karlsruhe, Germany, forthcoming_b.

Lynch, Kevin: *TheImage of the City*. MIT Press, Cambridge, Massachutes, 1960.

Malczewski, Jacek: *GIS and Multi-Criteria Decision Analysis.* John Willey & Sons, New York, 1999.

Strassert, Günther: *Das Abwägungsproblem bei multikriteriellen Entscheidungen.* Verlag Peter Lang, Frankfurt am Main, 1995.

Mainstreaming Disaster Risk Reduction:
A Trial of 'Future Search' on St. Kitts, West Indies

Tom Mitchell
Department of Geography, University College London, 26 Bedford Way, London, UK

Abstract

This paper details the use of the participatory method 'future search' to promote *disaster risk reduction mainstreaming* on the Eastern Caribbean island of St. Kitts. Results of the future search are presented, showing multi-stakeholder agreement on a number of concrete policies and programs. At present it is too early to make a full judgment on how successfully the future search process influences organizational approaches to reducing disaster risks, but early indications are promising.

Keywords: Disaster risk reduction, future search, St. Kitts

1 Introduction

Many countries, districts and urban areas have clear disaster management plans detailing the various responsibilities of each government department and agency in the case of a disaster. However, these plans have traditionally placed an emphasis on responding efficiently to an emergency, and this 'reactive' mode of planning for disasters is embedded in organisational structures. Far fewer plans adequately address the task of reducing the risk of disasters and if the plans do exist they are often limited to a small handful of agencies. In addition, many organisations have neglected to consider or implement proactive disaster risk reduction techniques and strategies. There are many reasons for this imbalance, including inadequate funding, lack of education and the compartmentalisation of disaster management agencies outside of larger, more powerful government organisations. The emerging project of 'disaster risk reduction mainstreaming' attempts to rectify this situation by introducing disaster risk reduction practices to all sectors of government and non-governmental organisations. 'Mainstreaming' in this context can be understood as the 'gradual adaptation of an individual, organisation or institution to new ideas and behaviours, leading to sustainable change'.

2 Development

Actors in the disasters and development fields are currently employing a variety of approaches at different scales to promote disaster risk reduction. One such international strategy is through the World Conference on Disaster Reduction (WCDR) in Kobe/Hyogo Japan in January 2005. This UN International Strategy for Disaster Reduction event 'is expected to motive and guide governments and their policymakers to pay more attention, identifying practical and concrete ways to incorporate disaster risk reduction into poverty reduction' (UN-ISDR, 2004). The conference is based on the idea that while 'commitment to the reduction of disasters has been growing – demonstrated through several existing international initiatives, agreements and declarations – actual materialization is still slow' (UN-ISDR, 2004). In preparation for the conference, over 600 academics, policy-makers and practitioners from around world took part in an online-conference called 'Priority Areas to Implement Disaster Risk Reduction: Helping to Set a New International Agenda'. This broad consultation exercise is designed to enable a wide-range of experts to contribute to the documentation being prepared for the meeting.

A second method being developed employs a series of 'frameworks' or 'indexes' to articulate best practices in disaster risk reduction (see UN-ISDR 2003; Mitchell 2003; IDEA 2003; Parker 1999; Ascanio 2003). According to the UN-ISDR/UNDP online conference (25th August – 30th September 2003) 'a universal and internationally endorsed [disaster risk reduction] framework can provide an organising tool to aid our understanding and guide action in disaster risk reduction' (Schlosser and Aysan 2003).

A third approach to broadening disaster risk reduction, encourages mainstreaming through multi-stakeholder participatory processes at the national, district and community scales. A number of participatory methods are being employed to help facilitate organisational learning (see for example Disaster Mitigation Institute 2003; UN-ISDR *National Platforms* project). This paper outlines one such method, *future search*, and details some of the experiences and results of testing it on the Eastern Caribbean island of St. Kitts. The goals of the research and the focus of this paper will discover if future search can help promote a national mechanism for mainstreaming disaster risk reduction and whether it can aid organisational learning in a diverse group of agencies[1]. With this in mind, participants were drawn from a wide range of governmental and non-governmental stakeholders. The two-day process was facilitated by members of the Department of Geography at University College London[2]. To the knowledge of the author, this is the first time future search has been used for disaster risk reduction planning.

3 Context: Participatory Planning in the Caribbean

With a single metropolitan area characterising the majority of Caribbean territories, an 'urban-led', 'centre-out', 'top down' planning structure was traditionally associated with colonial rule. This colonial model is resilient and still dominates in many islands. However, Pugh and Potter's (2003) edited volume traces the rise of participatory planning in the Caribbean, indicating that since the 1970s both donor agencies and grassroots' organisations have been pivotal in the movement towards more 'bottom-up', participatory practices in the region. However, by assessing case-studies from the region they find that the state still dominates policy-making despite the fact that 'western' participatory processes, such as planning workshops or community working groups are being embraced.

Pugh and Potter (2003, 204-214) identify that in many cases participatory planning is interwoven with 'the political'; meaning that socio-political hierarchies dominate these processes and strongly shape their outcomes. In turn, the use of participatory methods by the political elite allows them to present the results of participatory planning as 'the consensus' because a degree of consultation has taken place. While in reality, the process has been an exercise in containment and exclusion. For many in the political elite, the employment of a 'participatory' method is enough to present the process as fair and representative, while the

---

[1] The research forms the central fieldwork component of Tom Mitchell's PhD, supervised by Prof. Jacquie Burgess, Department of Geography, University College London (UCL) and sponsored by the Economic and Social Research Council (ESRC).
[2] The authors would like to thank the following organisations for their generous support of this research: The ESRC; the Environment and Society Research Unit (ESRU) and the Department of Geography at UCL; the Graduate School of UCL and the Benfield Hazard Research Centre, UCL.

'negotiations' over the invitation list attempt to render the process anything but representative.

On many Caribbean islands, the political divisions are strong, and any participatory approach must be aware of and sensitive to the possible political bias of participants. The complexities and repercussions of involving competing viewpoints must be delicately handled. In addition, any conclusions and concrete outcomes of a process are acutely vulnerable to changes of government or individuals in key positions. As Pugh and Potter suggest, the selection of certain individuals to join the future search process is based on political views, with key political stakeholders not wanting dissenting voices. The presents a challenge for mainstreaming disaster risk reduction, as it is an issue that must be addressed at all levels within a nation, from the governing elite through to the individual.

A successful participatory approach for disaster risk reduction mainstreaming must facilitate organisational learning. For example, the education, health, public works, media and many other sectors must embrace the principals of risk reduction, whether through school syllabi, community health programmes or infrastructure development if disasters are going to be prevented. A participatory process should bring all these organisational elements together in one room, so a wide variety of actors can explore what is meant by disaster risk reduction and work towards developing concrete programmes, policies and actions for their own organisations. Although somewhat idealised, the hope is these individuals take the learning from the process and apply it to activities in their own fields.

4 Methodology A: What is Future Search?

'Future search' is an action-oriented meeting, designed to help a wide range of people with a stake in an issue find common ground. The participatory process focuses on creating an environment for stakeholders to develop a shared vision for their organisations or communities (Weisbord and Janoff, 2000). Future search was developed by two Americans, Marvin Weisbord and Sandra Janoff in the early 1990s. Since the publication of their first edition in 1995, future search has been used in every continent.

Weisbord and Janoff suggest that future search has an underlying framework, which facilitators must adhere to if they are going to conduct a successful process. The first element of the framework is having **'a whole system'** in the room, which places emphasis on a diversity of stakeholders being involved in the process. This 'pre-condition of a successful meeting' stems from the work of Schindler-Rainman and Lippitt (1980) who organised numerous community conferences. They found that bringing together a wide cross-section of citizens into a single space created a problem-solving dynamic when facilitated skilfully. Through this technique, Schindler-Rainman and Lippitt were able to breakdown people's narrow views of an issue and create a new foundation for decision-making. Most importantly, everyone had to pool their understandings and then experienced a 'whole-room' outlook. Future Search terms this process 'touching the whole elephant' (2002, 52), a phrase coined from a folk tale.

The second element of the framework is **thinking globally and acting locally**, one of the many building blocks of future search drawn from the work of Emery and Trist (1964, 1978). Trist emerged from a long association with Wilfred Bion in the post-war years, where their work focused on group dynamics (Bion, 1961), to work with Emery on 'open systems thinking'. This concept suggests that everything is connected to everything else, no matter at what scale it operates. The 'open systems' approach models the 'global-local' linkage necessary for a future search conference. In its most simplified form, the process should address external issues beyond the local system; the interactions within the local system; the history and aspirations of the local people and agencies; the local impact of the external trends and how action planning might impact life outside. As Weisbord and Janoff state (2002, 70), 'to improve the whole system means to take all these relationships into account.

The third facet of future search is closely tied to Bion's **'self-management'** approach to group dynamics, where he found 'leaderless' groups organised themselves rather than needing an appointed leader to direct tasks. He discovered that such groups would not fight or split, and were often much less dependent on orders or other direct assistance. Through this 'self-management' process effective dialogue is fostered, as all parties talk about the same world and they develop a shared psychological field with common dilemmas and a shared fate (Asch 1952). So in a future search process, participants are expected to organise their own work in small groups, appointing a leader, a timekeeper, a scribe and a presenter with the roles being rotated at the end of each task.

The final aspect of the future search framework is for groups to **explore common ground**, rather than basing their work on problem solving or conflict management. This is partly achieved through placing a 'future focus' at the forefront of each task, where current trends and not conflicts help to build a shared scenario. While differences and disagreements are acknowledged and written down, they are not resolved in the meeting. As Wiesbord and Janoff (2002, 59) suggest, 'future search can take the focus off personality and political differences, problems and symptoms and instead focus energy toward building a sustainable community'.

To support this guiding framework, future search participants are also expected to follow a number of ground rules. These are: All ideas are valid; everything on flip charts; listen to each other; observe time frames and seek common ground and action – not problems and conflicts.

The 'ideal' future search design comprises of five segments spread over two to four hours each, with the meeting running over three days. The future search agenda would appear as follows (the five segments are in *italics*):

Day 1, Afternoon
- *Focus on the Past*
- *Focus on the Present* – External Trends

Day 2, Morning
- Present – External continued
- *Focus on the Present* – Owning our Actions

Day 2, Afternoon
- *Ideal Future Scenarios*
- Identify *Common Ground*

Day 3, Morning
- Confirm *Common Ground*
- *Action Planning*

(Weisbord and Janoff, 2000, 18):

5 Methodology B: Modifying Future Search to Work on St. Kitts

The Future Search process was modified considerably to 'fit' with some of the cultural and political pressures experienced in the tiny twin island federation of St. Kitts and Nevis. In a country of only 45,000 people (find statistics), a meeting of this nature draws significant attention. To reduce the complexity of the political dimension and to maximise the relevance of outcomes, it was decided to focus the Future Search process solely on St. Kitts. Unlike Nevis, St. Kitts is an island struggling under the burden of a failing sugar industry and an enormous national debt. Tourism is a fledgling industry in comparison with its near neighbours, Antigua or St. Maarten, and attractive beaches are rare. St. Kitts has also faced a number of damaging hurricanes in the last ten years, most recently Hurricane George in 1998.

The preparatory phase of the future search process took two months and involved many meetings with key stakeholders. In an ideal future search, facilitators would be approached by a community or organisation requesting assistance on a certain policy or issue. Facilitators would then begin by helping to establish a stakeholder executive group, who would determine the parameters of the meeting and the invitation list. Because the work on St. Kitts was driven by a research agenda, this initial approach was reversed. Effectively, the idea of a Future Search conference on disaster risk reduction was presented to the key stakeholders, some of whom requested a full proposal be sent from University College London to the relevant ministries of government. Once the authorisation had been granted, the invitation list was formulated by the principal investigators and then negotiated with individual agencies. Members of the St. Kitts National Emergency Management Agency gave final approval of the list about three weeks before the conference was scheduled to take place. Through meetings with stakeholders, the timings and culturally specific elements of the process were modified[3]. The main change was to reduce the meeting length to 1 ½ days rather than 2 ½ days. This allowed government and private sector employees to commit more easily to the process. In addition, it was deemed that 36 carefully selected participants (6 stakeholder groups, each with 6 people) were more manageable than the recommended 64 participants (8 stakeholder groups, each with 8 people). Invitees were drawn from a wide range of key agencies and organisations with a stake in reducing the risk of disasters on St. Kitts. Finally, the 'action planning' task element of the meeting was removed and incorporated in the 'common ground' task. This was due to time constraints and the fact that a stakeholder executive group had not been in charge of the organisation.

6 Results

The 'St. Kitts 2020: Building a Disaster Resilient Future - A Future Search Conference was held on May 5th and 6th, 2004 at the Ocean Terrace Inn, Basseterre. A broad range of stakeholder groups were represented. These included: Students; Public Services including the Emergency Services and Air & Port Authority; Local Disaster Committee Chairpersons; Education; Health; NGOs including the Red Cross; Public Works; The National Emergency Management Agency; Environment; Private Sector including Insurance and Banking

6.1 The Past Task

Participants, sitting in mixed (non-stakeholder groups) were encouraged to remember personal, local and global disaster events over the last 15 years and record them on a timeline. Groups were then asked to interpret the history shown on the each timeline and present a summary. General trends were observed, such as the concentration of hurricane

---

[3] For example, the first day of the meeting ended at 15:30 (so parents could collect their children from school). In addition, to conform to Caribbean customs, a prayer was said at the opening of the meeting and certificates were presented at the close.

events between 1995-2000 in St. Kitts and the fact that environmental hazards predominated. Conversely, comments were made about how man-made events were most prevalent on the global timeline, with specific attention being given to the rise of terrorism since 2001[4].

6.2 The Present Task

Mixed groups were joined to create three larger groups each with 13 people. The conference organisers facilitated a 'group' brainstorm exercise creating a 'mind map' of current trends in society – social, economic, technological, political and environment – that are currently shaping disaster experiences. When complete, participants were asked to view the other group's 'mind maps' and note similarities and differences. Finally the three 'mind maps' were placed on the floor together, and everyone was given 10 beans to vote which trends were the most important. The following issues received most votes (in no particular order):

- Economic fragility with particular focus on the magnitude of the national debt
- Sexual health issues, especially involved with HIV/AIDS and the associated cost of providing care
- The eroding social/moral fibre
- The need for more social education and training with a particular focus on public awareness related to a range of disasters and risks
- Poverty
- Physical development planning related to building code enforcement, fire and risk codes and improved infrastructure and shelters.
- The lack of provision for post-traumatic stress

6.3 The Future Task

Participants were asked to have a vision of what a disaster resilient St. Kitts would look like in the year 2020. Their visions had to be desirable, motivating and above all feasible. The whole room created a 'common vision'. The author has selected a few elements of the vision for inclusion in this paper.

- Behaviour and cultural change
- Better NGO co-ordination in event of disasters
- Annual simulation exercises, with institutions doing drills
- Information via internet for all families to access.
- Laws binded by crown
- Each household has own disaster plan
- Improved disaster budget
- Protection for volunteers and insurance for relief workers
- Improved community centres: bigger, better, with kitchens, bathrooms, food
- Proper coastal management
- Adequate water supply and health care
- Storm-water management project
- All buildings withstand cat. 3 Hurricane

---

[4] Due to limitations of space in this paper, full detailing of the timelines has been omitted.

6.4 Finding Common Ground

Participants were asked to formulate specific programs, policies and structures that would begin to help create the common vision of a disaster resilient St. Kitts. They were also directed to place their suggestions under the headings of economic, environmental, social and governance; indicate which proposals were their favourites and who should take responsibility for progressing the issue. From the results generated by this exercise, common ground appeared to be most prevalent with the following issues:

- The inclusion of disaster management education in school syllabuses at all levels, with responsibility being taken by the Ministry of Education.
- The creation of a database of information or a disaster information centre that can be accessed by everybody interested in disaster management, with responsibility being shared between the Ministry of Education, the Mitigation Council, the Ministry of Communication and the Chamber of Industry and Commerce.
- Schools should be encouraged to adopt their own disaster management plans and conduct simulation exercises with responsibility being shared between the Ministry of Education and the Schools and School teachers.
- Insurance companies should provide incentives (reduced premiums) for compliance with building codes and other mitigation efforts. Reduced mortgage rates should also be offered to those complying with strict building codes. Banks and the insurance industry would take the lead.
- NGOs should be involved in supporting long-term development programs through a newly-formed NGO umbrella organisation. NGOs themselves would take responsibility with assistance from communities and the Ministry of Social Development.
- The increased collection of revenue, with responsibility directed towards the cabinet and the Inland Revenue.
- Further diversification of the agricultural sector away from sugar, with the lead being taken by the Ministry of Agriculture.
- There should be an improvement of waste disposal from industrial sites, with responsibility taken by the Ministry of Health and the Environment, the Chamber of Industry and Commerce and NGOs.

7 Conclusions

On St. Kitts, the future search method has allowed a range of stakeholders with diverse political views, to come together to thoughtfully discuss disaster risk reduction. By finding 'common ground' at the end of the process on some directed policies and strategies, participants have a mandate to return to their organisations and effect change. In this regard, it is too soon to evaluate the success of the future search process in promoting disaster risk reduction mainstreaming across a range of agencies and organisations. In addition, organisational learning is notoriously difficult to measure, but the promise of future search should not be underestimated. At a time when many policy-makers, academics and practitioners are searching for a way to implement disaster risk reduction objectives, this methodology provides a relatively inexpensive way to highlight the issue among stakeholders. A subsequent research visit to St. Kitts in late 2004 will gauge the degree to which disaster risk reduction policies, programmes and learning have been introduced to both governmental and non-governmental agencies in the period since the future search conference was held in May 2004. Even this maybe too soon, but as more time elapses it will become increasing difficult to separate the success of the future search process from other influences.

References

Ascanio, C. 'Venezuela Mechanisms for Formulating Risk Prevention and Mitigation Programs: Local Management of the Caracas Metropolitan Area'. Available electronically at http://www.eird.org/ing/revista/revista.htm. 2003.

Asch, S. *Social Psychology,* Prentice-Hall, New York. 1952

Bion, W. *Experience in Groups*. Tavistock, London. 1961

isaster Mitigation Institute 'Action Learning for Disaster Risk Mitigation – Annual Report 2002-2003'. *Experience Learning Series 25*. Disaster Mitigation Institute, Ahmedabad, India. 2003

Emery, F. E. and Trist, E.L. 'The Causal Textures of Organizational Environments.' *Human Relations*. 18(1), 21-32. 1964

Emery, F.E. and Trist, E.L. *Towards a Social Ecology*, Plenum, New York, 1973.

Instituto de Estudios Ambientales, IDEA 'Information and Indicators for Disaster Risk Management' co-ordinated by the Universidad Nacional de Colombia, in Manizales – with financial support from Inter-American Development Bank. Available online at http://idea.unalmzl.edu.co, accessed 25/09/03. 2003.

Mitchell, T. 'An Operational Framework for Mainstreaming Disaster Risk Reduction'. *Benfield Hazard Research Centre Disaster Studies Working Paper 8*. Available online at www.benfieldhrc.org. 2003.

Parker, D.J. 'Criteria for Evaluating the Tropical Cyclone Warning System'. *Disasters* 23(3): 193-216. 1999.

Pelling, M. eds. Natural Disasters and Development in a Globalizing World. Routledge: London and New York. 2003

Pugh, J. and Potter, R.B. (Eds) *Participatory Planning in the Caribbean: Lessons from Practice*, Ashgate: Aldershot and Burlington USA 2003

Schindler-Rainman, E. and Lippitt, R. *Building the Collaborative Community: Mobilizing Citizens for Action*. University of California: Irvine. 1980

Schlosser, C. and Aysan, Y. 'Disaster Risk Reduction Framework – Synthesis of the UN-ISDR/UNDP online conference, 25th August – 30th September 2003'. Available online at www.unisdr.org. 2003

Twigg, J. (2004) Good Practice Review: Disaster Risk Reduction: Mitigation and Preparedness in Development and Emergency Programming. ODI Good Practice Review 9: Humanitarian Practice Network: Overseas Development Insitute, London.

UN International Strategy for Disaster Reduction (UN-ISDR) 'A Draft Framework to Guide and Monitor Disaster Risk Reduction'. Available online at www.unisdr.org/dialogue/basicdocument.htm, accessed 21/08/03. 2003

Weisbord, M. and Janoff, S. *Future Search: An Action Guide to Finding Common Ground in Organizations and Communities*. 2nd Edition. Berrett-Koehler Publishers, San Francisco. 2000

Flood Assignment Plans – A Successful Instrument for Avoiding Damages

Jörg Lotz
Lotz AG, Ingenieure, Schloss 3, D-63607 Wächtersbach, www.lotz-ag.de

Abstract

"The management of flood incidents must be improved."

Many reports and analyses of the last flood incidents can be summarized under this heading.

Measures for reducing the peak outflows as well as restoring or extending fixed local technical flood protection are the decisive basis - but ideal management at the interface between water engineering and catastrophe protection can also drastically and permanently reduce the damages from floods and the costs of response.

To improve this is principally not the task of catastrophe protection or the fire brigades. These organisations can expect that information and data will be provided for them in a form that is usable for their assignment.

Assessing the Latest Flood Incidents

is much the same. There are similar demands in many analyses regarding the sequences during flooding. Better coordination on the basis of the latest alarm and assignment plans is demanded. There were the same demands following the Oder floods but implementing these has not progressed very much.

These demands are based on different situational cases in various Federal Lands but can be reduced to three important criteria.

- overloaded assignment managements and responsible authorities,
- complicated paths of communication,
- unclear decision structures and decision bases.

Many people who have experienced the Elbe flooding live and on site for example, can tell you that there was a unique mixture of motivation, engagement, willing cooperation and courage to make decisions there.

Picture 1: There was no lack of motivation in the Elbe floods.

Everybody wants to help and there is agreement on the basic requirements; even the costs of the assignment measures during the work were no problem. What is the reason then, if the ideal situation is nevertheless not achieved?

Three causes are important here:

- There was no chance from the beginning. That happens - even if all sequences, decisions and measures ran very well during the floods. Everything possible was done correctly, but nevertheless the floods were all-powerful in many places.
- Those helping tried, but were not adequately trained. This refers to both areas: Fire Brigades and Catastrophe Protection must be more prepared and trained for flood assignment and learn to understand more of water engineering. Administration and water engineering on the other hand must learn more about fire brigade work and catastrophe protection. This applies to employees in the competent authorities just as to political decision-makers and government authority leaders. The work in the staff of technical assignment management also has to be learned. A political appointment alone does not automatically qualify a person for this.
- If everything is fulfilled, or motivated, trained helpers can implement adequate means and forces, then there is still a decisive instrument missing: data, information and strategies that have previously been worked out: a plan - which allows exactly the right thing to be done at the decisive moment with the defined strategy and detailed application hints.

And what happens in many places when there are floods?: work is started - and everywhere the wheel is discovered once again. There are exceptions however. The more often a town is confronted with flooding, the more often is each person clearer what is to be done. It is well-established. But when flooding occurs only seldom is there chaos.

A tool is to be presented that reduces the decision stress, provides a large degree of organisational safety and can thus lower the hectic activity during assignment. A tool that enables forces and means to be used efficiently and effectively. The technical language of water engineering will have to be translated into the technical language of the fire brigade for this.

Flood assignments are, from their structure, very much easier to plan in advance than, for example, fires, earthquakes or storm catastrophes. With an adequate data base it can be seen in great detail what happens with a certain water level and what measures are then necessary.

This observation must not stop at a certain flood level however, but must extend to the probable maximum flood level.

There is a large difference between floods and fires for the fire brigades for example. With fires the situation to be fought can be seen and appropriate action taken. If a flood wave rolls towards the assignment teams they must first ask about the situation they are to fight. But not only the water level expected is necessary here but also very much more information, e.g. what areas are then flooded etc.

Compare a flood assignment with the large damages of a forest fire: the fire brigade unit there will receive the order to respond: fire fighting in section x, extinguishing water extraction at y pond. These are the important items. A fire brigade team leader can then start his work. Flooding is completely different on the other hand. There is very much more precise information needed and reporting back and technical information is also required.

In addition it cannot be seen what measures are the correct ones, for example with a saturated dike, and how much risk is there of the damages spreading. A good knowledge of hydrology or hydrogeology is usually necessary for this. But in a rural district for example, there are insufficient water engineers in many places who can be available for assignments over several days and around the clock.

Even when damages have already occurred to a dike for example, the assignment team leader on site can often not decide himself what are the most suitable measures for combating this. He needs material for this that does not belong to the equipment of his vehicle and the availability of which he does not know, and he needs information that cannot be seen on site. On the other hand the assignment management do not know how much material (e.g. sand sacks) is needed. This requires a continual flow of information, permanent reporting back and thus a high degree of communication. These facts inevitably lead to the assignment management and communication paths being overloaded.

This can only be changed by allowing units on site to be more independent. The basis for this is naturally also that the unit leaders on site are suitably trained.

Information is also important that the unit leader has to be given so that he is independent in his section and can introduce the necessary measures there. Information and data can be given to him that was already available before the flood. This must be made in a form suitable for use. If it is shown that from a water level of xy metres a dam of sand sacks is to be erected from here to there at a height of 1.20 m and 600 sand sacks are needed for this, then the unit leader can arrange the appropriate work. If the same plan is also available to the assignment manager, then he knows how many sand sacks are needed on site.

A lot of data relevant for this is mostly available and only has to be supplemented by tactical fire brigade planning. There is data with various technical authorities that a team leader can gather during an assignment, but not with the required care; there is just not enough time available to do this.

How can this be Improved?

Acquiring water engineering knowledge and requirements in tactical assignments of the fire brigade are decisive for this. There is a lot of data in water engineering - but plans for flood protection and open waters are always only made by water engineers for water engineers.

A longitudinal section of a dike system to a high scale for example cannot necessarily be used by the fire brigade. Forecasts of water levels are unusable if there is no reference height at the site of assignment. To level between sand sacks is not ideal; it would be easier beforehand.

Picture 2: Dike measuring as emergency measure in the Elbe floods

Dikes are not uniformly high and not constructed equally as good. It is rather important to know where the lowest part of the dike is and where problems have already occurred. This information must be so illustrated and prepared so that even non-water engineers can understand it.

After several days even the local fire brigade force will not all be completely available. For assignment teams from outside the local area it is even more important to be able to orientate themselves with the help of assignment plans.

Deficits in Water Engineering

Unfortunately some water engineers take the view that a catastrophe has occurred above the calculated flood level and nothing more can be done. That is quite wrong! It should become known that a calculated flood water level, once selected, would be exceeded in any case. Only when this will be the case is unknown as yet today.

Water engineers must then explain to the helpers how water engineering systems react when the calculated flood levels are exceeded. Water engineers must first know this themselves - an obligation that is often overlooked. But one must learn to see this situation as it is.

With a catastrophe situation decisions must be made that have a completely different background to normal life, such as for example the option of having to also sacrifice values to avoid larger losses. It is permissible to sacrifice a village in order to save a town. Such "atrocious acts" are imperative for a thoughtful strategy in a catastrophe and have to be included in flood assignment plans. This decision will have to be made by those responsible in catastrophe protection in each case, but for water engineers these are new horizons to be considered beforehand so that these options can be implemented or also discarded.

Before the contents of the flood assignment plans are explained, the following must be briefly mentioned:

There is no Scale for Describing the Results of Flooding

There is generally no uniform term for showing the effects of a certain flood.

An HQ500 flood can be an absolute catastrophe in one place and in the next village it was relatively harmless.

The probability of an occurrence is the only term at present however, that describes the intensity of the flood. This statistical characteristic is completely unsuitable to describe the results of a flood however.

On the other hand flood registration - alarm or warning stages - are different in each Federal Land or town. These have 2, 3 or 4 stages and each have completely different meanings. This is unacceptable.

These alarm or registration stages also do not enable the definition of damage occurrence on the basis of the damages however.

A Flood Intensity Scale must therefore be developed and introduced that imaginatively describes the practical results for laymen and can be transferred into general language.

To improve the general understanding in the population the flood intensity scale should have a close resemblance to other scales such as: Beaufort - Wind; Mercali, Richter - Earthquakes; Fujita, Torro - Tornados; Saffir-Simpson - Hurricanes. A scale in 12 divisions would then be suitable.

Such a scale is at present being worked out by LOTZ AG. A technically-minded person can also visualise this. "Wind force 12" is a term that many people understand and represents total destruction.

Flood force 12 in future means roughly "something bad".

The results of a certain flood in a limited area can be described with a flood intensity scale. The intensity refers to the results, or property damages, victims, expenditure and other parameters. The larger the area observed the lower the total force, because some areas are always affected less than others. These can be expressed by a collective term:

Flooding of force 6/11 → means: force 6 for the town or area described, largest damages of force 11 however.

An initial draft will be presented at the lecture and can be seen in www.hochwasserintensitaet.de.

Suggestions or cooperations on this are explicitly wished.

What Must the Flood Assignment plan Contain?

Much more different information that exceeds the mere water engineering data: information about building infrastructures, supply and disposal pipework and equipment. The datum level of every sewer manhole in Germany is known for example, which can be very helpful in transferring reference heights. Other installations that are especially worth protecting must be contained in this. Also the details of sewers that are endangered by floods or which can be dangerous themselves.

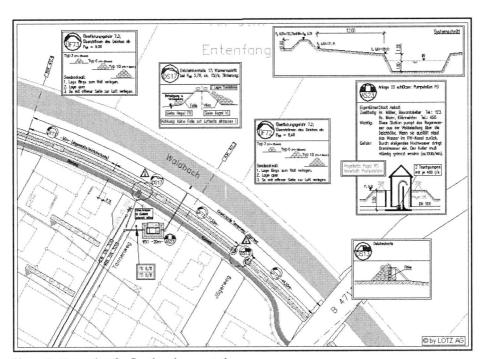

Picture3: Example of a flood assignment plan

Even the last stage in possible protection measures - the evacuation - must be planned in detail in advance. Exact ground levels are indispensable for this, to be able to proceed precisely.

At the latest here it is clear that water engineering will have to provide the basis for the plan, but they cannot develop all of the assignment tactics and therefore flood assignment plans must be agreed in close cooperation. The author of flood assignment plans must have experience in both areas, of water engineering and catastrophe protection.

On the contents:

- inventory data and weak points in the existing flood protection installations (exact heights, extent and location of known points of moisture penetration)
- possible measures for dike defence correlated to the local conditions (e.g. subsoil and construction of an existing dike)
- information on the necessary materials and personnel (e.g. from bridge A to point B there will be 2,800 sand sacks required for a water level of 9.73)
- information on other infrastructure (sewers, usable roads, special dangers in adjoining areas)
- information on areas especially worthy of protection (e.g. hospitals, telephone exchanges, computer centres, electricity supplies, sewer pumping stations)
- preparation of hydrological data, symbols, illustrations usable for assignment
- suggestions for assignment measures with appropriate tactical characteristics from catastrophe protection
- list of responsible persons, locations, supply installations (sand sack store, petrol stations), traffic route sketches, town plans. Prepared so that complete outsiders can find their way around.
- the basis for these plans is a detailed water engineering consideration of possible flooding occurrences with a risk study and process analysis of possible flooding occurrences up to pmf - the "assumed maximum flooding".

Can that be Paid for? And Who Should Pay?

Preparing flood assignment plans is an extensive task requiring a lot of data. The authority responsible for flood protection must have these plans prepared. Those who are to finance the building measures for flood protection should also finance the assignment plans. On the one hand the technical data is available there and on the other hand the flood assignment plan is the logical continuation and supplementation of the building flood protection, and feedback is required and necessary.

Complicated assignment situations require farsighted planning, which is nothing new. A building owner must provide the fire brigade with the appropriate fire-fighting plans. For the whole of Germany this is governed by DIN 14095. This has proved to be reliable and is practised daily. A similar standard is required for flooding. Flood assignment plans should be prepared for assignment personnel to uniform rules for the whole of the Federal Republic.

The costs, in comparison to the use in flooding and the costs of possible damages, are rather low, because it is often overlooked that the costs for a flood assignment are extremely high. They are mostly not made public - and a lot is voluntary and without charge - but there is economic damage in any case.

The next assignment will be considerably more effective through flood assignment plans and the assignment costs sink drastically. Even a single large flood balances the costs for preparing assignment plans alone through the assignment costs saved.

The Author

Joerg Lotz of LOTZ AG Ingenieure, is the managing director of a successful engineering office for many years with competence in the area of water engineering. He is also an experienced assignment manager of a large fire brigade base. Joerg Lotz was responsible, as water engineer, on the side of the fire brigades of Main-Kinzig District, as professional adviser in the TEL "Kornhaus" for 4 kilometres of the Elbe dike in Dessau and managed the cooperation between the fire brigades and the water engineers.

References

LOTZ, J. (11/2003): Hochwassereinsatzpläne – Schnittstelle zwischen Wasserwirtschaft, Feuerwehr und Katastrophenschutz. Vortrag anlässlich der acqua-alta – Internationale Fachmesse mit Kongress für Hochwasserschutz und Katastrophenmanagement, München.

LOTZ, J.(11/2003): Hochwassereinsatzpläne – Erfolgsinstrument zur Schadensminimierung. Fachartikel in „Hochwasserschutz und Katastrophenmanagement", Special 3/2004, Verlag Ernst & Sohn.

LOTZ, J. (2004): „Hochwassereinsatz" Fachbuch für Feuerwehren. Erscheint in der Reihe „Die Roten Hefte" als Nr. 82 im Herbst 2004 im Verlag Kohlhammer.

LOTZ, J.; STAAB, M (2003-2004): Marktübersicht mobile Hochwasserschutzsysteme. Informationsschrift der LOTZ AG.

LOTZ, J.; STAAB; M (2004): Vergleich der Deichverteidigungsmaßnahmen Quellkade und Sandsackrost - Simulationsberechnungen mit FE-Modellen (Veröffentlichung, in Vorbereitung).

Robots for Exploration of Partially Collapsed Buildings

Olaf Fischer[1,2], Marius Zöllner[2], Rüdiger Dillmann[1]
[1] *Institute of Computer Design and Fault Tolerance, Universität Karlsruhe (TH), D-76218 Karlsruhe, Germany*
[2] *Forschungszentrum Informatik, Interactive Diagnosis and Service Systems, D-76131 Karlsruhe, Germany*

Abstract

The reconnaissance of damaged structures after hazardous events like earthquakes or explosions is a dangerous venture for all rescuers involved. One goal of many research groups all over the world is the development and realisation of scout robots which autonomously or semiautonomously enter the damaged facilities and simultaneously or with short latency inform the rescuers about their discoveries. The aim of this contribution is to give a concise survey of existing systems for this application and an introduction of the walking machine LAURON , which we want to use for teleoperated reconnaissance. In addition to a brief description of the current state of the art, we will propose concepts for an effective robot-based exploration and demonstrate these concepts with our current work on LAURON III.

Keywords: USAR, rescue robots, LAURON III, controller interface, situational awareness, RFID tags

1 Introduction

The retrieval of humans or objects in partially collapsed buildings is a dangerous task for every rescuer involved. A lack of knowledge of the current state inside the building forces the rescuers to make decisions based on uncertain knowledge. With some basic knowledge about the destroyed area rescuers could act on a higher safety-level for them and for the vicitms. Here robots can help with an exploration of the area and collect valuable information about the collapsed building and the survivors. For this purpose, safe locomotion methods and robust teleoperated robots for rough terrain are needed. The necessity for such devices is recognized by many research groups in Japan (see e.g. Tadokoro, 2004), USA, and the European Community.

This article is organised as follows. Section 2 will give an overview about a selection of existing rescue robots from other research groups. In Section 3 we will focus on our work and give a short introduction of LAURON III and the modifications we plan to adapt LAURON III to the exploration task. This will be followed by a controverse discussion about open problems with the use of robots as exploration units in Section 4. The paper will end with some conclusions in Section 5.

2 Rescue Robotic Systems

There exist numerous robots escpecially designed or modified for exploration tasks in dangerous environments like collapsed buildings. In this section several advanced robots for the exploration of rescue scenarios will be described. A comparison table of all presented systems will be presented in Section 4.

2.1 Helios VII

At Hirose & Yoneda Robotics Lab, Tokyo, a very robust tracked robot is developed, which is called Helios VII (Fig. 1(a)). Two tracks are connected to the main body with rotary joints. To the main body a manipulator is attached (Guarnieri et al., 2004). This manipulator is equipped with a number of passive wheels. The manipulator arm is specifically designed to support movement of the whole vehicle. When starting to climb stairs, the arm will first swing forward, then apply force to the first step and by this lift the front part of the robot. The tracks will then get grip on the steps and the vehicle can ascend the stairs.

By moving the manipulator arm, the center of gravity of the whole machine can be shifted. To climb higher obstacles, the center of gravity is shifted to the back of the vehicle. In this configuration, the front of the vehicle easily lifts up when moving into the obstacle. While the tracks generate friction at the obstacle's front, the arm is used to apply force onto the floor, thereby pushing the vehicle up the obstacle. Even cooperation of several Helios VII robots is considered to conquer even more difficult obstacles.

At the current state, Helios VII does not carry energy packs, external sensors, or a controller unit. Energy is supplied by a tether, which also connects the onboard motor controllers to the control station.

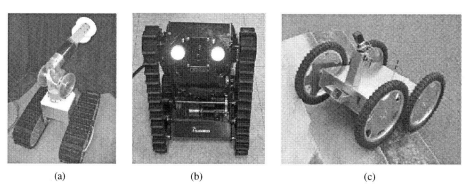

(a)   (b)   (c)

Figure 1: (a) Helios VII (Guarnieri et al., 2004), (b) modified Inuktun Micro VGTV at CRASAR in erected posture, (c) Orpheus (Zalud and Kopency, 2004)

2.2 Inuktun MicroVGTV at CRASAR

At the Center for Robot Assisted Search and Rescue (CRASAR), Tampa, FL, Murphy uses an Inuktun Micro VGTV for exploration purposes (Murphy, 2000). This impressive robot is one out of very few that have really been in action, e.g., at the disaster in New York City at September 11th 2001 (Blackburn et al., 2002). Like Helios VII, this robot has a very variable geometry. In standard posture, the Micro VGTV has the shape of a conventional crawler. However, the Micro VGTV can erect the front in that way that the tracks then take the shape of a triangle (Fig. 1(b)). In this marmot-like posture the cameras are lifted to a higher position, which results in a better overview about the inspected environment. In both extreme postures and all intermediate postures, the Inuktun Micro VGTV remains fully operable, in particular can the robot rotate in erected posture.

The Micro VGTV is connected with a tether to the power supply and the control station. This tether is robust enough to carry the whole weight of the Micro VGTV, and it is possible to let down the robot into the environment from a higher postition. It is equipped with a camera and lamps for illumination of dark environments. The camera and lamps are mounted on a tilt-unit which allows the operator to sweep them up and down. Bi-directional audio connection allows the user to hear what is going on in the remote location and even to communicate with a possible victim. Murphy added two lasers next to the camera, which project two parallel bars vertically into the scene. These bars of fixed and known distance support the operator's understanding for the inspected scene and helps him to estimate sizes in the remote location, thereby increasing the situational awareness.

A briefcase sized control station is available for the Micro VGTV. It consits of a display for the camera image and supplemental sensor information and several buttons and controller sticks for operation. The training time to operate Micro VGTV effectively is very short and is specified as several minutes.

2.3 Orpheus

At the Department of Control and Instrumentation at Brno University of Technology (CZ) the vehicle Orpheus (Fig. 1(c)) is especially developed for rescue teams, pyrotechnicians and firemen (Zalud and Kopency, 2004). One wheel on each side of Orpheus is driven directly and independantly, and a transmission belt connects the wheels on each side. With this simple configuration the vehicle can move forward and backward and turn to both sides. Orpheus is equipped with a color camera on a pan-tilt-unit, and two b/w-cameras on a tilt-unit, one pointing to the front and the other one to the back. The whole system is controlled by a heterogeneous set of 8 ATMEL microcontrollers, which are all stored together with the battery packs in the chassis. Controller and video data are transfered independently with two different wireless systems.

The operator station for Orpheus consists of a head mounted display (HMD) and a joystick connected to a notebook. The display is equipped with an inertial head movement sensor, which detects head movements and controls directly the pan-tilt-unit of the main camera-system. The robot's movement is controlled with the joystick. The display shows the camera image fullscreen, additional information can be projected transparently inscreen. The Orpheus system was introduced at the Robocup Rescue Competion 2003 in Padua (I) and placed first.

3 Walking Machine LAURON III as Exploration Robot

LAURON III (Fig. 2(a)) is a very advanced, six-legged walking machine and was developed by the Interactive Diagnosis and Service Systems Group at Forschungszentrum Informatik, Karlsruhe. Originally inspired by the stick insect, LAURON III has a height of about 50cm, length and width of about 70cm, and weighs about 18kg. The six legs each consist of three joints and can swing for- and backward, move up and down, and bend. All feet are equipped with 3-dimensional force-sensors. Each leg is controlled by one microcontroller, all microcontrollers are connected via CAN-Bus to an embedded PC-System, which is located in the trunk. There is a 3-dimensional orientation sensor also located in the trunk which provides the robot with its current orientation in the outer environment. This arrangement combined with a behavior-based control mechanism is sufficient for LAURON III to being able to stroll around freely, thereby stalking over medium-sized obstacles. LAURON III can walk conventional stairs and slopes of 30°-35° (depending on the surface), and climb obstacles of a height of approximately 28cm (Gassmann et al., 2001). Communication with LAURON III is established with a regular wireless ehternet connection.

3.1 Controller Interface for LAURON III

For enhanced perception LAURON III is equipped with a pan-tilt-unit which carries a stereo camera system and a set of microphones. We believe that vision is a keypoint to any successful

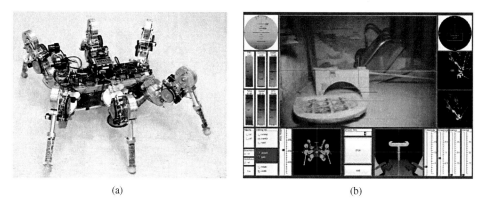

(a)	(b)

Figure 2: (a) LAURON III, (b) controller interface (Fischer et al., 2004)

teleoperation purpose, and without transmission of video images from the remote location it is very difficult for humans to understand the real scene. Additionally, transference of sound from the remote location to the user increases the user's ability to telecontrol LAURON properly.

To operate LAURON III remotely a control system which produces a high situational awareness to the operator was developed (Fischer et al., 2004). It is implemented as a multimodal controller interface (cf. Fig. 2(b)), which accepts input from mouse and keyboard. Output is presented visually and acoustically. The screen of the control panel displays the remote location on a large area. Additional information about the detected scene is projected directly into the image. Control and sensor elements are collocated at both sides and at the bottom of the camera image. The design of the interface implements high redundancy in input methods as well as in output generation, e.g., a critical body orientation will not only be indicated by a flashing of the artificial horizon but also by a warning sound.

Test runs with the described configuration were carried out. The overall impression of the system was good but still has potential for improvement. The low viewpoint of the cameras (about knee-height) seemed strange in the beginning of the tests. LAURON III moved at a leisurely speed, which resulted in impatient test persons. Critical is the narrow view angle from the 6mm lenses. Improvement is possible with the use of wider lenses, but this goes ahead with a quality loss for the stereo vision system. The use of mouse and keyboard as input devices and a regular display for the vision output is also not optimal. This combination proves useful in a laboratory environment, but outdoor tests show that it is difficult to find a suitable place where to place mouse, keyboard and screen. Therefore, it is further planned to use a head mounted display with earphones and a video game controller with force feedback as input and output devices.

3.2 Use of Landmarks for Autonomous Self-Retrieval

The main control mode for LAURON III as an exploration unit is teleoperation, i.e., a remote operator decides what the robot is going to do next, with only minor actions being performed autonomously. This is desired because the rescuers who are present outside the exploration zone want to be informed online about the exploration mission and decide at every moment what to explore next. However, it is always possible that communication to the robot breaks down for different reasons (e.g., technical failure, environmental noise, signal leaks). To prevent a loss of the machine in these cases we will use radio frequency identification (RFID) tags as landmarks for an autonomous self-retrieval effort.

RFID tags are passive transponders which are activated by radio waves emited from an RFID reader. When tags are activated, they use some of the absorbed energy to send back their unique identification number. On this way, RFID tags can be read from a distance which is determined mainly by frequency and antenna size; this distance ranges from several centimeters to a couple of meters. In our work we are using RFID tags which appear as 2mm coins with a diameter of 5cm. In our approach tags are placed by the robot every few meters along the path from the beginning of each exploration mission. When placing a tag, its unique number will be read and this information can be added to a topological map which is stored and updated in the robot. This map contains information about the sequence of tags as well as distances and directions between two neighbored tags. The track of tags can be detected and followed visually. Even if the machine is completely unaware of the correct direction to follow the track, the identification of a single tag by reading its unique number reveals the machine's location within the priorly built topological map. From this knowledge combined with the readings from the orientation sensor the machine will be able to pursue the track. Even branched tracks or loops can be followed with this approach. We believe that in this manner we can follow tracks even when one or more neighbored tags are missing.

Another positive effect which arises from the use of these tokens is that rescuers can easily follow the machine and automatically find all the places they have seen before in the presented images.

4 Discussion

In Sections 2 and 3 different robotic systems have been briefly introduced. In Table 1 some of their features are summarized. It is obvious that many different solutions are offered for a number of technological problems. In Section 4.1 these aspects will be discussed more closely. However, there is still another important issue to be discussed, which is of social nature. Section 4.2 will shortly discuss matters like necessity and acceptance of these machines.

	Helios VII	Inuktun	Orpheus	LAURON III
Size (WLH) in cm	62.5×71.2×?	16.5×31.7×6.5	43×54×112	70×70×50
Weight in kg	84.5	3.15	32.5	18
Locomotion	tracks	tracks	differential drive	different walking patterns
Control	tether	tether	wireless	wireless
Power Supply	external	external	onboard	onboard
Sensors	none	vision, sound	vision, sound	vision, sound, inertial

Table 1: Comparison chart

4.1 Technological Aspects

From the introduced robotic systems Helios VII and Micro VGTV are tracked, Orpheus is wheeled and LAURON III has legs. Tracks and wheels are by far the most common locomotion method used,because they have several advantages. First of all, they are easy to control, they are energy efficient, and capable to transport high payloads. Legs, on the other side, are very complex to build and to control. Energy is consumpted no matter if the machine is moving or standing still. Payload is a problem for all designers of walking machines. But nevertheless, walking machines have some advantages over tracked or wheeled vehicles, too. First of all, as weight is always an issue, walking machines are designed to be lightweighted. They can conquer very rough terrains and climb over obstacles that cannot be surpassed by conventional vehicels. Walking machines can decide thoroughly where to put their feet; no closed trajectory is needed. With this quality walking machines can pass through areas very carefully and with very few manipulations to the environment.

Size is another matter worth discussing. Small robots can enter small holes and explore areas where bigger robots cannot go. On the other hand, small robots cannot surpass big obstacles. Additionally, large scale robots can carry more sensors and equipment. Small robots often rely on external power supply and do not have enough computational power to act semiautonomously at least. Two of the robots introduced are connected with a tether to the operator station, while the other two rely on wireless communication. Tethers always limit the robot in mobility as well as in reach. Wireless communication can break down for several reasons, leaving the robot in a vulnerable position.

Many researches address the problem of teleoperation or telepresense (Zalud and Kopency, 2004). As long as exploration robots are not completely autonomous, the operator interface is of prior importance. The control station must be robust enough to be used in outdoor environments and resist water, dust and small mechanical impacts. The interface must be easy to control and present as much information about the remote location as possible. At the same time, the user must not be overloaded with unimportant or useless data.

4.2 Social Aspects

Until today, only very few robotic systems have ever been used in real disaster situations. Almost all systems still have research quality and are not ready for use by professional rescuers. What is needed by rescuers today are simple, reliable, and easy-to-use robots like the Micro VGTV from Inuktun. Only those systems have a chance of being accepted and utilized by rescuers. Many rescuers are willing to put themselves into danger as long as there is a chance to save a victim. The use of robots must at no time produce delays in the rescue operation and at the same time gather valuable information for the rescue process.

At the IEEE International Workshop on Safety, Security, and Rescue Robotics 2004, Bonn, Elger[1] reported about their robotic system for telemanipulating tasks. There, a robot is used to release pressure from barrels with excess pressure. This dangerous task is performed remotely controlled from a safe distance. He reports that almost all of the younger firemen commit a regular training with this robot, some of them even in their spare time, whereas many of the older firemen avoid the robot. This shows that acceptance of robotic systems may even be a matter of generation.

5 Conclusion

In this paper we have introduced the robotic systems Helios VII, Micro VGTV, Orpheus and LAURON III. As seen in Section 4.1, these systems differ extremely in form of locomotion, size, autonomy, and other aspects. Nevertheless, all systems are developed to perform exploration missions successfully. We believe that a great variety and availability of different systems is neccessary to fulfil future needs. There are still many technological problems, but most of them will be solved within the next few years. Additionally, there are also social matters that needs to be considered. We firmly believe that rescue robots can only be successful if they are widely recognized, accepted and utilized not only by rescuers for disaster management, but also by mean firemen. Only the equipment of many fire departments with exploration robots will lead to a large number of robots, which will still accelerate the development and improvement of rescue robots.

---

[1] BASF Factory Fire Brigade, Ludwigshafen

Acknowledgment

This project is funded by the German Research Foundation (DFG) and the University of Karlsruhe within the Research Training Group Natural Disasters.

References

Blackburn, Michael R., Everett, H. R., and Laird, Robin T.: After action report to the joint program office: Center for the robotic assisted search and rescue (CRASAR) related efforts at the world trade center. Technical report, SSC San Diego, 2002.

Fischer, Olaf, Wiersbitzki, Jörg, Zöllner, Marius, and Dillmann, Rüdiger: Teleoperating a six-legged walking machine in unstructured environments. In *IEEE International Workshop on Safety, Security, and Rescue Robotics*, 2004.

Gassmann, B., Scholl, K.-U., and Berns, K.: Locomotion of LAURON III in rough terrain. In *Advanced Intelligent Mechatronics* 2:959 - 964, 2001.

Guarnieri, M., Debenest, P., Inoh, T., Fukushima, E., and Hirose, S.: Helios VII a new tracked vehicle for rescue and search operations, In *IEEE International Workshop on Safety, Security, and Rescue Robotics*, 2004.

Murphy, R.R.: Marsupial and shape-shifting robots for urban search and rescue. In *Intelligent Systems* Vol. 15 (March-April), IEEE, 2000.

Tadokoro, Satoshi: Problem domain of Japan national project (DDT Project) on rescue robotics. In *IEEE International Workshop on Safety, Security, and Rescue Robotics*, 2004.

Zalud, L., and Kopency, L.: Teleoperated reconnaissance robotic system. In *IEEE International Workshop on Safety, Security, and Rescue Robotics*, 2004.

Optimizing Search Mission of Avalanche Dogs – GIS Based Simulation About the Availability and Transportation Time of Avalanche Dog teams to Rescue Buried Persons

Leopold Slotta-Bachmayr

Rescue Dog Unit, Austrian Red Cross, Minnesheimstr. 8b, A-5020 Salzburg, Austria

Abstract

If a person is buried in an avalanche, it is absolutely essential to react quick and effective. During the first 15 minutes the probability of survival is about 90 % and thereafter decreases dramatically. After 90 minutes only 25 % of the victims are rescued alive. Therefore, the quick and efficient action of rescue teams is an important prerequisite for the successful recovery of persons out of an avalanche.

For the planning and simulation of such rescue missions the new findings of avalanche research are very helpful. In this research, the actual avalanche danger scale and different landscape parameters are combined to give a risk factor. With this type of evaluation it is possible to calculate the risk of one skier setting off an avalanche over a large area.

Avalanche dog teams are mainly transported by helicopter to the scene. In a second calculation the helicopter flight time was simulated considering distance and terrain elevation. Four scenarios describe different types of availability of rescue teams.

The combination of the two simulations shows that the position of dog teams does not make a difference in transportation time. If the rescue teams are available at three different helicopter bases, the transportation time will be halved. In a second step, it was also evaluated, whether avalanche danger scale has any influence on transportation time. In this case no relation was detected.

The simulations have shown that availability of avalanche dog teams cannot be related to avalanche danger scale and it is the most efficient to have several helicopters and rescue teams ready for a mission at different places scattered over the county.

Keywords: avalanche emergency, avalanche dog, decision support, geographic information system, helicopter flight time, potential avalanche release site.

1 Introduction

After an avalanche emergency only quick response can save life. In the first 15 minutes the chances of survival for a buried person are about 90%. Afterwards the chances of survival decrease dramatically and after a burial time of 90 minutes it is only 25% (Falk et al., 1994). The quick and efficient response by rescue teams is an important requirement to recover buried persons out of an avalanche successfully. Therefore different rescue methods are available. Organized rescue teams mainly use avalanche dogs in addition to avalanche beacon and probing (Tschirky et al., 2000). With an avalanche dog team, which consists of a dog and a handler, the emergency site can be searched much quicker and more efficiently than with all the other rescue methods (Gayl and Hecher, 2000). As a disadvantage, avalanche dog teams have to be transported to the search site and this takes time. So optimal transportation and team siting are important factors to save time during an avalanche rescue operation.

Geographic Information Systems (GIS) are powerful tools to help with planning, mitigating, preparing for, responding to and recovery from disasters (Greene 2002). They allow a spatial impression of possible emergencies and the simulation of future events (Dash, 1997). The development of avalanche research during the last years to evaluate potential avalanche slopes has shifted more and more from the localized evaluation of a single slope based on stability of snow cover (Föhn, 1987) and the avalanche bulletin to a more generalized evaluation by combining the avalanche danger scale and landscape parameters like aspect and slope (Munter, 1999). This type of evaluation in combination with GIS makes it possible to calculate the risk of releasing avalanches for a whole region. Again these calculations and maps are important planning tools when deciding about preparedness and response to an avalanche emergency.

In a project to improve the efficiency of avalanche rescue teams in the Federal Province of Salzburg the recovery times of different rescue methods were investigated as well as the stress on rescue dogs during the search or the time the scent needs to diffuse through the snow. One module of this project will provide decision-makers with different scenarios of team siting and transportation to an avalanche. Therefore statistical data was analysed to show temporal distribution of avalanche emergencies. A model for risk of avalanche release and a model on helicopter flight time respectively was developed to show the influence of team siting on response time.

2 Data and Model Structure

2.1 The Temporal Distribution of Avalanche Emergencies

For the Federal Province of Salzburg there is a database which contains data about all the avalanche accidents between 1970 and 2003 (Niedermoser, 2000). Every accident is described by date, time of day, avalanche danger scale, weather, type of avalanche and position. To analyse the occurrence with regard to time the probability of avalanche emergencies in relation to day of the week, danger scale and weather was calculated.

2.2 The Spatial Distribution of Potential Avalanche Accidents

With this in mind a GIS based simulation was calculated. Basis for this model is the reduction method developed by Munter (1999). With this method every slope is evaluated by its aspect and its slope (Tab. 1).

danger scale	potential risk	slope	RF 1	aspect	RF 2
1 - low	2	>39°	1	NW-N-NE	1
2 – moderate	4	35-39°	2	NE-ESE	2
3 – considerable	8	30-34°	4	ESE-WNW	3
4 – great	16	<30°	8	WNW-NW	2

Table 1: Potential risk of causing an avalanche in relation to the actual avalanche danger scale and the values of slope and aspect (RF = reduction factor) for the respective slope following Munter (1999).

In combination with the actual avalanche danger scale the remaining risk for causing an avalanche is calculated following equation (1).

$$accepted\ risk = potential\ risk\ /\ (reduction\ factor\ 1\ *\ reduction\ factor\ 2) \leq 1 \qquad (1)$$

If the accepted risk remains below one, skiers are allowed through the respective slope. To calculate the remaining risk for the whole province the digital elevation model was used. There every 10m elevation is given. With the aid of GIS it is possible to calculate the remaining risk following the equation above. The remaining risk was calculated only for slopes outside woodlands and populated areas.

2.3 Availability of Transportation

In the Federal Province of Salzburg avalanche rescue teams consist mainly of volunteers. They are scattered over the whole province (Fig. 1). During the week they are available at home or at work (Slotta-Bachmayr, 2000). Avalanche rescue teams are transported to the site of the accident by helicopter. In the Federal Province of Salzburg there is one main base in the north of the province and two additional bases in the mountainous area of Salzburg (Fig. 1). An Ecureuil AS 355 helicopter is used at all bases. The parameters of the model are based on this type of helicopter. To calculate flight time from a helicopter base to a point elsewhere in the province a cost/distance model was developed. Basis for the model is also the digital elevation model described above, but now elevation data every 100m was used. Together with helicopter pilots the following values were defined:

- normal travel conditions: climbing rate 3 m/sec, travel speed 200 km/h
- maximal travel conditions: climbing rate 11 m/sec, travel speed 144 km/h

To date there are about 50 avalanche dog teams available (Slotta-Bachmayr, 2000). They are positioned at home or at the helicopter base, depending on the chosen strategy of availability (Fig. 1).

To provide decision-makers with different possibilities of team siting and transportation the helicopter flight time for four different scenarios was calculated.

Scenario 1: The helicopter takes off with a dog team from the main base in the north of the province and flies under normal travel conditions to the avalanche accident – take off time 2 min.

Scenario 2: The helicopters take off from all three bases with dog teams and fly under maximal travel conditions to the emergency – take off time 2 min.

Scenario 3: One helicopter flies from the main base to one of the other bases and picks up an avalanche dog team, unless it goes directly to the emergency, (normal travel conditions, take off and pick up time 2 min) and then flies under maximal travel conditions to the avalanche.

Scenario 4: One helicopter takes off from the central base, picks up the avalanche dog team nearest to the avalanche at home (take off and pick up time 2 min, normal travel conditions) and then flies with maximal travel conditions to the emergency.

To calculate the time it takes to transport an avalanche dog team to a potential avalanche the simulations about potential avalanche slopes and flight time were combined. With this overlay it is possible to simulate the influence of avalanche danger scale and team siting on the transportation time of a rescue team to an avalanche accident. To calculate the influence of team siting, danger scale 3 was chosen. Scenario 1 was used to show the influence of an avalanche danger scale on transportation time. To test the difference between the scenarios a Mann-Whitney U-test was used.

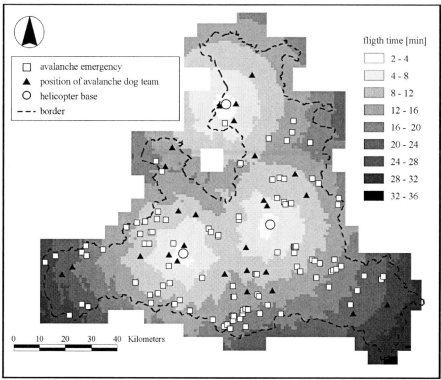

Figure 1: Flight time (scenario 2) and position of helicopter bases, avalanche dog teams and avalanche emergencies between 1970 and 2003.

3 Results

Figure 2a shows that avalanche accidents occur throughout the year, but mainly between December and March. An avalanche emergency can happen any day during the week, but mainly on weekends (Fig. 2b) when the weather is fine (Fig. 2c). To compare the accidents with the actual danger scale, the proportion of potential avalanche slopes for every class within the Federal Province of Salzburg was calculated. Figure 2d shows that during scale moderate and considerable, avalanche emergencies occur in a higher proportion than expected. During scale high, accidents occur as expected ($p < 0.05$).

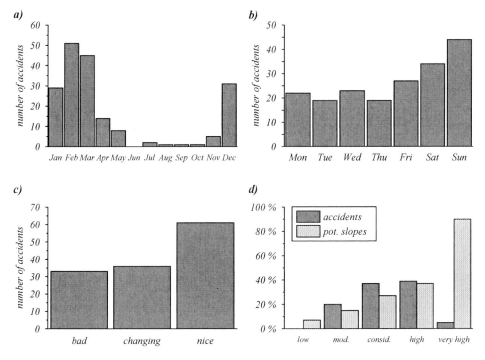

Figure 2: Temporal distribution of avalanche accidents in the federal province of Salzburg between 1970 and 2003.

    a) annual distribution

    b) weekly distribution

    c) weather on the day of the emergency

    d) emergencies in relation to avalanche danger scale (mod. = moderate, consid. = considerable) and availability of potential avalanche slopes respectively.

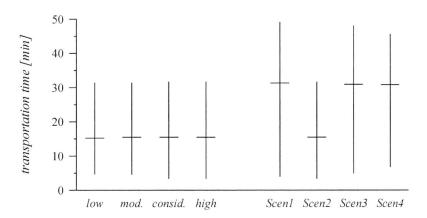

Figure 3: Minimum, maximum and mean transportation time depending on avalanche danger scale (see Fig. 2) and scenario of transportation availability (see 2.3).

It takes between 5 and 50 minutes to transport an avalanche dog team to an avalanche emergency per helicopter. Most of the time the helicopter needs 20 to 30 minutes to get there. Comparing the distribution of flight time in relation to avalanche danger scale, there is no significant difference. Avalanche danger scale does not influence the transportation time of a rescue team to the emergency (Fig. 3, $p > 0{,}05$). There is a clear difference in relation to availability of transportation (Fig. 3, $p < 0.01$). If scenario 2 is excluded from the 4 calculated scenarios, there is no longer any significant difference ($p > 0.05$). Figure 3 shows that the number of helicopters and not the siting of rescue teams influences the transportation time to a search site. For scenario 1, 3 and 4 it takes about 30 minutes to get there, whereas for scenario 2 it takes only 15 minutes.

4 Discussion

Geographic Information Systems have been used to calculate the risk of an emergency e.g. wildfire or landslide (Amdahl, 2001) and also for optimising recovery time or recovery route (List and Turnquist, 1998). GIS are also important tools to support decisions during an emergency (Wybo, 1998). In terms of avalanches GIS were used to document avalanche tracks (Gruber, 2001; Stoffel et al., 2001) and to manage avalanche hazard (Tracy, 2003). There are also models available which calculate potential avalanche release sites (Weetman, 1996; Maggioni et al., 2002). All these models evaluate slope and aspect and combine these two parameters with some other surface characteristics. The model used in this project is common practice in the Alps and is used out in the field to evaluate a single slope (Munter, 1999). Its usefulness is widely discussed but scientific investigations are missing (Pfeifer and Rothar, 2002). However, by calculating the accepted risk (Munter, 1999) for all the avalanche emergencies occurred in the Federal Province of Salzburg between 1970 and 2003, it was always ≥ 1, which means that an avalanche release on these slopes was very likely.

It is also known that during danger scale moderate and considerable, avalanche emergencies occur in a higher proportion than expected (Harvey, 2002; Nairz, 2002). This means that tour skiers are not aware of the risk which still exists even in lower danger scales. It is clear for everyone not to go out in a very high avalanche danger scale but only a few people are aware that avalanches may also be released in a low danger scale. That avalanche emergencies occur mainly at weekends and fine weather is mainly a consequence of many people being out in the country. This means a high probability of an emergency.

The analysis and simulation above have shown that avalanches are released mainly between December and March, at weekends with fine weather and avalanche danger scales of moderate and considerable. But during the winter avalanche emergencies can occur at any time.

The spatial distribution of avalanches depends mainly on surface characteristics and is modified by the avalanche danger scale. The danger scale has no influence on transportation time. So it is not necessary to change team siting and availability in relation to danger scale.

Transportation capacity has a great influence on transportation time. One helicopter at each base in the province halves transportation time. This means that with one helicopter rescue team arrive at the search site chances of survival are about 25 % whereas with three helicopters the chances of survival are still 60 %. Team siting does not influence transportation time. For transportation time it makes no difference if avalanche dog teams are available at the helicopter base or at home. Logistically it is much easier to have the search teams at the base because the pilot does not have to search for a proper landing area near the home of the team. It also saves pick up time which is possibly underestimated in this simulation. On the other hand the volunteer organisations have to provide at least 3 avalanche dog teams every weekend between December and March with a danger scale of at least moderate. This means that a lot of people hang around a lot of time. Maybe the tricky thing is to find a balance between optimal provision with avalanche dog teams without decreasing the motivation of the volunteers.

Acknowledgements

I want to thank N. Altenhofer and B. Niedermoser for their help with the data and S. Werner, D. Traweger and D. Fölsche for her help with the manuscript.

References

Amdahl, Gary.: *Disaster Response: GIS for Public Safety*, ESRI Press, 2001.

Dash, Nicole: The use of geographic information systems in disaster research. *International Journal of Mass Emergencies and Disaster*, 11: 135-146, 1997.

Falk, Markus, Brugger, Hermann. and Adler-Kastner, Liselotte: Avalanche survival chances. *Nature*, 368: 21, 1994.

Föhn, Paul M.B.: The Rutschblock test as a practical tool for slope stability evaluation. *IAHS publication*, 162: 223-228, 1987.

Gayl, Albert and Hecher, Hugo: Organisierte Rettung bei Lawinenunfällen. In *Lawinenhandbuch*, Land Tirol, Tyrolia Verlag, Innsbruck – Wien, pp. 173- 186, 2000.

Greene, Robert W.: Confronting Catastrophe: A GIS Handbook, ESRI Press, 2002.

Gruber, Urs: Using GIS for avalanche hazard mapping in Switzerland. In *21st ESRI International User Conference*, pp. - San Diego, 2001.

Harvey, Stephan: Lawine & Bulletin – Facts aus Schweizer Datenbank, *Berg & Steigen – Zeitschrift für Risikomanagement im Bergsport*, 11: 48-52, 2002.

List, Georg F. and Turnquist, Mark A.: Routing and emergency response team sitting for high-level radioactive waste shipments. *IEEE Transactions on Engineering Management*, 45: 141-152, 1998.

Maggioni, Margherita , Gruber, Urs and Stoffel, Andreas: Definition and characterisation of potential avalanche release areas. In *22st ESRI International User Conference*, pp. - , San Diego. 2002.

Munter, Werner: *3x3 Lawinen – Entscheiden in kritischen Situationen.* Agentur Pohl & Schellhammer, Garmisch Partenkirchen. 1999

Nairz, Patrick: Lawinenlagebericht, 1-2-3-4-5. *Berg & Steigen – Zeitschrift für Risikomanagement im Bergsport*, 11: 35-40. 2002.

Niedermoser. Bernhard: Lawinenunfälle im freien Gelände seit 1970', *Winterbericht 1999/2000*, Land Salzburg, Salzburg, 2000.

Pfeifer, Christian, and Rothar, Verena: Die Reduktionsmethode zur Beurteilung der Lawinengefahr für Skitourengeher aus statistischer Sicht. *Sicherheit im Bergland*, Jahrbuch 2002, Innsbruck, pp. 199- 207, 2002.

Slotta-Bachmayr, Leopold: Lawinenhunde im Land Salzburg, Winterbericht 1999/2000, Land Salzburg, Salzburg, 2000

Stoffel, Andreas, Brabec, Bernhard and Stöckli, Urs: GIS applications at the Swiss Federal Institute of Snow and Avalanche research. In *21ˢᵗ ESRI International User Conference*, San Diego. 2001.

Tracy, Leah: Using GIS in Avalanche Hazard Management. Available electronically at http://www.uah.edu/student_life/organizations/ASCE/AlumniNews/avalanche.pdf, 2003.

Tschirky, Frank, Brabec, Bernhard and Kern, Martin: Lawinenunfälle in den Schweizer Alpen – Eine statistische Zusammenstellung mit den Schwerpunkten Verschüttung, Rettungsmethoden und Rettungsgeräte, In *Durch Lawinen verursachte Unfälle in den Schweizer Alpen*, pp. 125- 136, Eidgenössisches Institut für Schnee- und Lawinenforschung, Davos, 2000.

Weetman, Georg: Avalanche Hazard Modelling Using GIS. Available electronically at http://www.geog.ubc.ca/courses/klink/g472/class96/gweetman/project.html, 1996

Wybo, Jean L.: FMIS: a decision support system for forest fire prevention and fighting. *IEEE Transactions on Engineering Management*, 45; 127- 131, 1998

Flood Protection Along the Elbe River: Simulation of the Flow Process of the Elbe and its Main Tributaries Regarding the Operation of Flood Retention Measures

Robert Mikovec
Institute for Water Resources Management, Hydraulics and Rural Engineering, University of Karlsruhe, Germany; Postgraduate College Natural Disasters

Abstract

The extreme flood event of 2002 affected a large part of the Elbe river basin. In a study in 2003, the efficiency of potential flood retention measures (polders) to mitigate the effect of extreme events along the German river course was analysed. According to this study, the maximal water level can be reduced by 20-50 cm depending on the retention volume and the event-specific operation of the polders. Since the operation of the polders will primarily depend on the hydrological and forecasting conditions in the upper catchment, mainly in the Czech Republic (CR), the influence of large reservoirs on the downstream flow conditions (German river course) are investigated. Therefore, data from Czech gauges and reservoirs have been collected and analysed for consistency. Further analyses revealed a significant impact of the reservoir installation in the 20th century on the long-term flood development.

Keywords: flood management, Elbe river, extreme flood events, Vltava cascade, polders.

1 Problem Definition

Flood events are caused by heavy rainfalls which are transformed by the catchment characteristics. Beside the precipitation, the initial hydrological conditions in the catchment play a decisive role for the flood extent. For flood protection purposes, the understanding of extreme events (return period of 100-year- or higher) is important. Their estimation requires the application of statistical methods and/or rainfall-runoff models. Both methods have their disadvantages due to the natural variability of processes and the data and model limitations. For example, flood records are often limited to observation periods less than 100 years. Extreme events have to be derived from these records, i.e. extrapolated from smaller events. In fact, the application of these methods is closely related to the data of gauging stations.

To reduce the risk of extreme events two type of flood protection measures are possible. One option is the diking along the river course with a sufficient safety against damage or even against overtopping in an extreme case. Consequently this leads to a loss of natural flood plain area along the rivers and thus to a reduced natural retention. Moreover, by this measure the flow peak is not reduced, but also the flow process is accelerated. The result is an increased flood risk for the downstream river sections. The second option is the use of controlled retention measures like reservoirs or polders upstream the interesting location.

In the case of the Elbe River, extreme events can be reduced by reservoirs operation and polders. In the Czech Republic, there are many large dams having a considerable storage volumes for flood-control. In Germany, polders can be located in areas where the terrain conditions are suitable, e.g. in the states of Saxony-Anhalt and Brandenburg. For the efficiency of the retention measures, the event-specific operation is very important.

2 Study Area and the Starting Point of the Research

The target area is the Middle Elbe river between Torgau and Dessau in the federal state of Saxony-Anhalt (East Germany). In fact, this study is a continuation from earlier research activities (compare Helms et al., 2002, Merkel et al., 2002, Nestmann and Büchele, 2002) funded by the German Ministry of Education and Research (BMBF), and a recent study by Ihringer et al. (2003) on behalf of the state authorities concerning flood retention measures.

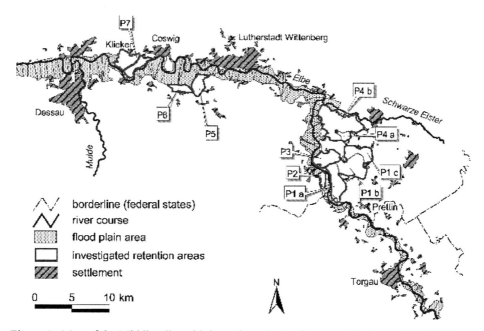

Figure 1: Map of the Middle Elbe with investigated retention areas (Ihringer et al., 2003).

The goal of the work of Ihringer et al. (2003) was to investigate the possible reduction of the flood peak by controlled retention measures, especially regarding extreme events which may lead to dike failure and uncontrolled flooding of building areas. The investigated retention areas are shown in Fig. 1. For this, the potential polder areas were identified and analysed in a geographical information system (GIS). The delineation of the potential polder areas was oriented on different criteria like protection of settlements and infrastructure, road connections in an event case etc. Emphasis was given on the terrain conditions in terms of retention volume and the hydraulic properties for filling. For every potential area, different areal extensions with their storage capacity were determined.

For the flood-routing along the river a hydrological model is used. The model is based on the established forecasting model for the Elbe (compare Merkel et al., 2002). In the study, the model discretisation was adapted consisting of nodes at gauging stations, at the mouth of tributaries, at important localities, locations of dike failures 2002 and last but not least for the potential retention areas. The present flood-routing model considers the river network starting from Usti in CR until Geesthacht near Hamburg. In future, it is the intention to extend the areal coverage to upper reaches in CR. As the magnitude of the flood event 2002 exceeded all records, the model parameters had to be revised and partially adapted (for the peak), too. In order to simulate the effect of the retention measures, first the event 2002 had to be reconstructed regarding the effects of several dike failures.

To assess the efficiency of the retention measures, Ihringer et al. (2003) simulated different flood scenarios (events) for different polder variants. The most important event is the flood 2002. In addition, the extreme event 1890 and five synthetic events (generated by a stochastic model) were considered. The results show the effect of the measures on the downstream reaches. The best retention effect can be carried out by an optimised polder operation in combination with an appropriate reliable forecast. Four polder variants were studied, from "3 polders with low storage volume" to "all polders with maximum filling". The simulations for the various scenarios show that the reduction of the maximum water level of about 20-50 cm (depending on the situation) along the river is achievable. The largest effect is promised by the first polders upstream of the river section (see Fig. 1: P1) which could provide up to 2/3 of the total potential retention volume. Thus, Ihringer et al. (2003) suggested to concentrate further investigations on this location while other areas should be treated with a lower priority, especially in terms of the complexity of operation.

These results represent the first steps of the planning of flood retention measures in hydrological terms. In the next steps, the studies have to follow two main directions. Of course, one is the further hydraulic and constructive design of the measures. The other is the question of operation for different types of extreme flood scenarios (exceeding the order of magnitude of a 100-year event).

3 Present and Future Research Activities

With regard to the overall catchment of the Elbe, the above results show that a more detailed understanding of the flow process, especially of the flood situation, is required. This includes a funding of flood statistics and an improved flood forecast for an optimised use of the polders in the cases of different kinds of extreme events. In an interdisciplinary context our current research is mainly focussing on:

- research on extreme flood events of the upper reach on Elbe in the CR, and
- the further hydraulic and constructive design of the polders (planned by the state authorities of Saxony-Anhalt in continuation of the study of Ihringer et al. 2003).

The following sections deal with first point and thus with the hydrological boundary conditions with regard to an appropriate polder operation. In respect to this, the (natural) hydrological processes in the basin upstream of the polders and the reservoir operation in the Czech basin part has to be taken into account.

The following main aspects have to be considered:

- characteristics of extreme flood events in terms of peak values and event shapes and their statistical evaluation;
- anthropogenic impact on the course of flood events (especially damping of peaks in large reservoirs on the Czech territory);
- Forecast conditions for extreme flood events with regard to an optimal adaption of the polder operation during flood events.

The final target is thus the development of a forecast and control system for the polders in Saxony-Anhalt taking into account the hydrological input characteristics and options of flood control in the upstream part of the Elbe basin. For this purpose the "Flussgebiets-modell – FGM" will be applied. A node plan of this modelling system already existing for the German part of the Elbe river has therefore to be extended to the Czech part. Using this extended modelling system it will be possible to couple models for the basin, for the Czech reservoirs and for the flood protection system along the river in Germany in order to derive recommendations for an appropriate flood management in the overall basin. However, the development of such a modelling system requires preparatory analyses.

With respect to this, the following investigation steps have been carried out in the Czech basin part:

In the first steps the extreme flood event of the year 2002 was excluded from the investigated period, since its analysis and simulation implies a high degree of uncertainty due to several reasons. The investigations presented here rather intend to contribute to a reliable evaluation base for the flood event 2002.

Beside the analysis and understanding of the large-scale and long-term hydrological situation the results described below mainly focus on the impact of large reservoirs installed during the 1950/60-ies on the flood situation. The following chapters give a summary of essential results and conclusions being achieved till now.

3.1 Data Collection

In collaboration with the Czech Hydrometeorological Institute in Prague (CHMU) and with POVODI-Vltavy in Prague long-term series of daily flow data at 23 gauges in the Czech part of the Elbe basin and informations on the characteristics of the reservoirs of the Vltava-cascade were collected (see Fig. 2). The selection of the gauges corresponds to the aim of a large-scale evaluation of the flood situation taking into account the impact of the above mentioned reservoirs.

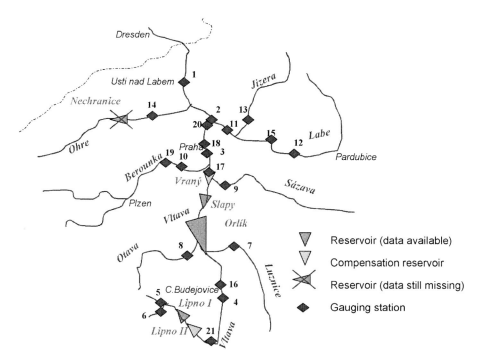

Figure 2: Survey map of the upper Elbe catchment area in the Czech Republic.

3.2 Consistency Analysis

In order to analyse the consistency of these series methods similar to those applied in the scope of the joint research project „Morphodynamics of the Elbe river" on German Elbe gauges were chosen (compare Helms et al., 2002, p. 101-104). For this, the hydrographs were plotted in a first step. From these plots appropriate flow events could be identified. For these events the flow volumes were determined and balances among adjacent gauges or gauge groups were calculated (Fig. 3). These investigations started from the (most reliable) German Elbe gauge of Dresden in order to ensure also the consistency with the data and analyses for the German part of the Elbe river. Altogether, the balances were plausible. The data base is therefore suitable for further analysis steps.

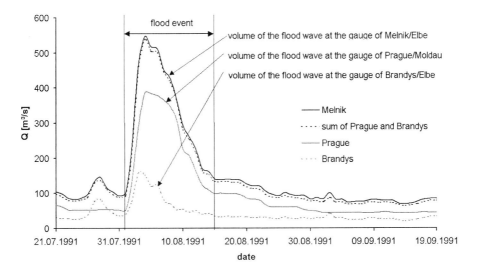

Figure 3: Flood event volume of the gauging stations Melnik, Prague, Brandys.

3.3 Analysis of the Long-Term Development of the Large-Scale Hydrological Situations Using Series of Mean Annual Flow Values

For these analyses the parameter MQ (mean annual flow values) was chosen, since it describes the overall properties of the flow process and since it was only marginally influenced by the reservoir effect. Therewith, the MQ-series may also serve as an appropriate reference series for the identification of the resevoir impact on the flood situation in further analysis steps (see below).

The MQ-series and the 5-year moving averages of all gauges were plotted and compared among each other (Fig. 4). It can be recognized that the grouping effect of wet and dry years identified by Helms et al. (2002) for the German Elbe gauges exists also at the Czech gauges. These long-term developments were similar at all gauges of the Czech basin part.

For most of the gauges including Praha-Chuchle (series 1924-2001) a statistically significant trend could not be identified for the overall series. This series can thus be used as a homogeneous reference series in the analysis of the long-term development of the flood situation (see below).

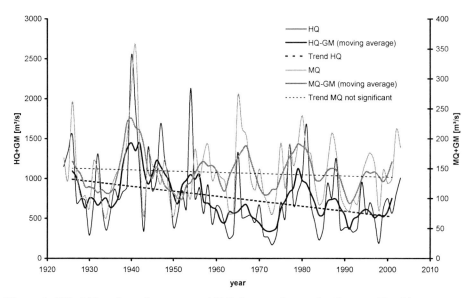

Figure 4: HQ, MQ and moving average (GM) for gauging station Prague-Chuchle.

3.4 Investigation of the Long-Term Development of the Flood Situation with Focus on the Impact of the Vltava Reservoir Cascade

The analysis was carried out based on the series 1924-2001 of annual peak-flow values HQ of the gauge Prague-Chuchle which is about 80 km downstream of the cascade outlet. In a first step the HQ-series was plotted together with the reference of the MQ-series (Fig. 4). It can be observed that the HQ-series includes a grouping effect, too. As opposed to the MQ-series a statistically significant trend was found. It can be assumed that the trend was mainly caused by the installation of the reservoir cascade and its damping effect on flood peaks. However, the installation of the reservoirs of the cascade concentrated on the 1950-ies/60-ies. Thus, it can be assumed that their impact corresponds to a shift rather than to a continuous long-term change like represented by the linear trend (see Fig. 4). This can be recognised in the plotted HQ-series, if it is compared to the (uninfluenced) MQ-series. An adequate method for the identification and proof of this effect is the double mass analysis with cumulative HQ values plotted versus the reference of cumulative MQ-values. In the resulting double mass curve a development similar to a shift like mentioned above can be recognized as a break point. In order to objectively identify the time and magnitude of the break point a break-point analysis was applied. For this, regression analyses for each two periods separated by all possible break points were carried out and the corresponding overall standard errors were calculated. The time of the break point was assumed for the year with the minimal standard error. In this way the break point of the gauge Prague-Chuchle was identified for the year 1953. In fact, the operation of the main reservoirs of the cascade began only some years later (Slapy: 1957, Lipno I: 1960, Orlik: 1963). However, due to the above mentioned grouping effects of wet and dry years the exact identification of the break point is hardly possible in this way (compare Helms et al. 2002, p. 125). In comparison to the gauge of Dresden the break point at the gauge of Prague-Chuchle is much more pronounced. This is a plausible result, because the position of the latter gauge is much nearer to the cause (the reservoir cascade).

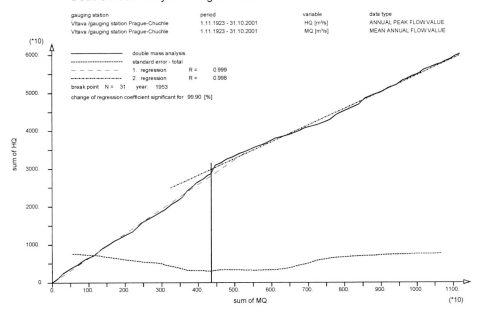

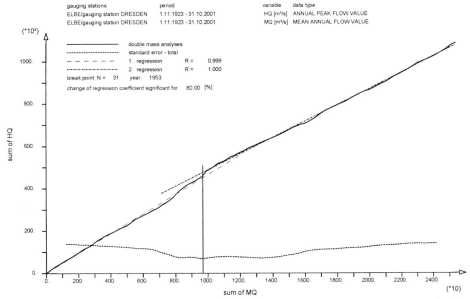

Figure 4: Double mass analyses at the gauging stations Prague-Chuchle and Dresden for the times series HQ (annual peak flow) and MQ (mean annual flow) of the period 1924-2001.

4 Conclusions

The analysis reveals significant changes in the flow regime, especially in the flood characteristics caused by the impact of the Vltava cascade. It can be concluded that for the calculation of long-term flood statistics a homogenisation of the flow series is required in order to avoid biased and misleading results. The homogenisation may be carried out using simulation models for the operation of the reservoir cascade and for the flow process in the river channel.

Together with the above mentioned development of simulation tools for forecast purposes in the scope of the modelling systen FGM this will be the topic of our work in near future .

Acknowledgements

The author thanks the Postgraduate College Natural Disasters and the German Research Foundation (DFG) for the financial support. For multiple suggestions and support the thank is directed to my tutor Jürgen Ihringer and the colleagues Martin Helms and Bruno Büchele.

References

Helms, M., Ihringer, J., Nestmann, F.: Analyse und Simulation des Abflussprozesses der Elbe. In: Nestmann and Büchele [Eds.], ISBN 3-00-008977, 91-202, Karlsruhe 2002.

Helms, M., Büchele, B., Merkel, U., Ihringer, J.: Statistical analysis of the flood situation and assessment of the impact of diking measures along the Elbe (Labe) river. *J. Hydrology*, 267: 94-114, 2002

Ihringer, J., Büchele, B., Mikovec, R.: Untersuchung von Hochwasserretentionsmaßnahmen entlang der Elbe im Bereich der Landkreise Wittenberg und Anhalt-Zerbst: Grundsatz-studie bezüglich der hydrologischen Wirkung. IWK-Report HY3/09, on behalf of the Landesbetrieb für Hochwasserschutz und Wasserwirtschaft Saxony-Anhalt, Institut für Wasserwirtschaft und Kulturtechnik (IWK), University of Karlsruhe, 2003.

Merkel, U., Helms, M., Büchele, B., Ihringer, J., Nestmann, F.: Wirksamkeit von Deich-rückverlegungsmaßnahmen auf die Abflussverhältnisse entlang der Elbe. In: Nestmann and Büchele [Eds.], ISBN 3-00-008977, 231-244, Karlsruhe 2002.

Nestmann, F., Büchele, B. [Eds.]: „Morphodynamik der Elbe". Schlussbericht des BMBF-Verbundprojektes mit Einzelbeiträgen der Partner und Anlagen-CD. Institut für Wasserwirtschaft und Kulturtechnik, University of Karlsruhe, ISBN 3-00-008977, 2002.

Risk Reduction in Industrialised Societies

Restructuring Urban Society for Mitigation: Risk Sectors in "The Earthquake Master Plan" of Metropolitan Istanbul

Murat Balamir
Disaster Management Center METU Ankara

1 Background

Following 1999, EQ risks were taken with greater caution by individuals and official bodies in Turkey than ever before. In response to the established likelihood of a major EQ in Istanbul, The Metropolitan Municipality of Istanbul (MMI) cooperated with the Japanese JICA teams in an analysis of hazard probability distribution in the region, and the preparation of microzonation maps. This study identified the extent of potential damages throughout the metropolitan area. Having obtained a 'diagnosis' of the hazard, the following step for the MMI was to obtain a 'prescription' for action to avoid or minimize the impacts of the EQ.

This prescription entitled as the "EQ Master Plan for Istanbul" (EMPI) aimed to identify all possible lines of action in the metropolitan area for the mitigation of losses. For the purposes of obtaining such a plan, MMI made a tender in late 2002 where four universities in pairs handed in propositions. Politically conscientious MMI preferred to hire them all in collaboration. The EQ Master Plan for Istanbul, a road-map for action, was prepared in eight months and submitted by the end of July 2003. It was the METU-ITU approach that rested on an Urban Risk Analysis methodology in which the natural hazard risk distribution together with the conceptualizations of 'Urban Risk-Sectors' led to the structuring of a comprehensive line of action, or the 'Mitigation Plan'. The approach thus considered hazards of natural and human origin in combination, within a framework of 'risk sectors', and proposed lines of action in each to involve all related factions of the urban society. The purpose was to bring together and activate, in every risk sector, related components of public administration, business and industry, NGOs and local community representation in the long-term management of urban risks, to draw mutual agreements of conduct and control, and to run various sub-project packages.

13 distinct Risk Sectors were identified, followed by the formulation of procedures and methods for risk mitigation. The purpose was to encourage stakeholders and enable parties exposed to risks in each sector, devising complementary regulatory tools in risk management as sub-components of an umbrella Social Contract. A reassessment of existing city administration procedures, new powers of implementation and tools for physical planning, encouragement of partnerships and private investments in comprehensive rehabilitation complemented the EMPI approach. Thus a comprehensive technical methodology for risk management and a complementary line of social action are simultaneously recommended. This assumes a new understanding of governance on behalf of MMI, and participatory arbitration processes on behalf of social agents and NGOs for the transparent management of risks, a step towards the reflexive society.

Distinct sub-project packages were also described in each sector that could be independently tendered and implemented. EMPI indicated areas of highest risks where immediate Action Plans and special projects of risk management are necessary. Action Plans recommended in high risk areas are for the total transformation and upgrading of such areas in physical and social terms, methodology for which is identified in detail. Comprehensive rehabilitation operations demand appropriate planning powers and tools, calling for regulatory measures with supplementary methods of financing and management. Such amendments are currently under consideration by the central government.

2 Natural and Urban Hazards, Approaches to Risk Management and Planning

Finance, insurance, management, engineering, medicine, politics are some of the areas in which concepts and methods of risk analysis have proliferated during the past two or three decades. On the other hand, sociological reinterpretations of risk, provides a broader understanding today to contemplate about social developments and policy decisions. Based on large scale catastrophes, it is only very recently that implications of risk management are considered in the context of urban planning and administration. There is every reason to believe that this area of concern will grow more extensively in the near future, due both to the global environmental factors and the severity of local hazardous conditions.

Urban seismic hazards are one of the most harmful causes in the built-environment. Earthquakes do not only have direct destructive impacts but could also induce equally harmful secondary natural effects as landslides, liquefaction, and faults. Further deleterious effects could be generated as tertiary impacts, when human-made sources of hazards are destroyed or interrupted. Such incidence may be large scale fires, spread of chemicals and toxic material, break-downs in dams, industrial complexes, energy-generators and other life-lines, as well as in infrastructure like hospitals. It is not coincidental therefore that attempts to promote risk management were enthusiastically initiated in attempts of mitigating EQ impacts.

Current approaches to urban seismic risk management could be broadly considered in two groups. Often seismic properties of individual buildings are investigated by geophysical and engineering analyses, and recommendations for retrofitting/removal made according to technical and economic feasibility criteria. Alternatively, recommendations could target the improvement of ground-bearing capacities for specific buildings. A second set of management efforts focuses on urban systems vulnerabilities due to natural hazards, and undertakes scenario analyses. The propositions of this latter approach are usually technical measures to be implemented and conducted by urban authorities by means of land-use control and other tools of urban planning.

Another 4-part classification is conceivable if forms of planning considered. Planning for the post-disaster reconstruction stage may overwhelmingly involve the conventional land-use and physical planning procedures. These may cover a wider range of work including compensations and programs devised for social and economic healing, and entitled in general as *'recovery planning'*. Planning for the purposes of post-disaster rebuilding activities in urban areas is a widely practiced form of public or private service (Spangle, 1991). There are no pre-determined principles and most of such work is carried out on an ad-hoc basis.

Planning for emergency preparedness is another common form of planning, often formally undertaken by local administrations as a legal obligation. This public responsibility is given to all provincial administrations in Turkey by the so called 'Disasters Law' which describe methods and standards of *'preparedness planning'*. The manner this task is to be carried out in the various branches of preparation activities is described in general by the mandates of central authorities.

It is the third type of planning activities that deserves elaboration. This aims to reduce the overall impact of hazards in the urban area. The approach can be identified as *'mitigation planning'* and described as an attempt to 'avoid, minimize, and share' the risks (Kreimer, et.al, 1999). This activity is necessarily based on the identification and analysis of risks, estimation of impact magnitude and costs of the most likely disaster, and demands intensive collaboration of the disciplines, orchestrated preferably by planners in the urban context. Although the potential role of urban planning in the mitigation of earthquake damages is often expressed, examples of such plans and implementation are very rare. Methods for such professional and administrative interventions remain largely underdeveloped. 'Scenario analyses', employed for this purpose are only <u>predictive</u> models whose merits depend on the built-in assumptions, rather than operational tools for local authorities or communities. Methods and tools for urban risk management are also scarce commodities within the <u>prescriptive</u> and action-oriented body proper of urban planning theory and practice.

Still further, a fourth category could be described as *'resilience planning'* that aims to monitor the dynamics of the economy and society for robust and sustainable development and in-built safety. It involves the long-term structuring of agents and legal systems, shaping the cultural background for greater awareness of hazards, and improving the capacity of communities in the risk and emergencies management. This multi-task operation requires the integration of programs prepared in distinct disciplines and aims to generate a lasting synergy based on coordination.

EMPI largely falls into the third, but with all the aspirations for the fourth category. Functions attributed to forms of urban mitigation planning concerning seismic risks are only a few and recent. Coburn (1995) has provided a description of various technical requirements at the urban scale, and in a previous work the subject was given a limited treatment (Coburn and Spence, 1992, 149), allocating most of the attention to the task of 'improving earthquake resistance of buildings'. A concern for the development of a methodology for mitigation at the urban scale is largely omitted.

"Elements of a Disaster Preparedness Plan"
(Columbia University 2001, p. 87)

- Hazard Identification (microzonation)
- Assessment of Critical Assets, Fragilities and Activities at Risk (infrastucture and lifelines, critical facilities, industries)
- Loss Estimation (economic modeling)
- CBA for Optimal Mitigation Strategy
- Risk Reduction (zoning, early hazard warning, improve codes, give incentives, reduce fragilities, increase resilience)
- Training Response Teams
- Communication and Education
- Distribute Risks by Insurance

The work carried out under the UN program of IDNDR during the last decade also promisingly concentrated on cities as entities subject to earthquakes and disasters. Some of this work did provide methods of investigating the probabilities of disasters and simulating their effects. These have not necessarily generated a comprehensive framework, and described channels within which prescriptive response and action for mitigation could take place. Another work that deserves attention is that of the Columbia University (2001). In their study entitled 'Disaster Resistant Caracas', planners' efforts were concentrated on geological information and geographical analysis to start with, identifying the likely spatial distribution of the intensity of EQ hazard. This enabled the determination of damages and losses upon which alternative courses of action could be evaluated. A response agenda is then developed to include public training and preparedness operations as well. The subtitles to explain the logic of work is given in the information box.

This work is significant in its treatment of the urban entity in its totality as an object of planning and management. However, it considers 'disaster preparedness' largely as a technical task to be accomplished and implemented by the authorities. It stands in between preparedness planning and mitigation planning. Secondary social processes likely to be triggered with the recommendations of training, education and communication are the implicitly expected mechanisms for the sustainability of this approach. No proposition is made for the direct restructuring of social processes for sustainable mitigation and city management. Urban planning that aims at earthquake mitigation and extends beyond conceptions of 'preparedness' is yet to be devised.

Availability of social and spatial information is a major constraint in the development of plans for mitigation. Most of the existing studies had to employ the geographical and cartographical information, and had to rely in their spatial analyses on cellular subdivision of the urban area. In general, it is assumed that the smaller the cells, more precise or accurate the analyses are. Existing approaches are constrained by available data, in their capacities to identify risks and risk management. Almost all depend on geological and geophysical information in their spatial description of earthquake risks, and assume all urban risks occur in linear variation with the natural properties of ground conditions. Yet risks do not solely depend on natural phenomena, but also on the human-made elements of the urban environment.

3 Risk Society and the Organization of Risk Management

The claim that disasters are human products is not recent. In the exchange of views between Voltaire and Rousseau on the 1755 Lisbon earthquake, Rousseau points to the fact that most of the adverse consequences of natural events are socially constructed (Dynes, 2000). It is highly significant that at an era when nature is materially and culturally considered external to society, major responsibility of the disaster could be attributed to the social decisions. This is more readily perceived today since in interactions with the environment 'the social and the natural have become inseparable from one another, (and) neither the social nor the natural can be seen as independent or self-contained' (Irwin, 2001, 50). The issue is attended from a more comprehensive perspective in the recent works of Beck (1998, 1997, 1992) where the sources of current social and environmental degradation are seen at the very nature of the modern society. Late modernity experiences numerous processes of destruction of 'nature, society and the individual' determined by the scientific/technological development, which have irreversibly increased the risks. These risks could not be controlled by means of the economic and political institutions of the modern society itself, insurance included. It is for this reason that the contemporary social formation is described as the Risk Society, a transitional stage in the world social history of development. The material forces and conditions that shape the Risk Society are leading to a stage of a second Enlightment and the Reflexive Society. In this context, although not explicitly addressed by Beck, EQ hazards also transform into major disasters as science and technology assist social processes in the pooling of larger and larger risks in space. The Beckian concepts of 'organized irresponsibility' and 'manufactured uncertainty' perfectly apply to the processes of land use planning and building construction in the urbanization of societies (Balamir, 2001a, b). It was on this background that EMPI was conceived as a road-map for action, and organized according to the nature of existing material risk structures in the society, besides the natural givens. Explicit structures of collaboration in each risk sector of the risk generators and those affected are expected to initiate reflexive processes of devising collective risk assessments, mitigation agreements, and risk supervision mechanisms.

4 The Priority of Mitigation Planning

Mitigation, rather than emergency preparedness could be far more vital for some countries than others. This is particularly true in the case of Turkey. The Anatolian traditions of building construction that evolved over the centuries have optimized resources and safety in this most seismic part of the world. The use of timber and infill materials did not only spare life in the event of major earthquakes, but also maintained recycling of construction materials. With the macro-economic changes of population growth and high rates of urbanization, this delicate ecological balance could not be sustained any longer. With the introduction of reinforced concrete in Turkey (late 1940s), a new era in constructional activity ruled the day, and despite low levels of capital accumulation, a phenomenal rate of building stock formation and urban growth took place, new ownership relations in property facilitating the process. Another major disadvantage has been the deceptive ease of concrete technology. Almost one third of the urban stock in urban centers today is unauthorized, deprived of architectural or engineering projects, and constructed by builders who learned the trade by observation. This implies building production processes with little or no supervision, and therefore the formation of a stock of high vulnerability. Under the circumstances, most of this growth corresponds to town expansions and on seismically vulnerable locations. The 1999 earthquakes are the first wide-scale trial of this unchecked physical growth. Istanbul with highest rates of unauthorized stock in the country represents even greater vulnerabilities.

5 The Earthquake Master Plan of Istanbul

The METU-ITU approach is distinctly based on the concept of risk, the sociological and philosophical tenets of which are in the expositions of Ulrich Beck (1998, 1997, 1992) and others. This does not confine the work and the analyses of risk to an academic exercise, but provides a methodology for action and a framework for the democratic involvement of the whole society in 'risk analysis and management'. This proactive approach exclusively describes '**risk sectors**' in the Istanbul metropolitan area, for which independent risk analyses could be conducted, methods of which are described in detail in the main report. Secondly, parties involved in each risk sector are identified with a description of tasks of risk management (risk avoidance, risk minimization, and risk sharing) attributed to each. This demands agreements and protocols between these parties on collective and organised action. Stake-holders in each risk sector are thus to be activated in relation to a general 'road map' that combines all action in independent risk sectors. Altogether 13 risk sectors have been described within the EMPI framework. These are selectively outlined here, in terms of 'Scope, Observed Problems, Possible Risk Management Methods, Responsible Bodies, and Proposals for Action'. The expectation is that the city administrations lead the way to bring the stakeholders together in the management of each risk sector, draw the necessary protocols in which responsibilities and tasks are identified with reference to the overall **Contingency Plan** (*Sakınım*: Precaution Plan) that integrates all risk sectors. The identification, assignment of roles within a framework, and encouragement of various bodies in the city has also generated alternative courses of action. The professional chambers have been highly receptive in leading the NGOs in the city and generate further activism since they are organized at the national level. **Action Planning** is recommended in the high risk areas of Istanbul, as a set of activities to take place especially by means of comprehensive local rehabilitation and redevelopment projects directly involving the residents of the area. Such processes are expected to be initiated by means of a few local projects guided by local and national expertise in pilot areas. EMPI also contains recommendations on revisions in legal provisions and devices, methods of procurement and use of financial resources, formation of 'Local Community Administrations', and public education and local community training.

The over-all purpose of EMPI is to enhance safety and total quality of life in the city by reducing infrastructural deficiencies, gradual elimination of the unauthorized stock, the integration of city management processes, protecting natural and historical assets, reclaiming urban quality and identity, participatory local community management, comprehensive rehabilitation of high risk areas, retrofitting or removal of buildings according to the local revision plans.

Since the task was an unconventional one, special terminology and spelling of principles were necessary during the preparation phases of EMPI to keep the work within track. In this context EMPI is not:

- An operation confined to the 'retrofitting' of specific buildings in the metropolitan area; Rather, the urban environment is considered in its totality, with its life-lines, emergency facilities, land uses and management processes.

- A conventional 'development plan' describing simply some future physical state, employing solely the devices of methods of physical rearrangements and standard land-use planning apparatus; Rather, EMPI has to generate tools to monitor organizational and economic tendencies and processes.

- An exercise in strict confines of existing 'legal and administrative constraints'; Rather, proposals are made for the development of new methods and tools of enforcement, and the revision of existing legal frameworks.

- A 'one-shot' undertaking; Rather, sustainable mechanisms and institutions for a safer and more robust city and resilient communities are to be introduced.

- An excuse to allow further expansion of the city, generating new waves of demands over the forests and water basins; Rather, it is a comprehensive methodology for upgrading the existing built-up areas in safety and quality, and protecting the natural assets.

- A program for post-disaster activities or a form of crisis management plan, but a comprehensive plan to cover all forms of action for the long-term minimization of damages or loss in the city.

- A simple exercise of diagnosis, but a scenario of action and steps to be followed.

6 Risk Sectors of EMPI

A selective description of the Risk Sectors is given here in a specific format. The thirteen sectors specified include: 'macro-form risks', 'risks in urban texture', 'risks related to incompatible uses', 'risks of productivity loss', 'risks in special areas', 'open space scarcity risks', 'risks related to hazardous materials and uses', 'risks in lifelines ', 'vulnerabilities in historical and cultural heritage', 'risks in buildings', 'risks related to emergency facilities', 'external risks', 'risks of incapacitated city administrations'.

6.1 Risks in Urban Texture

Scope: Independent of the building safety, determination of Risks in the differential formation of urban fabric comprising plots, building coverage and density, access roads and car-parking, ownership pattern, and other environmental properties;

Problems: Great disparities of risk between various types of urban pattern, and unauthorized changes in due course are observed;

Risk Management: Differentiation of urban texture zones in development plans, and long-term physical policies for redevelopment, collective or singular buildings; differentiated property taxation and obligatory insurance enforcements;

Responsible Bodies: MMI and municipalities, LCAs;

Proposals: Formation of an inter-municipality working committee, functioning with improved powers of municipalities in development, supervision of construction, differential property taxation, municipal assessment in the determination of obligatory insurance; Legal changes in Municipalities (1580), Development (3194), Property Taxation (1319) Laws, and modifications in the Obligatory Earthquake Insurance Draft Law.

6.2 Risks of Productivity Loss

Scope: Investigation of seismic sensivity of industrial enterprises and Risks of productivity losses in the industrial establishments, in the case of earthquakes, based on their size, location, building and facilities robustness, technology employed, materials processed, and dependencies on infrastructures, access, input-output relations, etc.;

Problems: Many industrial enterprises are extremely vulnerable in terms of location and building quality; Resilience of the city is largely dependent on the sustainability of the productive potential of the city in many direct and indirect ways;

Risk Management: Carrying out essential research on vulnerability classes of the industry; Building a data-base and developing methods of mitigation; Promoting local and sectoral cooperation between industries; Provision of credits for different types of mitigation; Imposition of obligatory insurance; Compulsory early warning systems; Training for emergency; Information dissemination;

Responsible Bodies: MMI, ICI, ICT, Ministry of Industry, Universities, Business Associations, UCAET, and other related NGOs;

Proposals: A 'Safe Industry' Committee to be established by the representatives of responsible bodies, with information and inspection teams, facilitating the special mitigation measures each enterprise has to take; Supervision functions of the Committee; Enabling with provision of information, and credits for retrofitting and other safety measures; Building up a technical information pool, with standards and regulations of safety;

6.3 Risks Related to Hazardous Materials and Uses

Scope: Urban uses that process, store, and distribute combustible, explosive, poisonous and pollutant materials are sources of further risks, the location, environment and routes of which should be separately investigated;

Problems: Unauthorized and ignorant operators; Ineffective regulatory devices and standards; Disregard of the need for contingencies, waste management and responsibilities; No supervisory system; Uncontrolled spatial spread and levels of concentration;

Risk Management: Survey and determination of enterprises that deal with hazardous materials, development of a comprehensive data base; Classification of enterprises according to the potential risk contained, and their spatial distributions; Developing a unified permit system, periodical inspections and warnings to neighboring uses;

Responsible Bodies: MMI, Governorate, Ministry of Energy, LCAs, environmentalist NGOs.

Proposals: MMI-Governorate protocol for comprehensive control over the Province; Instituting a permits and inspection system in line with EU standards and procedural constraints; Access to the spatial data-bank and transparencies in management and information; Proficiency requirements in the sector; Lists of hazardous materials as used by international organizations; Enterprises processing and/or distributing hazardous materials to share insurance costs of neighboring uses; Obligation of warning the neighbors by the enterprise dealing with hazardous material; Preparation of a special regulation by IMM (3030); Obligatory earthquake insurance of the enterprise; Changes necessary in Property Taxation (1319), and Environment (2872) Laws.

6.4 Risks Related to Emergency Facilities

Scope: Hospitals, schools/ dormitories, communications centers, fire-stations, police-quarters, major commercial centers and storage facilities, banks, and other public and private buildings that are expected to provide emergency services after the earthquake are investigated for their satisfactory functioning; Their malfunctioning imply further risks for the city;

Problems: Structural risks of the emergency facilities are beyond tolerable limits; Facility management is not geared to emergency conditions; Locational and spatial risks are high; Disregard of an integrated planning approach prevails;

Risk Management: Structural safety of emergency facilities as part of an integrated emergency plan is a first step; A second aspect is the intra-risk management within each facility; Thirdly, inter-facility management has to be reviewed as part of the integrated emergency city-plan; Fourthly, safety of location and spatial distribution of facilities with respect to predicted emergency service demand has to be evaluated, complementarities and substitutive nature of facilities verified;

Responsible Bodies: Governorate, MMI, municipalities, Ministry of Education, Ministry of the Interior; Ministry of Health, SHAT, infrastructure managing corporate enterprises; private enterprises, NGOs, media enterprises;

Proposals: A joint 'Emergency Risks Committee' of Governorate and the MMI to develop a comprehensive plan of mitigation measures for emergency; The 'Emergency Facility' status should provide priorities as in services and special infrastructural support, and special pecuniary and non-pecuniary benefits or exemptions as of tax or insurance costs; Emergency Facility status could be granted to public, or if necessary to private buildings; Production of emergency facilities system map of the city and its dissemination to citizens; Preparation of a special regulation by the MMI, specifying mitigation standards and prerogatives of the municipalities in Law 3030; other prerogatives to be provided in Property Taxation (1319); Changes in Disasters Law (7269), Law of Municipalities (1580), empowering the joint Committee; provisions necessary also in the National Health, Education, and Civil Defense Laws.

Abbreviations

MMI: Istanbul Metropolitan Municipality; MPWS: Ministry of Public Works and Settlement; SPO: State Planning Organization; SIS: State Institute of Statistics; SHAT: State Highways Administration of Turkey; SAWW: State Administration of Water Works; NGO: Non Governmental Organizations; ICI: Istanbul Chamber of Industry; ICT: Istanbul Chamber of Trade; UCAET: Union of Chambers of Architects and Engineers of Turkey; CAT: Chamber of Architects of Turkey; CET: Chamber of Engineers of Turkey; LCA: Local Community Administrations (as proposed by EMPI)

References

Balamir, M. (2001a) *Disaster Policies and Social Organisation*, unpublished paper presented at the Disasters and Social Crisis sessions of the European Sociological Association Conference, Helsinki 27 August-3 September, Finland.

Balamir, M. (2001b) Shaky Grounds for Architecture in Oblivion: Whose Agenda is Earthquake Anyway? Paper presented at ACSA Conference, Istanbul.

Balamir, M. (2001c) Recent Changes in Turkish Disasters Policy: A New Strategical Reorientation?, in *Mitigating and Financing Seismic Risks in Turkey*, eds. M. Sertel, P. R. Kleindorfer, NATO Science Series, Kluwer Academic Publishers, 207-234.

Balamir, M. (1999) Reproducing the Fatalist Society: An Evaluation of the Disasters and Development Laws and Regulations in Turkey', in *Urban Settlements and Natural Disasters*, *ed.* E. Komut, International Union of Architects and Chamber of Architects of Turkey, 96-107.

Beck, U. (1998) Politics of Risk Society, in *The Politics of Risk Society*, *ed.* Jane Franklin, Polity Press, Cambridge UK, 9-22.

Beck, U. (1997) The Reinvention of Politics: Rethinking Modernity in the Global Social Order, Polity Press, Cambridge UK

Beck, U. (1992) Risk Society: Towards A New Modernity, Sage, London.

Coburn, A.(1995) Disaster Prevention and Mitigation in Metropolitan Areas: Reducing Urban Vulnerability in Turkey, in *Informal Settlements, Environmental Degradation, and Disaster Vulnerability: The Turkey Case Study*, eds. R. Parker, A. Kreimer, M. Munasinghe, IDNDR and the World Bank, Washington, D.C.

Coburn, A., Spence, R. (1992) *Earthquake Protection*, John Wiley & Sons.

Columbia University (Spring 2001) *Disaster Resistant Caracas*, Urban Planning Studio of Graduate School of Architecture, Planning and Preservation, and Lamont-Doherty Earth Observatory, report photocopy.

Dynes, R. R. (2000) The Dialogue between Voltaire and Rousseau on the Lisbon Earthquake: The Emergence of a Social Science View, *International Journal of Mass Emergencies and Disasters* (18: 1) 97-115.

Gülkan, P., Balamir, M., Sucuoğlu, H. et.al. (1999) 'Revision of the Turkish Development Law No. 3194 and Its Attendant Regulations with the Objective of Establishing a New Building Construction Supervision System Inclusive of Incorporating Technical Disaster

Resistance-Enhancing Measures', 3 volumes of reproduced research report supported by the World Bank and submitted to the Ministry of Public Works and Settlement.

Irwin, A. (2001) Sociology and the Environment: A Critical Introduction to Society, Nature and Knowledge, Polity Press.

Kreimer, A., et. al. (1999) Managing Disaster Risk in Mexico: Market Incentives for Mitigation Investment, The World Bank, Washington D.C.

Severn, R. T. (1995) Disaster Preparedness in Turkey and Recent Earthquakes, in Erzincan, in D. Key editor, '*Structures to Withstand Disaster*', The Institution of Civil Engineers, Thomas Telford, London.

Schwab, J. et.al. (1998) Planning for Post-Disaster Recovery and Reconstruction, FEMA and APA, Planning Advisory Service Report 483/484, Chicago.

Spangle, W. and Associates (1991) *Rebuilding After Earthquakes: Lessons From Planners*, Spangle and Associates Inc., CA.

Analytical Approaches for Critical Infrastructures Protection

Walter Schmitz

IABG, Industrieanlagen Betriebsgesellschaft mbH, Einsteinstrasse 20, 85521 Ottobrunn, Germany, schmitz@iabg.de

Abstract

Our current societies are fully dependent on large complex critical infrastructures (LCCIs). LCCIs are large scale distributed systems that are highly interdependent. Failures, accidents, physical or cyber attacks can provoke major damages which can proliferate by cascading effects and can severely affect vital functions of economy, governmental services and society. This paper aims at providing analytical modelling and simulation (M & S) approaches for analysis and assessment critical infrastructures protection. M & S helps the planner to recognise how susceptible the system is and where the risks lie. This teaches him how he can improve system stability.

Keywords: Critical Infrastructure Protection (CIP), large complex systems, networked thinking, interdependency.

1 Risk and Prevention

Our world has become more dynamic, more complex, more dependent and more vulnerable. Daily news reports of many risks, natural catastrophes, technical disasters, international crime and terrorism including cyber terrorism. Risks imply the possibility of considerable damage, but they are also characterised by a great uncertainty. Therefore scenarios have to be considered to study also "unthinkable" events and their primary and follow-on effects. An important reason for such cascading effects is that our daily life is networked via information and communications technology (ICT). Although infrastructures have a considerable criticality due to this ICT penetration, their interdependencies are hardly known and are only insufficiently investigated. Figure 1 gives a first impression of the complexity of interdependent infrastructures [2]. It will be used in this paper to discuss the different aspects in modelling and simulation of CIP.

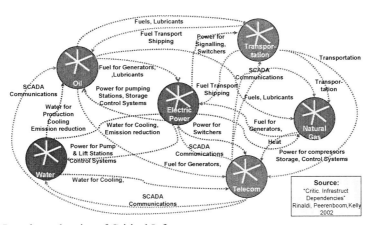

Figure 1: Interdependencies of Critical Infrastructures

2 Handling of Complex Systems

The purpose of this paper is to introduce a practicable process model for CIP that will help to understand critical infrastructures as a cybernetic system and to derive decision support instruments suitable for improving their survivability.

2.1 Fundamentals of Complex Systems

Knowledge of the individual parts of a system is not enough to be able to assess a complex system. It is also important to know their cross-linking. Intervention into the network changes the relationships between the parts and consequently the character of the system. Ecological systems for example are open systems and remain viable through permanent exchange with their environment. Such an exchange causes characteristics like feedbacks and self-regulation that are not contained in the individual components of the system. Therefore the survivability of a complex system can not be derived alone from the survivability of its components. Primarily survivability depends on the fact that the organisation of the network follows cybernetic principles.

2.2 Prevalent Shortcomings

According to Dörner [4] we make consistently strategic mistakes in dealing with complex systems. Typical mistakes are for example:

- Incorrect definition of objectives
- Inadequate modelling.

2.2.1 Incorrect Definition of Objectives

Sub-optimisation and selection of inappropriate objective functions are often observed in dealing with complex systems. Instead to focus on survivability of the whole system, planners often follow repair service strategies or select shareholder value as objective function. The consequence is that sustainability, stability and robustness of the system are not furthered. In the long run sub-optimization of individual system components leads to inefficiency and also often to irreversible erroneous trends.

2.2.2 Inadequate Modelling of the System

Often the aggregation levels of system components are not adequate to the problem. Consequentially too many details lead to an information overload and important relationships and interactions will be overlooked.

Systemic analysis means first of all to recognise the interactions of details on a suitable aggregation level. Without knowledge of the network with interdependencies between the components the performance of the system cannot be assessed even if the individual components are studied in detail. Interrelationships or disturbances of a complex system can reveal surprising effects which are seldom manifested by a direct cause-and-effect relationship between neighbouring elements. Instead of simple extrapolation hidden feedback loops and self-regulation mechanisms have to be identified.

2.3 Modelling Principles

We have to think in networks in order to recognise the cybernetic rules of a complex system. Survivable systems contain control loops that enable the system to absorb disturbances without external interventions. Faults may happen, but the system does not collapse.

The lack of knowledge concerning indirect effects with their time delays leads to the fact that we normally realise the impact of interventions or disturbances too late. Therefore extrapolations are not qualified except for a short time horizon. Policy-tests have to be carried out. The results of these policy-tests deliver important hints for the solution development. The forecast will refer less to the fact which events when occur, but to the fact how the system behaves and how it reacts to certain events. That means we need a capability to simulate the behaviour and response time of the system.

3 A Possible Process Model

The Challenge of a holistic approach is to avoid shortcomings discussed above and to find answers on questions like

- How does the system react to certain events?
- How robust and flexible is it?
- How can its behaviour be improved?
- What are suitable leverages for control?
- What cybernetic rules as for example self-regulation or fault tolerance can be exploited?
- What are the critical and uncritical areas of the system?

The knowledge of the individual parts of a system is not enough to answer these questions. First of all we need knowledge of the cross-linking of the parts. Vester, Probst and Gomez ([5], [6], [7], [8]) showed, that cybernetic interpretation can indeed produce quite accurate descriptions of system behaviour, even when working with rough data provided the cross-links are realised. Therefore this paper wants to introduce a practicable process model that will help to understand critical infrastructures as a cybernetic system and to derive decision support instruments suitable for improving their survivability. The working steps of this cybernetic approach are given in Figure 2. These steps will be described as recursive process. After each step we can return to one of the previous steps and improve its content.

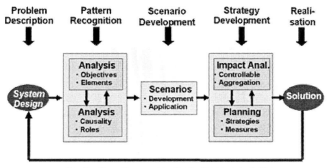

Figure 2: CIP Process Model

3.1 Step 1: Objectives and Modelling of the Problem Situation

The correct description of the problem situation is decisive for a successful problem solution and it is also important to recognise the true objectives which should guide us to the problem solution.

Critical infrastructure description covers at least four hierarchy levels representing different levels of critical infrastructure relevant decision making with different objective functions:

Level 1 represents the "System of Systems" level. This is the level of the economy as a whole, the international community and the organisations like EU and the national governments. CIP at this level includes the definition of interests of the society, the achievement of national and international awareness of risks, preparation, administration and legislation, and development of a CIP framework. Responsible actors are EU, national governments, and trade associations. Objective function is the survivability of the complex system of critical infrastructures.

Level 2 represents the level of individual critical infrastructures. This is the level of the economy, the EU, the national governments, and the stakeholders of the individual infrastructures. Objective function is to minimize the risks of an individual critical infrastructure.

Level 3 is the level of systems. Systems are represented by elements belonging to an individual critical infrastructure, single enterprises or a group of co-operating and competing enterprises. Actors are the stakeholders of the individual infrastructure systems, management of enterprises and trade association. Objective function may be to improve the shareholder value.

Level 4 is the level of technical components. At this level technical simulation algorithms, vulnerability analysis, sustainability and maintainability calculations and experimentation may be applied. Actors are the management and technical experts responsible for security tasks. Objective function is to maximize the technical functionality.

Survivability, risk minimizing, shareholder value and technical functionality are different objective functions, where shareholder value and risk minimizing can be contradictory. The decision process on each hierarchy level can be supported by decision support tools such as socio-economic models, scenario techniques, gaming, systems dynamics, empirical modelling, cost-effectiveness analysis, simulation, optimisation algorithms, risk analysis methodology, human behaviour models, cost-effectiveness models and others. In analogy of the approach of the networked thinking a critical infrastructure can be described as control model (see Figure 3). Elements of critical infrastructures are

- Actors for example operator, data administrator, user and others
- Controllable factors as computer, network, switches etc and
- Criteria or indicators that indicate how the objective of the system will be fitted.

Criteria indicating how well the considered system fulfils its mission can be for example: integrity, safety, reliability and others. Actors control the controllable factors and vice versa controllable factors can influence the behaviour of actors. Controllable factors determine the indicators and vice versa the indicators regulate the controllable factors (f.e. refrigeration will be activated if the temperature is too high), as well as the behaviour of the actors. As critical infrastructures are not isolated, so called non-controllable factors influence the system. Non-controllable factors are factors that can not be influenced by critical infrastructure itself. Examples for such non controllable factors can be liability, international standards and others. Non-controllable factors influence or disturb actors and controllable factors of critical infrastructure. External and internal factors influence the indicators that determine the value of the objective function as for example technical functioning. The value of the objective function causes actors to change controllable factors, if the value of the objective function lies outside of normal sector. That means each critical infrastructure can be treated as classical control model with feedback loops.

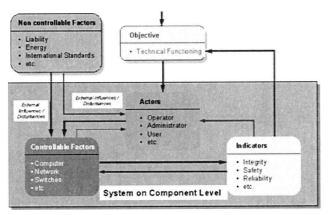

Figure 3: CIP Control Model

3.2 Step 2: Analysis of Causality

Tools are needed to investigate interrelationships, influences, time periods and changes in order to get a comprehensive understanding of the problem. Networks allow us to describe the causality of the relationships and to analyse their characteristics provided we have all relevant relationships registered with respect to their intensity, impact direction and time aspects. According to Figure 4 the objective function of a particular system of critical infrastructures is to improve the survivability of the whole system. Survivability is a complex function and may be measured in terms of technical operativeness, acceptance, and environmental compatibility. Actors are the individual critical infrastructures such as electricity, telecommunication, water, gas, oil, transportation.

- If element A influences element B then the question comes up whether A will have a reinforcing or diminishing impact on B. The plus and minus sign at the top of the arrows indicate whether element A influences element B in direct or reversal direction. For example
- the improvement of the electricity can lead to a better water supply, therefore "+"
- and the increase of transportation can lead to lower environmental sustainability, therefore "-".

The thickness of the arrows indicates the degree of influence. Not all relations have the same effect. Therefore the relations have to be assessed in a quantitative or qualitative way.

Criticality of elements:

These influence intensities help us to categorise the networked elements as active, reactive, critical or buffering elements (see Figure 5):
- Elements, which influence strongly elements in the network without being influenced strongly by others, are called "active" or "driver".

- Elements, which influence faintly others and are influenced strongly by others, are called "reactive" or "passive" or "driven element".
- Elements, which influence and react strongly, are called "critical".
- Elements, which neither influence nor react strongly, are called "buffering".

The network system GAMMA [9] considers all elements with their relations and classifies them in drivers, driven, critical and buffered elements only taking into account the relations with their strength. This portfolio analysis shows that our demonstration network for Critical Infrastructures is fairly sensitive concerning disturbances. Almost all critical infrastructures reside in the yellow field for critical elements. That means they are both driver and driven elements. This snapshot on a high aggregation level indicates that first of all electricity, then telecommunication, oil, transportation and gas should be investigated for survivability purposes. It may be asked what must be done to shift these infrastructures into the green field? Detailed analyses with higher resolution will provide the answer.

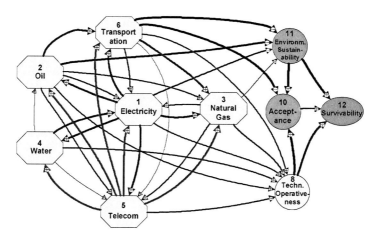

Figure 4: System of Interdependent Critical Infrastructures

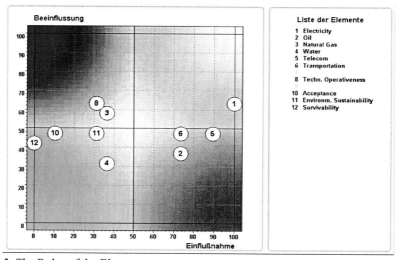

Figure 5: The Roles of the Elements

3.3 Step 3: Scenario Development

Of course, the future can not be predicted exactly in complex problem situations. But we can devise possible scenarios for specific parts of the network and simulate the consequences. In practice it has been successful to develop a basic scenario and some alternative scenarios, for example an optimistic and pessimistic scenario. Scenario development requires the following work steps:

- Determination of the necessary timeframe
- Identification of the influencing factors within the network
- Selection of the relevant scenario areas
- Development of the basic scenario
- Development of alternative scenarios
- Interpretation of the scenarios.

3.4 Step 4: Impact Analysis

In this step control possibilities should be identified. In doing so, we have to distinguish between controllable elements, non-controllable elements and indicators. Controllable elements are to be considered for steering tasks as well as disturbances. Non-controllable elements are to be monitored with respect to preventive actions. Indicators notice the degree of success of a steering measure or the degree of impairment caused by disturbances. Our main task is to improve the survivability of the system of critical infrastructures. So, the question comes up which elements influence the survivability of the whole system.

3.5 Step 5: Planning of Strategies and Measures

Planning of strategies and steering measures for survivability improvement is a creative and challenging process. Viable strategies have to consider very carefully aggregation level and system characteristics like reinforcing loops, feedbacks, control cycles, etc. But the exemplary network of Figure 4 comprises only reinforcing feedbacks, self-regulation loops are missing. That means, this specific system is instable and not fault tolerant. A possible stabilizing measure on this aggregation level is to introduce a regulation agency for electricity for example (Figure 6). At least two self-regulation loops can be identified (Figure 7):

(1) The regulation agency observes decreasing technical operativeness due to insufficient electricity and tightens the standards for electricity. That will lead to an improved technical operativeness.

(2) The regulation agency observes decreasing environmental sustainability and tightens the standards for electricity. That will lead to an improved environmental sustainability.

These self-regulation processes will be iterated until survivability is achieved. This simple example should show how changes of the system topology can stabilize the system behaviour.

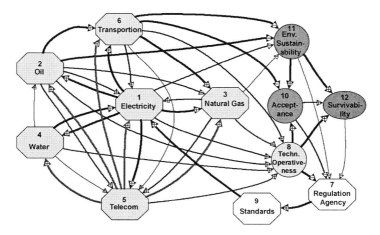

Figure 6: Regulation Agency and Interdependent Critical Infrastructures

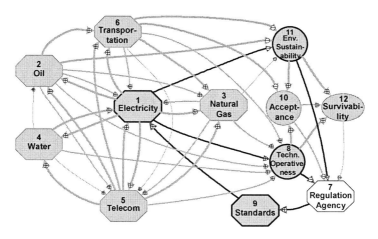

Figure 7: Self-Regulation Processes

3.6 Step 6: Realizing of Robust and Adaptable Problem Solutions

Problem solutions should be realized in such a kind that they endure also in adverse circumstances and that they are able to adapt to changed situations. Thereto it is necessary to control progress and to accomplish respective corrections. It is important to define early warning signals that indicate deviations and changes as early as possible. To identify early warning indicators, the time response of the system must be studied. Sensitivity analysis supported by simulation helps to find out suitable early warning indicators and to test robustness and adaptability of the strategies in all considered scenarios.

4 Dynamic Analysis of Networked System

The simulation of critical business and public areas of a highly developed technical society is based on the entities, relations and sensitive parameters as developed in seminar sessions with GAMMA. These relations are transferred into a logic structure based on the systems dynamics methodology. The methodology permits the dynamic representation of the most sensitive parameters and interactions of the areas under consideration.

The simulation progresses to the first intrusion event (see Figure 8). During the undisturbed time period every variable is showing the normal level of 0.9. After this period a terrorist action takes place and reduces the productivity of the control centre of the production of energy in a disastrous manner for some time. A crisis management team will have to react on this development and to plan and act according to crises response planning. After the crisis management team has found the right means to repair the problem, the value of reduced parameter is slowly approaching its initial level.

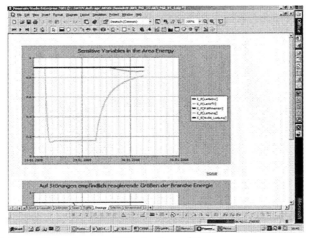

Figure 8: Simulation of Intrusions and Elimination of them

5 Recommendations

When developing decision support tools for CIP planning, one should therefore look for an approach that cannot only simulate the pattern of interactions but also allows the user to interpret and evaluate the cybernetics thereof. Only such a model will allow to decide whether interferences in a system will lead to alternative chances of survival and what subsystems may not have to be viable alone because of given cross-linkages. The purpose of such a model is to recognise the stability of the structure, the ability to adapt, the onset of irreversible trends, the risks of dissolution and the actuating elements that allow the planner to steer the system in the desired direction.

In addition to these more static features a simulation model is necessary to study the dynamics of the system to understand how feedback and amplification suddenly bring strong reactions into moderate interferences.

Modelling and simulation (M&S) should support the planner to understand the actual as well as the nominal status of the system. M&S helps the planner to recognise how susceptible his system is and where the risks lie. This teaches him how he can improve system stability, what precautions are necessary to avoid risks, how to safeguard critical points.

Sensitivity tests show which variable should be changed to achieve a desired effect. Sensitivity analysis gives hints to the planner where steering interventions are successful or not in respect of overall system behaviour.

In this way, the planner can continually improve his model and is thus part of the system, also in the sense of cybernetic reality.

References

[1] ACIP project homepage, http://www.iabg.de/acip/

[2] Rinaldi, S.;Peerenboom, J.; Kelly,T.: Complexities in Identifying, Understanding, and Analyzing Critical Infrastructure Interdependencies", invited paper for special issue of IEEE Control Systems Magazine on "Complex Interactive Networks", December 2001

[3] Roßnagel, A.: Sicherheit für Freiheit?; Nomos Verlagsgesellschaft Baden-Baden, 2003

[4] Dörner, D.: Problemlösen als Informationsverarbeitung, Kohlhammer Verlag Stuttgart 1976

[5] Vester, F.: Die Kunst vernetzt zu denken: Ideen und Werkzeuge für einen neuen Umgang mit Komplexität, DVA Stuttgart, 2000 ISBN 3-421-05308-1

[6] Probst, G.J.B./ Gomez,P: Vernetztes Denken – Ganzheitliches Führen in der Praxis; Gabler Verlag , Wiesbaden, 1991

[7] Ulrich, H./ Probst, G.J.B.: Anleitung zum ganzheitlichen Denken und Handeln, Bern und Stuttgart 1988

[8] Vester, F.; von Hesler, A.: Sensitivitätsmodell; Umlandverband Frankfurt, 1988

[9] UNICON Management Systeme GmbH: GAMMA Das PC Werkzeug für Vernetztes Denken

Terrorism as Perceived by the German Public

Michael M. Zwick

University of Stuttgart, Department of Sociology of Technologies and Environment, Seidenstr. 36, 70174 Stuttgart, Germany, zwick@soz.uni-stuttgart.de

Abstract

Due to the attacks of September 11, 2001 and the following political reactions, terrorism became a premier topic, a long runner in media reporting. With the attacks in Madrid the international terrorism reached European terrain and made the subject even more explosive. Although it meets high attention in the public, the empirical research on the public's perception of terrorism and the assessment of its risks is rather sparse in Germany.

Between December 1st, 2003 and February 20th, 2004, 43 semi-structured interviews with very different interviewees throughout Germany were carried out. The intention was to gain differentiated insights into the present perception of terrorism, its risks and its regulation.

Keywords: Terrorism, empirical study, risk perception, risk management.

1 Introduction

»Due to the recent incidents, the Instanbul attacks and of course September 11th 2001, terrorism has gained a complete new dimension... I am now in the early twenties, and have known terrorism only with regard to the German RAF, only by narrations, by books or in TV news. Now I am realizing terrorism as a pressing matter of current interest.« (A01.1.008)[1]

The history of terrorism, attempts of ostracized people (Bude, 2003: 96) to change prevailing political conditions by means of spectacular use of violence (Hoffmann, 2003: 53ff), goes far back. The beginnings are ascribed to the Zealots in the first century AD (Helmerich, 2003: 14, Laqueur, 2001: 16), although the notion of terrorism developed not until during the course of the French Revolution. (Hoffmann, 2003: 16, Badiou, 2002: 63p) From the beginning, terrorist acts of violence occurred in different ways: Mostly it were activities of nationalistic, ethnic-separatist or anti-colonial liberation movements, often joint with ideologically motivated, radical or fundamentalist groups. (Hoffmann, 2003: 30ff.)

Two changes seem to be worth mentioning as it is an extremely explosive subject today. When on July 22, 1968 activists of the PFLP, a branch of the PLO, captured an EL-AL plane flying from Rome to Tel Aviv, terrorism entered the international arena. (Hoffmann, 2003: 85) And it became a mass phenomenon, an everyday media event: (Waldmann, 2001: 56pp) Colin Mac Lachlan counted just in the years from 1976 till 1995 more than 9.800 attacks that can be ascribed to international terrorism (Mac Lachlan, 1997: 112). It is worth mentioning that the internationalisation did not at all take the place of national terrorism.

The most momentous attack up to now happened on September 11, 2001. Many authors believe to see there a metamorphosis, a sign for a historic transition: (Simon, 2002: 24, Bolz, 2002: 84, Gumbrecht, 2002: 100, Hitzler and Reichertz, 2003: 7) "Terrorism, in the 20th century in its significance often overestimated, threatens to become one of the main

---

[1] A01.1.008 indicates interview number 8, tape side 1, and tape counter 008.

hazards for the humanity at the beginning of the 21st century" resumes Walter Laqueur the new edition of his monograph. (Laqueur, 2001) Ulrich Beck classifies the modern terrorism of the 21st century among the boundless threats of the 'world risk society'. The incidents of September 11th, 2001 gave rise to a "globalized culture of fear", (Beck, 2003: 281), followed by global and long-term effects which went far beyond the direct damages: "It showed clearly to which (before unbelievable) extent economic systems… are vulnerable by terrorist acts". (Münkler, 2003: 19) Since September 11th there is a development towards an omnipresence of terror and a globalization of its psychic, political and economic consequences.

With that background the restraint of the German sociology concerning the research in the public's perception of terrorism is astonishing. Therefore, in 2003 the Department of Sociology of Technologies and Environment at the University of Stuttgart started a research project: Its objective was to gain differentiated insights into the public´s perception of terrorism, its reasons and risks, its risk prevention and risk management, by the use of qualitative methods.

2 Material and Methods

Between December 1st, 2003 and February 20, 2004 – even before the attacks in Madrid – 43 semi-structured narrative interviews of about 30 minutes each were carried out in all Germany. Such qualitative methods promise particularly detailed insights in the perception of terrorism. The relatively high number of interviews (for qualitative research strategies) allows deep and broad analyses when the samples are composed with high variability. (Flick, 1995: 89) Admittedly, only theoretical but no statistical generalizations are allowed.

Selecting the interviewees, a direct field access was chosen, whereby the interviewees should differ regarding gender, age, socio-economic status, but first of all regarding the characteristics relevant for the expected attitudes towards terrorism. For example persons with very conservative, state-supporting or protest-orientated, anti-American attitudes, with inclination to Islam or Judaism, with xenophile or xenophobe orientations were chosen. Additionally interviewees of professional groups that might be affected by the consequences of terrorism were selected. These are people working in the security or tourism branch, but also in fire brigades, rescue services, police, army, politics and insurance branch.

The semi-structured interviews began with an open question: "What do you think about terrorism?", followed by questions about the assumed reasons, the perceived threat and the risk semantic of terrorism. In each phase of the interview the respondents had the opportunity to explain their attitudes exhaustively. The interviews were concluded with two rating-scales on which the personal and the societal threat of three risks had to be marked: traffic, nuclear power and terrorism. Due to the topic of this article, mainly reactions to the initial question and answers to the rating scales will be displayed subsequently.

3 Towards the Perception of Terrorism by the German Public

The question for the perception of terrorism evoked by most of the 43 interviewees associations with international terrorism or general statements that did not relate to countries. However, national terrorism played a marginal role: Concerning Germany, only three of the interviewees remembered the terrorism by the "Red Army Faction" (RAF).

In Fig. 1, the structure of coding is demonstrated. First, for each interact, the subject is identified: Do people speak about terrorism in general, about international terrorism or national ways of terrorism? Second, it was examined how elaborate the respondents expressed their points of view: Therefore all aspects and single arguments referring to each subject were coded. Aspects describe, on what is spoken about terrorism. They encompass, for instance, the suspected reasons of terrorism, trials to define or valuate terrorism, its risk perception, strategies, organization, means and aims, as shown in Fig. 2. Each aspect can attract one ore more arguments. For instance, considerations about the terrorism's causes may yield different arguments: the role of social inequality between the first and the third world, political or cultural colonization, religious fundamentalism, and so on. So, Fig. 1 and 2 give us a synopsis on what is relevant to the public, which points of view give rise to different descriptions, and which subjects remain marginal in our data.

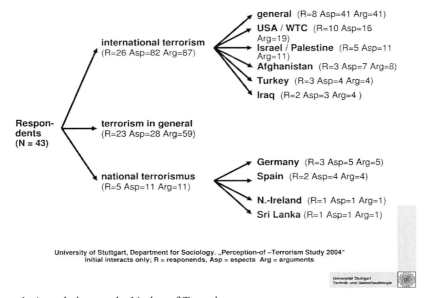

University of Stuttgart, Department for Sociology. „Perception-of –Terrorism Study 2004"
Initial interacts only; R = responends, Asp = aspects Arg = arguments

Figure 1: Associations to the Notion of Terrorism.

Figure 1 indicates a particular inclination towards international terrorism. 26 conversation partners dealt with this subject and developed 82 aspects and 87 arguments. The respondents mainly reflected generally on international terrorism, on the other hand they pointed on the USA respectively the attacks of September 11th, 2001.

Third comes the conflict in the Middle East. 5 respondents with 11 aspects and arguments each turned to that, whereas other types of international terrorism remained marginal.

The respondents talked in very different ways about "terrorism in general". 23 of them worked out 28 aspects and 59 arguments, whereas, as already mentioned, national terrorism played in their narrations only a minor role. So what is spoken about? Subsequently, the most important findings are presented.

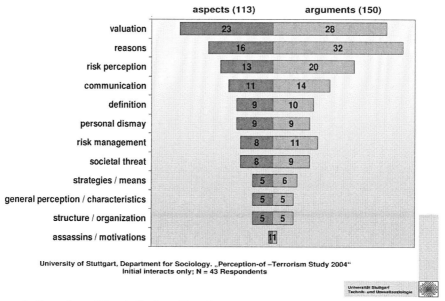

Figure 2: Terrorism - What is Spoken About?

3.1 What is Terrorism? Attempts to Find a Definition

Terrorism is understood mainly as a violent strategy to communicate problems respectively to push through objectives. This strategy is particularly chosen by politically weak groups to which legitimate means to realize their objectives are denied: Terrorism is the attempt to *"push violently through political or religious ideas.... in politically weak... or in third world countries that do not have good opportunities, like in Germany where we can bring it into the parliament. But they push it through with violence."* (A06.1.005) Some of the respondents point to the normative content of the term terrorism. Accordingly it is a question of the power of definition if it is possible to stigmatize the opponent with the pejorative label 'terrorist': *"I think that 'terrorism' is to a certain extent a matter of opinion."* (A40.1.006) *"It always depends on the definition and who defines terrorism. I give just one example: During the cold war the Afghan Mudjahedin were called freedom-fighters, because they were on the side of America. Now the tables have been turned, now they are terrorists. And because they resist America. The question always is how the political and the historical situation use this term."* (A47.1.010)

3.2 Assessment of Terrorism

More than half of the respondents gave value judgements already in the beginning of the interview: *"Terrorism is a threat for the whole civilized world."* (A40.1.006) All value judgements are negative. Use of violence is refused, particularly when 'innocent people' are affected: *"I absolutely condemn terrorism, because mainly those people suffer from it who are really innocent. If the right people were affected, then it would be different."* (A54.1.004) In some cases the judgement is connected with reasons of terrorism. Interview 50 refers to differences in wealth and power that makes the terror to be the strategy of the marginal: *"Terrorism is a means that is not legitimate at all, but is a means of the poor against the rich,... against the military superior, the more powerful ones."* (A50.1.072) With respect to the deprived life situations of the terrorists, often disapproval is combined with understanding for the people who exercise terror: *"On the one hand I condemn terrorists, on the other hand I can understand that someone who is pushed to the wall must take that as a least resort."* (A53.1.155) Therewith we enter the much discussed field of the assumed reasons of terrorism.

3.3 Reasons of Terrorism

This subject evokes the second most interacts and the most arguments. On the one hand, this underlines that it plays an important role in the public debate about terrorism. On the other hand it is an indication for the distinct variety of opinions. In spite of all variety, the economic inequality and the corresponding intolerable life situations dominate among the assumed reasons for terrorism. Terrorism is mainly represented as a phenomenon of poverty and inequality; almost half of the mentioned reasons follow this pattern of argumentation: *"Poverty in the world surely is one of the main things that lead to the formation of such terrorist groups, because there they find their breeding ground. That has to do with oppression of people, and on the other hand poverty is the reason for religious fanaticism that leads to terrorism* (A03.1.190) Economic deprivation is accompanied by political deprivation, which hardly allows to pursue ones objectives legitimately: *"Terrorism emerges from ... an absolute helplessness. These are people who do not have other possibilities to push through their objectives or ideas.* (A07.1.010) From the economic or political point of view modern terrorism can also be interpreted as a consequence of globalization: *"Since three or four years it is often called globalized terror... And globalization is demonstrated to the terrorists accordingly: Economy is globalized, capital is globalized, politics are globalized.... , and I think this is just the beginning: Globalization is at the expense of those that are already put at a disadvantage, and opening markets are no step towards equal distribution but rather a step to unlimited access for the industrial countries. And I think .... it will get even more radical, more globalized."* (A08.1.077)

There are different opinions – like by the way in literature (Krieg, 2002, Krönig, 2001, Mohn, 2003, Ostendorf, 2003) – about the function of religion: For the ones in Islamism itself are the reasons for terrorism: *"For me the reason is fanaticism. There are people who are fanatic, in each direction: there are religious fanatics, there are other fanatics; and when someone is fanatic he can not see things in an unbiased way and he does unpredictable things that are absolutely senseless.* (A21.1.033) In addition, religion can fulfil the function to legitimize actions and to recruit activists: *"The worst thing is that young people are used and intimidated, that they are given big promises and are lied at: for example the Palestinians or the Arabic people are told that they would come to paradise where they will be received by 72 virgins – really senseless things – and then they are prepared to sacrifice themselves for any idea.* (A21.1.016) Also sense of mission and missionary motives were mentioned, arguments that refer to cultural disparities between Islamic and rational western societies: *"The reason is a too strong belief in Islam.... I just think that, with their faith that is too strong, the Islamic people want to convert the bad, seedy western world, that means Europe and America. And they think that everything that is not Islamic is bad. And I believe that their faith is simply too fanatic."* (A33.1.07)

A small majority of respondents is doubtful. They guess that religion is just harnessed to recruit people to push through political and economic interests: *"I think that religion is harnessed here by people who have other interests, political, ideological and surely, if one wants to include the psychological aspect, there are fears."* (A50.1.133) Some of the interviewees hint, *"that particularly Mr. Bush instrumentalizes religion as well, ... for his personal crusade."* (A53.1.434)

Closely connected with the debate about religious disparities is the interpretation of terrorism as a cultural conflict, respectively attempts of the west to colonialize Islamic cultures: The problem is that the *"western culture of US provenience often acts in a very imperialistic way,... that Americans wage war against the backdrop of 'we bring democracy and prosperity to your country',...but with really imperialistic affectation, and with economic interests."* (A50.1.060)

The reflection of reasons signals, that the majority of interviewees interpret terrorism and the terrorists´ motifs rather rational than irrational. And, only two respondents describe assassins in psychiatric terms.

3.4 Terrorism and Communication

Terrorism is connected with communication in three ways. As we have already seen, it is first perceived as a (violent) strategy to communicate problems. Second, modern terrorism bases on a functionalization of the public. That requires a global communication of their actions by the mass media. In this way mass media represent a link between inequality, terrorism and globalization: *"Mostly the purpose of a terrorist act is to attract attention and to get something with the help of the attention of the population or the whole world. Everywhere are mass media. Everything is broadcasted by TV, and on the other hand there are always weak, groups that present themselves and become the focus of attention via the mass media - if necessary by terrorist attacks."* (A41.1.055)

The third variant of communication is the spiral of violence: "I just think that the conflict parties have become set on each other so terribly that they are unable to see beyond the end of one's own nose." (A53.1.067) Thus the spiral of violence and counter violence hints to an exclusive form of bilateral 'communication' that has lost the ability of self-reflexivity and meta-communication.

4 Risk Perception and Risk Management

The extent to which people feel menaced by terrorism differs widely: People living in big cities respectively close to symbolic places, females, and people regarding themselves as particularly exposed – e.g. Jews – tend to feel more frightened than others. Becoming involved in a terrorist attack is regarded as a fatal event. However, most of the respondents take it as an omnipresent but still highly improbable risk in Germany: *"The probability that I am in Germany just at that time at the place where an assassination happens, is extremely small."* (A03.1.257) *"Everywhere where there are a lot of people, where there is a lot of public, is a potential objective... One can never be safe, it can affect everybody everywhere."* (A24.1.319) Other respondents think that symbolic places are particularly exposed: *"They try to hit the opponent at a sensitive spot... Terrorists attack rather the symbolic.* (A071.070). *"They always search for spectacular objectives... spectacular places that have a certain symbolic value. That one on September 11, that was the greatest symbol of all which was attacked."* (A50.1.180) In a respondent's opinion the World Trade Center signals *"a bit of megalomania and a bit of missing empathy. It is a symbol of megalomania shown outwardly."* (A50.1.270)

Almost all interviewees perceive terrorism as an increasing risk: "In Germany we have not been very much affected by terrorism, compared with other parts of the world, but it will come up to us. What then will be, we can not predict it." (A04.1.024)

Still, the risk of terrorism is thought to be rather a "switching risk" than a "pervasive risk", (Zwick 2002) that needs special opportunities to be remembered, but does not affect people for a long time in everyday life: *"Normally, one hardly thinks of terrorism, except, there are news in TV. Everybody has enough serious problems and does not look after terrorism... in my own life it plays only a marginal role, since I have never been directly affected by terrorism."* (A35.1.038) Indeed, most people, asked for personal consequences regarding terrorism; think about travelling and answer like this: *"I avoid to make holidays in trouble spots."* (A43.1.060)

4.1 Risk Semantic

The perceived risk semantic of terrorism is similar to a sword of Damocles. In contrast to smoking, traffic or nuclear power, terrorism is seen as an uncertain risk. It is *"a big unknown... It is something that smoulders and then really happens, as an incident. Fortunately it does not happen often, but it is a time bomb ticking away."* (A08.1.315) Not only the probability but as well the extent of possible damages are uncertain, since terrorism can aim at the infrastructure and then develop multiplying destruction.

In the eyes of most interviewees, terrorism is an intentional, forced on, individually almost not controllable risk, a particularly pejorative risk, in the face of which one is helpless. *"I do not have any influence on terrorism and therefore... I am helpless in the face of it and feel extremely menaced."* (A05.1.100) Nearly all of our conversation partners believe that terrorism is a risk which is uncontrollable, inevitable and which makes safety to be an illusion: *"If it shall affect someone you can not do anything against... There is no safety."* (A12.1.045)

If one compares individual and societal threat of traffic, nuclear power and terrorism, the latter shows symptoms of a systemic risk: Regarding traffic and nuclear power the societal threat – measured on a 100-points-rating-scale – exceeds subjectively felt threat by 10 points maximal. However, regarding terrorism the difference is more than 35 points! For this, two reasons are given in the data.

First, the global damages are made responsible. For example, in the perception of several interviewees, the direct damages of the WTC-attack are exceeded by far by damages that influence the world's economy up to today. The second reason affects the societal dealing with terrorism.

4.2 Risk Management and Risk Prevention

A couple of interview partners argue, that the risk management fails in the establishment of safety and becomes a threat to the society itself: *"The rights guaranteed by basic law are.... extremely restricted... The missing personal freedom,... is a very severe threat for our society, for our democracy."* (A39.1.148) *"A world like in the book of George Orwell '1984', a totalitarian power, in order to eliminate terrorism and to establish absolute security for the people, is coming up."* (A25.1.085) Partly hidden agendas are suspected behind the state's risk provisions: *"I think that threat scenarios can be misused by the state to stir up fear among the population in order to push through certain measures easier and quicker, without much resistance."* (A49.1.173)

Almost all respondents refuse the US strategy fighting terrorism by war: "It is a big mistake to wage war against terrorism. It would be better to try to regulate it in a political or diplomatic way. Because war... will probably cause terrorism again. Therefore, I think that it is madness to declare war on terrorism." (A55.1.306)

The best precaution against terrorism is therefore seen in a cautious foreign policy and – in the long term – in the elimination of its causes: *"Concerning each of the two wars, Iraq and Afghanistan, German politics behaved very cautiously. I think, this was just the right way to deal with it!"* (A56.1.217) *"As long as the world does not do anything to create fair conditions everywhere, and self-determination of the peoples, and a certain minimum of justice, ... there will exist terror."* (A08.1.066)

Acknowledgements

Special thanks to Ortwin Renn for the granted scope to plan and carry out such projects, which meet sociology of technologies and environment only indirectly, and for his support in conceptual and methodical questions. Thanks to my students who carried out most of the narrative interviews, and to the University of Stuttgart which financed this project. Sabine Mertz was a great help with the transcription of the interviews, with the translation of this article and the improvement of its legibility: Many thanks!

References

Badiou, Alain: Philosophische Überlegungen zu einigen jüngsten Ereignissen. In: Baecker, D., Krieg, P. and Simon, F.B., editors: *Terror im System*, 61-81, Carl Auer, Heidelberg, 2002.

Beck, Ulrich: Weltrisikogesellschaft revisited: Die terroristische Bedrohung. In: Hitzler, R. and Reichertz, J. editors: *Irritierte Ordnung*, 275-298, UVK, Konstanz, 2003.

Bolz, Norbert: Die Furie des Zerstörens: Wie Terroristen die Kritik der liberalen Vernunft schreiben. In: Baecker, D., Krieg, P. and Simon, F.B., editors: *Terror im System*. 84-99, Carl Auer, Heidelberg, 2002.

Bude, Heinz: Die Rache der Überflüssigen. In: Hitzler, R. and Reichertz, J., editors: *Irritierte Ordnung*, 95-102, UVK, Konstanz, 2003.

Gumbrecht, Hans-Ulrich: In eine Zukunft gestoßen: Nach dem 11. September 2001. In: Baecker, D., Krieg, P. and Simon, F.B., editors: *Terror im System*, 100-109, Carl Auer, Heidelberg, 2002.

Helmerich, Antje: Wider den Etikettenschwindel. Ein politikwissenschaftlicher Erklärungsversuch des Begriffs »Terrorismus«. In: Bos, E. and Helmerich, A., editors: *Neue Bedrohung Terrorismus*, 13-32, LIT, Münster, 2003.

Hitzler, Ronald and Reichertz, Jo: Die gesellschaftliche Verarbeitung von Terror. Einleitung. In: Hitzler, R. and Reichertz, J., editors: *Irritierte Ordnung*, 7-9, UVK, Konstanz, 2003.

Hoffmann, Bruce: Terrorismus - der unerklärte Krieg. Fischer, Frankfurt/M., 2003.

Krieg, Peter: Ewige Gerechtigkeit oder anhaltende Freiheit? Subtext und Logik eines Konflikts. In: Baecker, D., Krieg, P. and Simon, F.B., editors: *Terror im System,* 160-174, Carl Auer, Heidelberg, 2002.

Krönig, Jürgen.: Jihad versus McWorld. *Aus Politik und Zeitgeschichte*, B41-42, 2001.

Laqueur, Walter: Die globale Bedrohung. Econ. München, 2001

Mac Lachlan, Colin M.: Manual de Terrorismo Internacional. Mexiko, 1997.

Mohn, Jürgen: »Clash of Religions« - der 11. September 2001 als Rückkehr der Religion im Gewand der Gewalt? In: Bos, E. and Helmerich, A., editors: *Neue Bedrohung Terrorismus*, 69-92, LIT, Münster, 2003.

Münkler, Herfried: Grammatik der Gewalt. In: Hitzler, R. and Reichertz, J., editors: *Irritierte Ordnung,* 13-30, UVK, Konstanz, 2003.

Simon, Fritz B.: Was ist Terrorismus? Versuch einer Definition. In: Baecker, D., Krieg, P. and Simon, F.B., editors: *Terror im System*, 12-31, Carl Auer, Heidelberg, 2002.

Waldmann, Peter: Terrorismus, Provokation der Macht, Gerling, München, 2001.

Zwick, M.M.: Risk as Perceived by the Public: Disparities of Qualitative and Quantitative Findings. Meeting Paper for the SRA Europe annual meeting, Berlin, Humboldt-University, Berlin, 2002.

A New Model of Natural Hazard Risk Evaluation

Thomas Plattner

Chair of Forest Engineering, Swiss Federal Institute of Technology (ETH Zürich), CH-8092 Zürich, Switzerland

Abstract

Since risk mitigating measures have to be accepted by the society, the risk evaluation by laypeople has to be considered in risk management. Due to time and financial constraints, it is impossible to gather this information on a case-by-case basis. Therefore, a so far theoretical model of risk evaluation that accounts for the public risk perception and evaluation is suggested. The model is based on the assumption that the perception affecting factors PAF and the objective risk R_{obj} are important for the perception of the risk and the evaluation criteria EC are relevant for the decision about the acceptance of risk. Together, the PAF, the EC and the R_{obj} allow to predict the accepted risk R_{acc}. According to the different types of human behavior, the prospect theory is applied to obtain relevant information about the evaluation and acceptance of risk for different types of persons.

Keywords: natural hazard, risk evaluation, risk perception, risk aversion, perception affecting factors PAF, evaluation criteria EC, prospect theory

1 Introduction: A Risk-Based Approach for the Management of Natural Hazards

Dealing with natural hazards has recently shifted away from being hazard-oriented towards a more risk-based approach that can be described with the slogan 'From the avoidance of hazards towards a culture of risk' (PLANAT, 2004). According to Hollenstein (1997) such a modern proceeding comprises three components (see Fig. 1).

On the one hand, *Risk Analysis* is the process of quantification of the probabilities and expected consequence for identified risks (SRA, 2004). Consequently, it provides information about the extent and frequency of the expected damage and answers the question 'What can happen?' (Hollenstein, 1997). In its simplest formulation, the objective risk R_{obj} may be defined as the product of the frequency (or probability) of occurrence F of an event and the extent E of the associated consequences, i.e.,

$$R_{obj} = F \cdot E \tag{1}$$

On the other hand, *Risk Evaluation* is the socio-politial and moral-ethical component in which judgements are made about the significance and acceptability of risk (SRA, 2004). It addresses the question 'What may happen?' and produces the information that is needed to determine the acceptability of risks and thus the acceptable residual risk (R_{acc}). Within the framework of economical (increasing marginal cost of protective measures) and environmental reasons (negative environmental impact of the measures) *Risk Management*, finally, aims at reducing risks that are found to be unacceptably high and prevent other risks from becoming so by maintaining a safe state. It combines the results of risk analysis (R_{obj}) and risk evaluation (R_{acc}) in a political process and implements measures based on economic and technical principles (Heinimann, 2002).

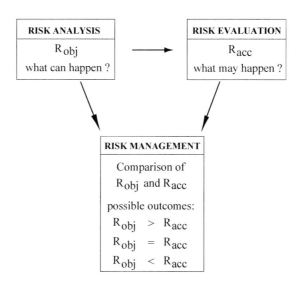

Figure 1: The three components of a modern, risk-based approach of dealing with natural
         hazards: risk analysis, risk evaluation, risk management.

2 State of Knowledge: Current Representation of Risk Evaluation

Currently, two common formal risk evaluation procedures are the approaches of the *Boundary
Line* (see Farmer, 1967) and the *Aversion Term* (Schneider, 1985), (Bohnenblust and Troxler,
1987), (BUWAL, 1999). Both procedures are based on the theoretical assumption that *Risk
Aversion* (α), depending on the extent E of the direct damage of an event, is the driving
evaluation criteria

$$\alpha = f(E) = \alpha(E) \tag{2}$$

2.1 Boundary Line

The two procedures differ in the way of including α. The Boundary Line represents constant
risk on a double-logarithmic chart (plotting the frequency F and the extent of damage E of
events) by a line with the gradient $g(B)$

$$g(B) = \frac{\Delta(log(F))}{\Delta(log(E))} = -1 \tag{3}$$

and risk aversion can be accounted for by changing the slope of the boundary line such that

$$g(B) * \alpha = g^*(B) = \frac{\Delta(log(F^*))}{\Delta(log(E))} < -1 \, , \tag{4}$$

i.e., if the extent of damage doubles, the acceptable frequency is no longer half as high, but
usually less than half. This approach is often used in regulatory applications, e.g., in the Swiss
Ordinance on Technical Hazards (see EDI, 1991). However, it is often not obvious if and to
what extent risk aversion has been integrated.

2.2 Aversion Term

The use of an Aversion Term (a factor or an exponent in the representation of actual risks) is an explicit way considering risk aversion. Often, the outcome of this approach is called perceived risk R_{perc}, although only the extent of damage is considered and none of the other factors affecting the perception and evaluation of risk. Consequently, the result should be called aversion-corrected risk R_{AC}

$$R_{AC} = F \cdot E \cdot \alpha(E) \qquad (5)$$

or

$$R_{AC} = F \cdot E^{\alpha(E)} \qquad (6)$$

In both cases, the relations

$$\alpha(E) > 1 \qquad \text{and} \qquad d\alpha/dE > 0 \qquad (7)$$

hold, i.e., risks with a greater extent are artificially and progressively increased.

3 Four Principles in the Development of a New Risk Evaluation Model

The development of a new evaluation model is based on four principles:

1. Mathematical formalization of the new approach

2. Correct and complete representation of current knowledge about risk perception and risk evaluation due to the fact that there are significant differences between the (phenomenological) characteristics of natural and technological hazards. The above mentioned approaches have their origins in the field of technical hazards and though provide a wealth of knowledge.

3. Role and representation of the perceiving person

4. Design for easy implementation and application

4 The Implementation of the Principles

4.1 Mathematical Formalization of the Risk Evaluation Approach

Under the assumption that the perceived risk R_{perc} does not comply with the objective risk R_{obj}

$$R_{perc} \neq R_{obj} \qquad (8)$$

but that R_{obj} contributes to R_{perc}, the perception affecting factors PAF are used to define the perceived risk R_{perc} according to

$$R_{perc} = f(PAF, R_{obj}) \qquad (9)$$

The acceptable risk R_{acc} depends on the evaluation criteria EC, i.e.,

$$R_{acc} = f(EC) \qquad (10)$$

and the decision about the acceptability Acc of a certain risk R_i can now be made according to

$$Acc = \begin{cases} 1 & \text{if } R_{i,perc} \leq R_{i,acc} \\ 0 & \text{if } R_{i,perc} > R_{i,perc} \end{cases} \qquad (11)$$

Under the assumption that for a given combination of a risk and its environment i the acceptable risk $R_{i,acc}$ is usually more or less constant, it follows that

$$Acc_i = f(PAF_i, R_{obj,i}) \,, \tag{12}$$

i.e., the objective risk R_{obj} is not a sufficient basis to judge the acceptability Acc of a risk even if the limits of acceptability R_{acc} are known.

4.2 Current Knowledge about Risk Perception and Risk Evaluation

4.2.1 Perception Affecting Factors (PAF)

An important step in the model development is the determination of i) the relevant risk perception factors PAF driving the perception of risks due to natural hazards and ii) the evaluation criteria EC used for defining the acceptable risk R_{acc}.

The determination of the PAF is based on a survey of perception literature and an adjacent qualitative selection process. The chosen set of factors was submitted to a group of experts from technical, administrative and social institutions dealing with natural hazards. The experts assessed the relevance of the individual factors and the results were compiled. From this procedure resulted a smaller list of factors that was tested on a qualitative basis for collinearity, such that redundant factors could be grouped without significant loss of information (see Tab.1, left column).

PAF/EC:	Represents/Includes
Voluntariness	Voluntariness
Reducibility	Reducibility, Predictability, Avoidability, (Controllability)
Dread	Controllability, Number of people affected, Fatality of consequences, Spatiotemporal distribution of victims, Scope of area affected, Immediacy of effects, Directness of impacts
Experience	Familiarity, Knowledge about risk, Manageability

Table 1: List of relevant perception affecting factors (PAF) and evaluation criteria (EC). The factors in the column on the right are those known from the literature. *Note:* The suggested selection of PAF and EC is work in progress and subject to change.

4.2.2 Evaluation Criteria

As a working hypothesis, it is assumed that the evaluation criteria EC are similar to the PAF (see Table 1), i.e., a factor is not only relevant for the perception of a risk, but also for the limits of acceptability after

$$R_{acc} = f(EC) = g(PAF) \,. \tag{13}$$

From this it follows that

$$EC = h(PAF) \,. \tag{14}$$

Additional criteria according to

$$c \in EC \wedge c \notin PAF \tag{15}$$

will only be defined if this appears to be required for the model building.

The next step is the weighting of the relevant PAF and EC considering the following methods:

- Analysis of the risk perception literature in order to gather information about the weighting of several PAF and EC.

- Breakdown of the results of an ongoing risk perception survey in selected swiss areas (University of Zurich, Sozialforschungsstelle).

- Analysis of current risk evaluation case studies (using the approaches of the Boundary Line or Aversion Term) in order to gather information about the consistency of the used aversion terms.

- Analysis of pre- and post-event data of case studies regarding the expenditures for the protection of natural hazards in order to gather informations about the considered PAF and EC.

4.2.3 Risk Aversion

Risk aversion should be taken into account using an improved definition and understanding of it. Not only the extent of direct damage (e.g., number of fatalities or monetary costs) has to be considered but also the indirect effects of a hazardous event (e.g., costs of enacted laws, costs of psychological support of affected people, cost of recovery actions, etc.). Both, indirect and direct effects, can be included in the term expected consequence C_{exp}. Furthermore, such an approach must also consider what those affected and the society think about the relevance of these consequences from a sociopolitical, an environmental and an economic point of view.

Based on this considerations and the comments of Page (1982) and Bähler (1996), one can now develop a new definition of risk aversion and an associated concept for estimating its extent. Aspects to be included in this approach are:

- Objective risk R_{obj}, particularly the expected extent of consequences C_{exp}.

- Evaluation criteria EC. Since the magnitude of the potential damage is relevant, the scale for judging this magnitude is also important. The aversion against an event i may be stronger than that against an event j even though all the relations $R_{obj,i} < R_{obj,j}$; $R_{perc,i} < R_{perc,j}$ and $C_{exp,i} < C_{exp,j}$ may hold. This can, e.g., occur when i results from an involuntary and j from a voluntary activity.

- Effects for the system considered (stability, regenerability).

- Speciality of a risk (extraordinary vs. normal events).

One of the potential pitfalls of such an approach is the mingling of risk perception and risk aversion. Evaluation criteria EC are closely related to the perception affecting factors PAF. A complete delineation between risk perception and aversion is anyway impossible if the validity of the relation $R_{perc} = f(PAF, R_{obj})$ is assumed, since it follows that there is also a relation

$$PAF = g(R_{perc}, R_{obj}) \tag{16}$$

because the processes of perceiving a risk and developing an aversion are influencing each other.

4.3 The Role and Representation of the Perceiving Person

The perception of a certain risk also depends significantly on factors that have no immediate relation to the risk itself, but to the perceiving person:

- Economic perspectives

- Social environment (structural and cultural properties of the community)

- Values and world views (e.g., technocratic vs. naturalistic, progressive vs. conservative, individualist vs. collectivist)

- Psychological and behavioral characteristics (e.g., risk seeking vs. risk averse)

These factors must be accounted for, but it is probably not possible, and also not desirable, to do so in a parametrized way. On the one hand, it is not likely that a useful representation of the individual perceiver can be derived, on the other hand, this is also not desirable because the collective perspective is of interest. The suggested evaluation model should have as simple a structure as possible and be easily appliable but still allow a causally correct representation of the reality. Thus, the application of a model of human behavior seems to be a promising way of integrating the personal factors. A large number of individual risk perception processes can collectively be represented by a prototype behavior.

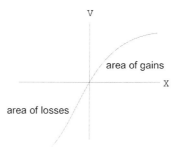

Figure 2: The asymmetric and non-linear value function of the Prospect Theory. The losses are weighted stronger than thze numerical equal gains (loss aversion)

Due to the fact of the irrationality of risk evaluation, a model of man has to be elected that allows modelling human behavior without neglecting all the irrational aspects of it. People usually do not use a value system in order to evaluate an action alternative, but the amount of benefit or loss due to a decision for a specific action alternative. Within the prospect theory (Kahneman and Tversky, 1979), the value of this amount is calculated using the value of the reference point as a basic value (and not using the final state of an outcome). The reference point allows to model two main characteristics of human behavior: i) a variable reference system and ii) a asymmetric and non-linear value function, i.e., there is more emphasis on losses (convex value function: loss aversion) than on gains (concave value function; see Fig. 2). This effect has also to be considered in the scale of the aversive and non-aversive effect of PAF and EC (see Fig. 3).

4.4 Design for Implementation and Application

Based on the above mentioned principles, the operationalization of the risk evaluation model, considering a simple design, can be achieved. A main task is the integration of the redefined risk aversion. This is part of the ongoing step of quantifiying and weighting the perception affecting factors PAF and the evaluation criteria EC (see chapter 4.2.2). A possible method of resolution for this problem is shown in Fig. 3.

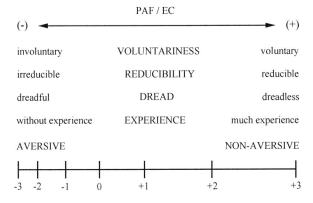

Figure 3: A possible approach of considering risk aversion and judging the magnitude of the risk perception affecting factors PAF and the evaluation criteria EC. *Note*: This figure shows only a possible example of scaling the PAF and EC. The quantification is work in progress.

5 Discussion: Shortcomings of the Chosen Approach

The most critical points in the model development process are the many simplifications and assumptions required to create an applicable model. But there are several issues that warrant more detailed discussion.

First, the idea to derive a normative formulation from social and psychological factors originally derived in a non-normative context and applying this formulation within an engineering context may give raise to questions. However, managing risks involves making decisions, and while it is obvious that a perfect representation of the real risk evaluation will never be possible, one should nevertheless try to improve the basis for the decision making. The goal is not to replace the evaluation process, but to facilitate the selection of 'good' options (i.e., those that are likely to be in line with the evaluation) within the framework of risk management.

A second issue is the definition of PAF and EC components as suggested above (including their weights and scales). It is obvious that laypersons play a vital role in evaluating risks. They should be represented in the definition of the factors that drive the model. This was not possible due to time and budget constraints. This is a shortcoming of the chosen approach, but one that will be diluted using the outcomes of a currently ongoing inquiry about the perception of natural hazard risks in affected Swiss regions as an evaluation and calibration tool of the model.

A third issue is the inclusion of risk aversion in the evaluation process. While aversion was the only factor considered so far in formal evaluation procedures, it may almost disappear in the new model since it is expressed by the perception affecting factors PAF and the evaluation criteria EC. This would be a clear change in practice, and it remains to be seen if such a change will be endorsed by the risk management community.

6 Conclusions: The Importance of a Risk Evaluation Model for a Culture of Risk

In a culture of risk based thinking, decision making must not be limited to scientific findings and methods, but it must also include societal aspects and concerns as brought forward in the evaluation process. As it is usually not possible to directly operationalize and quantify this public evaluation, the decision makers are left with the obligation to consider the evaluation appropriately. Up to now, this has happened in an intuitive and subjective manner, which is unsatisfactory for both the decision makers and the public. A model as suggested above would allow this process to be formalized, ensuring that the public evaluation is taken into account in a consistent manner and at the same time relieving the decision makers from an obligation that most of them did not really seek.

However, while the knowledge about the objective risk R_{obj} and the perceived risk R_{perc} is a valuable input for the allocation of resources, it is by far not the only factor relevant in risk management. Other issues such as the comparison with other risks or regulatory provisions (legal limits etc.) are in reality often at least as important as the socioeconomic aspects that are addressed by risk analysis and risk evaluation results.

References

Bähler, Fritz: *Ein Beitrag zur Strukturierung und Abgrenzung der Risikoaversion.* Eidgenössisch-Technische Hochschule ETH Zürich, unpublished, 1996.

Bohnenblust, Hans and Troxler, Christoph: Risk analysis - Is it a useful tool for the politician in making decisions on avalanche safety ? In *Davos Symposium: Avalanche formation, movement and effects*, pp. 653-664, Davos, Switzerland, 1987.

Bundesamt für Umwelt, Wald und Landschaft (BUWAL): Kosten-Wirksamkeit von Lawinenschutz-Massnahmen an Verkehrsachsen. Bern, 1999.

Eidgenössisches Departement des Innern (EDI): *Verordnung über den Schutz vor Störfällen (Störfallverordnung, SFV). Entwurf*, Bern, 1989.

Farmer, F.R.: Siting criteria - a new approach. In *Containment and siting of nuclear power plants*, pp. 303-329, Wien, Austria, 1967.

Heinimann, Hans Ruedi: Risk Management - a framework to improve effectiveness and efficiency of resource management decisions. In *23rd session of the European Forestry Commission's working party on the management of mountain watersheds*, pp. 16-19, Davos, Switzerland, 2002.

Hollenstein, Kurt: Analyse, Bewertung und Management von Naturrisiken. PhD Thesis, Swiss Federal Institute of Technology (ETH), Zurich, Switzerland, 1997.

Kahneman, Daniel and Tversky, Amos: Prospect Theory: an analysis of decision under risk. *Econometrica*, 47(2):263-297, 1979.

Page, Talbot: Definitions of risk aversion. *Transactions of the american nuclear society*, 41:445-446, 1982.

Plattform Naturgefahren Schweiz (PLANAT): Naturgefahren: so wehrlos sind wir nicht. Available electronically at http://www.planat.ch, 2004.

Schneider, Thomas: Ein quantitatives Entscheidungsmodell für Sicherheitsprobleme im nicht-nuklearen Bereich. In *Risikountersuchungen als Entscheidungsinstrument - Risk analysis as a decision tool*, pp. 113-143, Köln, Germany, 1985.

Society for Risk Analysis: Glossary of Risk Analysis. Available electronically at http://www.sra.org/resource_glossary.php, 2004.

Streamlined Permitting in Building Laws of the Federal Republic of Germany and its Effects on Mass Movements

Bodo Damm

Institute for Geography, University of Göttingen, D- 37077 Göttingen, Germany

Abstract

In recent years, the laws pertaining to construction have been simplified with the goals of reducing administrative procedures and decreasing government spending. On the one hand, the new regulations are intended to simplify construction through the elimination of obstacles toward approval and through the simplification of government oversight procedures. On the other hand, a simplified approval process shifts the responsibility for compliance with building codes to the owners, planners (architects, civil engineers, experts, etc.) and project managers. First experiences with damages show that streamlined construction approvals run the risk of neglecting basic geomorphologic and geologic fundamentals. Studies document errors and omissions in this context, which have led to failures at embankments and on building sites, causing considerable damages. Because simplified approval processes currently are allowed only in areas with development plans, the responsibility is shifted within the administration from the offices that approve individual construction processes to those who design general development plans.

Keywords: building laws, geomorphology, mass movements, low mountain ranges, Germany

1 Introduction

German construction law stipulates that the safety and order of private and public buildings and installations is guaranteed by the government. While there is a private responsibility by the author of the construction plans and by the owner of the building, the government has to make sure that legal requirements are met. As part of a general political goal to abolish, speed up or limit bureaucratic acts, German construction law has been simplified in recent years. The new law intends to facilitate construction by eliminating obstacles to the approval of plans, by simplifying oversight processes, and by reducing material requirements. For example, in Bavaria, the revised construction law was incorporated into state law in 1994. Between 1994 and 2002, more than 103,000 residential buildings were constructed without permits, saving owners about 110 million Euro in permit fees (BayBO Erfahrungsbericht, 2002, Bayerische Staatsregierung, 2004).

By now, most states have replaced the previous construction permitting processes with a combination of simplified permits, permit exemptions and simple reporting requirements (Hauth 2001). The most important regulations are based on development planning law and construction regulations law:

- The development planning law is federal law (rules in the "Baugesetzbuch" BauGB). The planning sovereignty of municipalities specifically is protected by law.
- The building regulations law is state law (e.g., HBO in Hessen state). It covers the construction approval process (Hauth, 2001, Moog, 1999).

Since the revised building regulations became law in June 1994, the new rules have been adopted into state law by numerous states, often revised several times (cf., MBO 2002). After a few years of experience, architects and also some building administrations have given highly favorable reviews to the new permitting processes (among others, SBUB 1998). However, as a consequence of the reversal of previous preventive administrative law, an increasing number of cases have been documented where basic technical rules have been neglected. In this context, faulty planning has caused significant economic damages (cf., MFBS 2003).

2 Simplified Permitting Processes as Exemplified by the Hessen State Building Regulations

The new Hessen State Building Regulations (HBO) is representative for the new rules. The simplified permitting process was introduced relatively late (October 2002), when the existing HBO was modified. The new HBO is based in its essential parts on the Musterbauordnung (MBO, sample building regulations). According to paragraph 2 (3), it divides buildings into five classes (fig. 1), which are the basic criterion for selecting the applicable permitting process. The new law provides for the following processes:

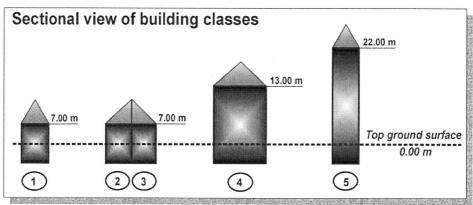

Figure   1:   Section   of   building   classes   after   the   Sample   Building   Regulations (Musterbauordnung MBO), status of November 2002.

- **Exemption from permitting (after paragraph 56)** applies to plans for residential construction and simple commercial buildings up to the "high rise" limit in areas with valid development plans that apply to the planned construction. These construction projects do not need approval. Only notification of the municipality is required. The exemption from approval does not cover special buildings and the demolition of buildings. The exemption process is the direct process to maintain the municipalities' planning sovereignty, because the municipality is informed immediately.
- **The Simplified Building Permitting Process (after paragraph 57)** applies to all construction projects that are not covered by the exemption from permitting and that are not special construction projects requiring permits. Thus, the Simplified Permitting Process is the standard procedure. The limited governmental preventive evaluation is reduced to examining whether the planned construction meets the legal planning requirements of the construction law

(BauGB). The owner of the building project may hire consultants as defined by paragraph 59 HBO (technical examinations of construction), including experts on substrate stability. The HBO excludes special buildings from this privatization because of safety concerns.

- **The classic Building Permitting Process** includes an examination by the department of planning and building inspection, which attempts to provide a comprehensive legal determination that the project poses no risks and meets all applicable laws. This process now is limited to buildings of classes 4 and 5 (except residential buildings up to the high rise limit), as well as special buildings after paragraph 2 (8). Because of their design or use, special buildings pose special risks. Examples are high rise buildings, retail or meeting buildings, commercial facilities, sports facilities as well as office buildings. The sometimes significant potential dangers and the increased need for coordination during the examination of these buildings continue to require a comprehensive permitting process by the department of planning and building inspection. In these cases, at least two government officials have to sign off on the construction plans.

With the new practice there are significant risks associated: Municipalities decide whether a construction project is exempt from permit approval. They have to examine whether their interests are concerned and whether the project meets the guidelines of their planning. However, it is uncertain whether smaller municipalities have the required personal and technical resources for this evaluation. If the construction supervision administration does not object to a project within four weeks, construction may begin. However, the risk of a faulty assessment of a construction project is borne by the building's owner. According to current law, a full building permit confirms that the project meets all applicable legal standards and rules. As far as the law is concerned, the permit is a completed process. This safety no longer exists for the streamlined processes allowed by the new HBO. A building's owner thus has to rely completely on the technical knowledge of the consultants and the quality of the involved firms. If it is determined after completion of construction that the development plan did not apply to this project, then there is no legal certainty. It is possible that the use of the building may be limited or completely prohibited or that the building has to be demolished. However, simplified permit processes also pose geomorphologic and geologic risks, as evidenced by several examples.

3 Damages through Mass Movements as a result of Simplified Permitting

A special danger in this respect risks associated with substrate conditions ("Baugrundrisiko"). Apart from problems that are caused by interactions between building and substrate (Boeck et al., 2003, Witt, 2002), this comprises especially large-scale problems such as the stability of slopes surrounding a building and other geologic dangers. In Bavaria, several years of experience with the new building law have shown that the legal requirements are not met in significant numbers of cases (BayBO Erfahrungsbericht, 2002). This applies especially to the proof of stability in connection with the load-bearing capacity of the substrate, which often is neglected. In certain cases, this has led to significant damages. Experiences with damages caused by mass movement show that streamlined permitting processes risk neglecting general geomorphologic and geologic concepts. Studies show numerous examples of poor judgment, which even on relatively unproblematic properties can cause failures with significant damages. Often, mass movement is caused by improper construction methods.

- In March 2000, a failure in a 10 m tall bedrock cliff of an old quarry in the Rheinhessen region released about 80 m³ of rocks. The roof and outside walls of a residential house suffered significant damages. Safety devices (catch fence, protective planks) erected by the owner were obliterated. Because of continuing danger to the building and its inhabitants through further rock falls, geotechnical measures were required. The cost of these measures (80,000 Euro) was paid by the building's owner. The construction of the single-family house below the cliff of the abandoned quarry had been permitted without special requirements. Both the permitting agencies and the building's owner had underestimated the reach and magnitude of potential rock falls. No expert geo-scientific evaluation was undertaken. Such an evaluation would have determined the risk (cf., Schroeder, 2003).

- A failure with relatively large damages occurred in Wolfstein (Rhineland-Palatinate), in the borough of Roßbach, in February 2003. After prolonged precipitation, 50,000 m³ of substrate started sliding above several newly constructed houses (fig. 2). Slope movements had occurred already in April 2002. Differential movement and pressure caused shear in the foundations and walls of buildings. Water and gas lines were damaged. Eight inhabitants had to be evacuated from three buildings. Two buildings were damaged. One (built in 1997) was written off and subsequently was demolished. The damage of 250,000 Euro was covered by natural hazard insurance. The slope was stabilized with measures costing 770,000 Euro (communication of the local government of Wolfstein community, July 2003).

Figure 2: Slope movement in Wolfstein - Roßbach (Kusel Department) in February 2003.

- In Balingen (Swaebian Alb), construction measures caused a slide in December 2001 (fig. 3). Various utilities and a telephone line were damaged. A dewatering trench was covered. During the excavation of a foundation pit, liquefiable soil had been stored in a pile, which begun to slide.

- In Hagen (North-Rhine Westphalia), water released from a flooded foundation pit caused soil movements on neighboring properties in January 2003. The resulting settlement of house foundations caused damages. During prolonged rainfall, the pit had not been pumped dry.

Figure 3: Mass movement in a pile of liquefied soils in Balingen (Swaebian Alb).

4 Recognizing Dangers During Development Planning - Examples from Southern Lower Saxony State

In Hann. Münden (southern Lower Saxony State), mass movements have caused significant monetary damages in recent decades (fig. 4, Damm, since 2000). As a result of damages, municipalities have been sensitized to the problem over the last few years. The planning agencies attempt to discover dangers and risks in development areas at an early stage and to ensure that neither the planned construction activity nor the structures themselves present dangers for inhabitants of the area. For the development planning on critical areas they try to realize whether the project is feasible and what slope stability measures were required (vgl. Damm & Pflum, 2004). However, the municipalities do not examine the suitability of individual buildings. Instead, they assume - as prescribed by the Lower Saxony Building Law (NBauO since 1995) - both for buildings that require permits and those that are exempt, that architects and engineers are able to plan buildings that are safe, and are able to evaluate projects with legal certainty. If construction results in damages, private parties are responsible for these damages. Therefore, municipalities generally are responsible only for damages that occur as a result of public construction projects.

The sensitization of planning agencies in Lower Saxony state to the special problems of slope instability in Hann. Münden was a result of scientific studies, direct and indirect communications (conversations with experts, press reports) as well as direct experience from damages to public and private property. The scientific studies and the resulting data have become an important factor in decisions for development planning. For example, land use planning locations with unstable slopes, which had been slated for future development, were removed from the municipalities. In this context, potential conflicts were discussed and weighed in the decisions, because prohibitions on constructions may run counter to the interests of the land owners.

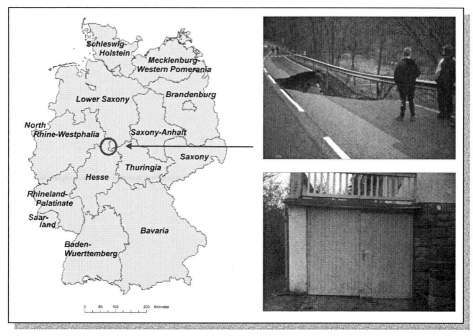

Figure 4: Examples of mass movement in southern Lower Saxony State.

5 Results

Several years of experience with streamlined construction permitting processes have shown that often fundamental technical rules are overlooked. This holds especially true for the evaluation of substrate stability and for the securing of embankments. Experiences with damages resulting from mass movements show that streamlined construction permitting processes carry the risk that general geomorphologic and geologic fundamentals are neglected. Because the streamlined construction law was enacted only a short time ago, there are no statistical records of damages due to mass movement at this time. Those results probably will take further 10 to 15 years to develop.

Determination of risks and education of owners are becoming increasingly important tasks for municipalities, as more and more unstable sites are developed with streamlined permitting processes. The new construction law in all German states no longer requires individual permits for most buildings up to the high rise limit, including small commercial structures. This shifts the responsibility for the safe planning and construction of buildings to the building's owners, the authors of construction plans and the project managers. Special importance thus falls on the qualified planning of new subdivisions, where risks and dangers have to be recognized at an early stage. Planning authorities have to ensure that neither the planned construction activities nor the proposed buildings pose any risks for residents. Special care also is required where buildings are added as infill on unstable sites.

In individual cases, municipalities have become sensitized toward geomorphologic natural hazards and mass movement. This was caused among others by education, but also by direct experience with damages on public and private property caused by slope movements. The increased risk awareness in some cases led to the abandonment of construction plans in locations prone to sliding. More often, it resulted in a more thorough evaluation of the substrate with respect to the stability of the entire site, as well as engineering requirements and limitations of use.

6 Consequences and Recommendations

When weighing public and private interests, geomorphologic hazards gain increasing importance as the conflicts become more focused. Considering that more and more construction projects are located on difficult sites, the decisions of planning agencies can have far-reaching consequences: for private citizens, this goes all the way to financial ruin. Governmental agencies thus should aim to optimize processes with respect to avoiding damages, not just to streamline permitting processes. Current laws offer sufficient possibilities for this. Based on the experience from southern Lower Saxony State, the following recommendations can be made:

- Sensitize employees of agencies toward geomorphologic and hydrologic natural hazards.
- Train employees in geomorphologic and climatologic-hydrologic processes and their consequences for the occurrence of natural hazards.
- Evaluate (historic) data on natural hazards, as well as collect these data in databases.
- Give more consideration to *geo-scientific* expertise in decisions on construction on unstable sites.
- Consider and include knowledge of local experts and local information sources.
- Define latitudes and guidelines for evaluations, based on political discussions.
- Ensure that planning and permitting agencies and external consultants offer targeted consultation to those owners affected by geomorphologic hazards, such as mass movements.

References

Becht, M. and Damm, B., Hrsg.: Geomorphologische und hydrologische Naturgefahren in Mitteleuropa. *Z. Geomorph. N.F., Suppl.- Bd.* 135, 2004.

BayBO Erfahrungsbericht: Erfahrungsbericht BayBO 1998 vom 6. Mai 2002, pp. 1- 50. Bayerisches Staatsministerium des Innern, München, 2002.

Bayerische Staatsregierung: Bayern reformiert - Maßnahmen zur Verwirklichung der Baurechtsreform, www.bayern.de, 2004.

Boeck, T., Itzeck, H. and Rührmund, K.: Gründungsversagen durch Fehleinschätzungen des Baugrundes. In *Ber. 14. Tagung Ingenieurgeologie, Kiel 26. – 29. März 2003*, pp. 221- 225, 2003.

Damm, B.: Hangrutschungen im Mittelgebirgsraum – Verdrängte „Naturgefahr"?. *Standort – Zeitschrift für Angewandte Geographie* 24/4:27-34, 2000.

Damm, B.: Rutschungen im Fulda- und Oberweserraum (Nordhessen/Südniedersachsen) - Ursachen, Auslöser und zeitliche Häufungen. In Fiedler, F., Nestmann, F. and Kohler, J., Hrsg., *Naturkatastrophen in Mittelgebirgsregionen.* Verlag für Wissenschaft und Forschung, Berlin, pp. 129– 147, 2002.

Damm, B.: Die „Altmündener Wand" am Rabanenkopf – 120 Jahre Hangrutschungen – Rutschungsgeschichte, Sanierungskonzepte, Hang- und Verkehrssicherung. *Göttinger Jahrbuch* 50:7-20, 2002.

Damm, B. and Pflum, S.: Geomorphologische Naturgefahren und Raumplanung – Bewertungsprobleme am Beispiel von Rutschgefahren in Südniedersachsen. *Z. Geomorph. N.F., Suppl.-Bd.* 135:127-146, 2004.

Hauth, M.: Vom Bauleitplan zur Baugenehmigung. Beck – Rechtsberater, 6. Auflage, München, 2001.

HBO: Hessische Bauordnung (HBO) vom 18. Juni 2002. *GVBL. L S. 274*, Wiesbaden, 2002.

MBO: Musterbauordnung (MBO), Fassung November 2002. Arbeitsgemeinschaft Bau, Bauministerkonferenz vom 8. November 2002, Wiesbaden, 2002.

MFBS – Ministerium für Finanzen und Bundesangelegenheiten des Saarlandes, HRSG.: Beitrag zur Rede des Ministers anlässlich der ersten Lesung des Gesetzes zur Neuordnung des saarländischen Bauordnungs- und Bauberufsrechts am 14.05.03 im Landtag des Saarlandes. Saarbrücken, 2003.

Moog, W.: Landesbauordnung Rheinland-Pfalz. LBauO 1999. Textausgabe, 4. Auflage, Düsseldorf, 1999.

NbauO: Landesbauordnung Niedersachsen, 7. Gesetz zur Änderung der NbauO vom 15. Juni 1995 in *Nds. GVBl. S. 158*. Letzte Novelle vom 11. Dezember 2002 in *Nds. GVBl. S. 796*, since 1995.

SBUB: Erfahrungsbericht Landesbauordnung Bremen, Teil I u. II, pp. 1- 62. Senator für Bau und Umwelt Bremen. Bremen, 1998.

Schroeder, U.: Geotechnische Schadensfälle als Ergebnis vereinfachter Genehmigungspraxis. Seminar „Geotechnische Untersuchungen und Berechnungen im konstruktiven Ingenieurbau und im Erdbau des Straßenbaus", 11. März 2003, Kaiserslautern, 2003.

Witt, K. J.: Das Baugrundrisiko aus geotechnischer und vertragsrechtlicher Sicht. Vortragskurzfassung 4 p. *VSVI-Seminar „Schadensfälle" 7. November 2002.* Emmelshausen, 2002.

Vulnerability, Perception and Reaction of Companies in the Case of Natural Disasters

Isabel Seifert

Postgraduate Programme "Natural Disasters", Institute for Finance, Banking and Insurance, Chair of Insurance, University of Karlsruhe, Kronenstrasse 34, D-76133 Karlsruhe, Germany

Abstract

High damage due to floods can be reduced by taking precautionary measures in time. A requirement for this reaction to flood events is the awareness of flood as a risk. This study investigates the factors influencing risk perception and the reaction to risk of chemical companies in Baden-Wuerttemberg, Germany.

Keywords: risk perception, risk reaction, damage potentials

1 Introduction

Floods occur irregularly in Germany but they have, together with storms, a high damage potential. Types and magnitude of damage depend on one hand on the flood characteristics (water level, duration of flooding, mechanical forces, water contaminations) and on the other hand on measures taken to prevent losses.

During flood events, chemical companies face not only the risk of damage on one own's property but also a great risk to contamination of the environment. Historically, companies have been aware of the business risks of chemical production thus they developed a lot of tools for assessing and reducing risks. Flood risk is counted as an external risk which means that only the damage potential can be affected but not its probability of occurrence.

Whether and how chemical companies recognize flood as a risk and what they do to protect themselves and to reduce their damage is the subject of the presented study. Factors are to be determined which influence the risk perception and the reaction to it and loss potentials are to be detected.

2 Perception of and Reaction to Floods

The reaction to flood risk depends on the awareness of risk and how risk is perceived. Risk perception in the context of this study is understood on one hand as an intuitive process and on the other hand as a scientific assessment of probability and impact of a flood. WBGU (1998) does "not consider it useful to blur the distinction" but subjects of this study are both big companies which have the capacity to do a scientific risk assessment and small up to middle size companies where risk perception is more like an intuitive process due to the lack of, e.g., personal capacities.

Factors which can influence risk perception of companies might be earlier experiences, source and kind of information on flood risk, application of risk analysis methods, fear of the negative consequences of a flood event and cultural background. Risk awareness includes the previous perception of risk and finds its expression in the reaction to the risk.

The reaction to flood risk itself can be separated into measures taken before (ex-ante), during and after (ex-post) the disaster (Mechler, 2003). What kind of and to what extent damage-preventing or -reducing measures are taken depends on the importance of the flood risk for the company which is again depending on flood risk perception.

3 Flood Situation in Baden-Wuerttemberg

The risk derived from flood is classified as a risk with low probability of occurrence but high extent of damage with a high reliability of estimation for both factors (WBGU, 1998). This definition is applicable to storm floods and river floods but not for flash floods which can occur anytime and everywhere but mostly show a minor damage potential compared to river or storm floods.

Storm floods happen only in coastal areas. They are caused by strong winds blowing inland and thereby pressing also the water inland. Their damage potential is very high but high damage has become rare during the last decades thanks to improved warning systems and coastal protection.

By contrast, flash floods can happen anywhere. They are caused by sudden heavy rainfall which cannot be absorbed totally by the ground. The consequence is a quickly increasing surface runoff which might be intensified by a surface covered by ice or sealed through impermeable materials like asphalt, e.g., in urban areas. Also an already saturated underground can intensify the surface runoff because the soil can not absorb additional water. Flash floods mostly last only hours. Short warning times in the range of minutes are the main problem of flash floods because they make direct damage-preventing measures nearly impossible.

River floods happen, as the name says, along rivers. They rise and fall slower than flash floods so that the warning times are at least in a time range of several hours. The magnitude of a river flood event depends on many factors: kind, location and quantity of precipitation, the recent precipitation in the area, the size and nature of the river catchment area, the shape of the river bed. An impermeable ground (e.g., water saturated or frozen soil) or long strong rainfall over a quickly draining catchment area with little water interception can shorten the formation time. A regulated river stream leads to increased flow velocity and therefore to higher mechanical damage.

Baden-Wuerttemberg is affected by flash floods and river floods. Flash floods happen every year but mostly only small areas are affected so there is hardly any press release to be read. Along the main rivers Neckar, Rhine and Danube, the last severe floods happened in 1995 and 1999. Tributaries of Rhine and Neckar show higher water levels, mainly after heavy rainfall (often in combination with flash floods).

4 The Chemical Industry in Baden-Wuerttemberg

Baden-Wuerttemberg, located in south-western Germany, together with Bavaria has one of the strongest economies of all states in Germany. Most people are working in mechanical engineering. In 2003 this branch, together with vehicle construction, accounted for nearly 50% of the total turnover of manufacturing industry in Baden-Wuerttemberg. The chemical industry represents only 6.2% of the total turnover of the state. But in relation to the number of employees, after vehicle construction, this branch has the second highest turnover per employee (Statistisches Landesamt Baden-Württemberg, 2004).

Within the branch of chemical industry nearly 50% of companies produce pharmaceutical, body-care and washing products. About 40% of the companies work in the area of basic or speciality chemistry, including the production of rubbers, plastics, glue, mineral oil,

Region	Rivers
Freiburg/Hochrhein	Rhine, Wiese, Elz
Ulm/Biberach	Danube, Ri
Mannheim/Karlsruhe	Neckar, Rhine
Stuttgart	Neckar, Enz, Nagold, (Fils, Rems)

Table 1: Rivers in the "centers of chemical production" in Baden-Wuerttemberg (source: Chemie in Baden-Württemberg (2003))

Function	Direct Damage/ Reason for Interruption	Indirect Damage (microeconomic scale)
Housing	Building damage	Business Interruption
Production	Damage to Factory Equipment (e.g. machines or computer)	Business Interruption
Transportation	Damage to Motor Vehicles	Discontinuous supply or delivery chain
	Damage to Transport Infrastructure	Employees cannot reach factory
"Natural Environment"	Contamination (e.g. soil)	
		Decrease in Turnover
		Loss of confidence of stake holders

Table 2: Example for classification of damage-types after floods

fiber, varnish and dye (Chemie in Baden-Württemberg, 2003). In the following, the terms "chemical industry" and "chemical companies" are used to name the whole branch.

Chemical companies are not distributed equally over the whole area of Baden-Wuerttemberg. There are four centers of chemical production: The regions of "Freiburg/ Hochrhein", of "Ulm/ Biberach", of "Mannheim/ Karlsruhe" and Stuttgart and its surrounding area. Rivers flowing through these areas are shown in Table 1.

5 Loss Potentials

Damage caused by flood events can have varied causes: the water itself, contaminations in the water, mechanical forces due to the velocity and transport capacity of the flowing water, further water damage which causes consequential damage. This damage is classified as direct damage because it is a direct result of the physical impact of the water to the infrastructure. An advanced classification of damage-types could be made after the function of the destroyed subject (Table 2).

Indirect damage or losses are a consequence of direct damage. On microeconomic scale indirect losses are mainly business interruptions. They could be classified more exactly according to the reason for the interruption (Table 2). Another type of indirect damage is the loss of confidence of stake holders which can have both economic consequences (e.g., declining stock price, decreasing sale) and social consequences (e.g., unsettled staff, lawsuits by citizens). Indirect losses on macroeconomic scale are the effects of disasters to the regional or national economy. These losses are hard to measure. Dacy (1969) mentions a decrease in tax revenue in industrialized countries.

6 Floods and the Chemical Industry

For all risks originating from the production of chemicals it is either possible to reduce the probability of occurrence or to limit the damage. Nevertheless, there is always a residual risk depending on the measures taken to reduce it.

Flood risks cannot be influenced in their probability of occurrence but to a certain degree in their magnitude of damage. Possible measures to reduce damage before a flood event are mostly constructional measures, like dyke construction, dislocation of critical infrastructure (e.g. computers) or goods, prevention of upwelling of buildings or tanks, sealing of doors and windows and backwater protection. Constructional measures are designed for a certain magnitude of flood. That means that a known residual risk is accepted.

Organizational measures can help to reduce damage shortly before the flood (e.g. after the flood warning). To guarantee a smooth course of action under time pressure the flood emergency must be planned and trained regularly. A flood contingency plan can include co-operation with relief organizations like fire brigades to keep pumps and sand bags ready and to move mobile goods to higher places. Insurance contracts are among ex-ante measures which help to reduce financial losses after a flood event and can help the company to come quickly back to their daily business. Due to big accidents in the past with personal, economic and ecological damage, today chemical companies have to fulfill a great deal of safety regulations to run their business. But compared with other industry branches the chemical industry still faces a higher risk of contaminating the environment with a wide range of chemical substances. Large companies in particular have devised methods for analyzing risk potentials, and have developed tools to minimize the probability of occurrence of hazardous incidents. With increasing safety they reduce at the same time their damage potential.

7 Approach of this Study

The goal of this study is to investigate the perception of and reaction to flood risk in chemical companies in Baden-Wuerttemberg. Is the risk awareness for external risks as flood as high as the awareness for internal business risks? How is the flood risk estimated and what are the consequences of this estimation? Further points of interest are how the decision about the acceptable residual risk is derived and what measures are taken to limit damage potentials.

I expect to find well prepared companies which can serve as examples as well as companies which totally ignore their flood risk. Here the question is what factors influence risk perception and reaction of companies.

Another aspect of this project is to collect data about the vulnerability against flood damage of chemical factories in Baden-Wuerttemberg and compare them with other data recorded in Saxony and Saxony-Anhalt after the big flood along the river Elbe in 2002.

References

Chemie in Baden-Württemberg: Branchenstruktur. Available electronically at http://www.chemie.com/branchen0.html, 2004.

Chemie in Baden-Württemberg: Branchenstruktur. Available electronically at http://www.chemie.com/region_schwerp.html, 2004.

Dacy, D. C. and H. Kunreuther: *The Economics of Natural Disasters: Implications for Federal Policy.* Free Press, New York, 1969.

German Advisory Council on Global Change (WBGU): *World in Transition: Strategies for Managing Global Environmental Risks.* Annual Report 1998, Springer, Berlin, Heidelberg, New York, 2000.

Mechler, Reinhard: Natural Disaster Risk Management and Financing Disaster Losses in Developing Countries. *Karlsruher Reihe II: Risikoforschung und Versicherungsmanagement*, Eds.: R. Schwebler, U. Werner. Karlsruhe, 2004.

Statistisches Landesamt Baden-Württemberg: Verarbeitendes Gewerbe, Bergbau und Gewinnung von Steinen und Erden in den Stadt- und Landkreisen von Baden-Württemberg 2003. Artikel-Nr. 3526 03001, Statistisches Landesamt Baden-Württemberg, 2004.

Local Experience of Disaster Management and Flood Risk Perception

Tina Plapp
Postgraduate Programme Natural Disasters, Institute of Insurance, University of Karlsruhe, 76128 Karlsruhe, Germany

Abstract

According to various studies risk perception is subject to many influencing cognitive, personal, situational and contextual factors. This paper focuses on the interrelation between cognitive factors of flood risk perception and local experience of loss and experience with disaster management in a particular social context. The observations presented in the paper originate from a field study on perception of risks from windstorm, flood, and earthquakes conducted in summer 2001 in six regions of Germany. To study risk perception, the so-called psychometric paradigm was applied among other approaches to risk perception. In this paper, the interrelation between risk perception and experience persons living in three affected flood areas are compared regarding their cognitive flood risk perceptions and loss experience. Then, the observed differences are contrasted with the respective local hazard experience.

Keywords: risk perception, flood, psychometric paradigm, empirical study.

1 Introduction

Risks from natural hazards (earthquake, windstorm, flood, landslides, avalanches, etc.) have not been the major topic of interest in risk perception research which has focused more on technical, chemical, environmental risk and risks from everyday or recreational activities (Slovic 1992, Peters/Slovic 1996, Vaughan/Nordenstam 1991, Siegrist 2000). Yet, increasing losses from disasters, rising risk potential due to climate change, and the growing vulnerability of societies require knowledge of the potentially affected peoples' risk perceptions to develop efficient risk management strategies.

In this paper, risk perception is referred to as the process of attributing risk to an object, a situation or an action. From a social science perspective, consequently, risk perception is a construction process. "Risk perception is about ideas, attitudes and beliefs." (Sjöberg 2000a, 408). Since our empirical experiences and also the way how we word our experiences are embedded into and determined by the society and culture we belong to, risk perception is as well socially constructed (Lupton 1999, 33). Risk is defined as a multidimensional concept that combines everyday, subjective quantitative assessments based on experience and information as well as risk characteristics that are perceived or attributed within a certain context (Renn 1995, 27-28).

Affected peoples' perception of risk is subject to many factors. According to various studies on the perception of technical and environmental risks, there are several factors which influence risk perception (Sjöberg 2000): the characteristics of the hazard itself that shape cognitive risk components (Slovic 1987), worldviews and social ways of life (Douglas/Wildavsky 1983, Dake 1991, Peters/Slovic 1996), trust in authorities (Siegrist 2000), ethnic-cultural and socioeconomic background (Vaughan/ Nordenstam 1991), and personal variables such as gender and profession (Barke et al. 1997). Because of its complexity and variability, it is very difficult to deduce general statements or a general theory of risk perception. Nevertheless, knowledge about the risk perception of potentially affected people is relevant whenever risk management strategies are developed or applied. Affected peoples' risk perception is, among others, a principal base for decisions and mitigation behaviour with regard to natural hazards (Mileti/Fitzpatrick 1992).

The main purpose of the paper is to focus on the relationship between cognitive factors of risk perception and the relevance of local experience in a specific social context. The empirical study to investigate risk perception is outlined in Section 2. In Section 3, respondents from three affected flood areas are compared regarding the cognitive components of their flood risk perceptions and their loss experiences. Afterwards the observed differences are contrasted with the respective local situation and experiences of disaster management (Section 4).

2 Methods

Risk perception from windstorm, flood, and earthquake were studied in a mail survey conducted in summer 2001 in six different areas. All six survey areas in West and Southern Germany have been affected by flood, windstorms and / or earthquake within the last 30 years: Cologne-Rodenkirchen, Passau, Neustadt / Donau, Albstadt, Karlsruhe, and Rosenheim.

In the research design, several approaches used in risk research to study risk perception from technical, chemical and environmental risks as well as risks from daily or recreational activities were used to approach risk perception from natural hazards, among them the psychometric paradigm (Slovic 1987, 1992). The psychometric paradigm was designed to explore cognitive components of risk perception and has seldom been applied in the field of natural hazards. The questionnaire covered the ratings of general dangerousness, the judgement of the hazard on the basis of nine predefined risk characteristics, and the experience from windstorm, flood, and earthquake. A total of 450 persons returned the questionnaire (for details of the research design see Plapp, 2004).

In the sample of 450, 273 persons live in areas that have been affected at least once by flood. Out of these 273, 114 persons live in Cologne-Rodenkirchen at the river Rhine, 96 persons live in Passau and 64 in Neustadt, both located at the river Danube. Cologne Rodenkirchen is a district in the flood area of the city Cologne (circa 1 Million inhabitants). The survey area in Passau (circa 50.000 inhabitants) covers the flood area of the old town centre which is located at the confluence of the rivers Dabube, Inn and Ilz. The study area in Neustadt (circa 12.000 inhabitants) covers the area which was inundated in 1999 due to a dam failure. Compared to the total population of Germany and compared to the respective communities' population, older persons are slightly over represented in the samples of Cologne-Rodenkirchen and of Neustadt. In the Passau sample, younger persons (most of them students) are over represented compared to the total population of Passau. The education level is higher compared to the total German population in all three area samples. Both the high proportion of older people and the high education level are a typical feature of the self-selection effect of mail surveys, but the effects of age and gender are considered as having no major influence on risk perception compared to other factors (Sjöberg 2000b).

In Section 3, the significant differences between the three groups are presented. The groups were compared regarding their general assessment of dangerousness of flood which was measured on a scale from 0 to 100 (*general risk assessment*). The significant differences between the groups were tested by analysis of variances (ANOVA). Then, the groups were compared regarding the perception of risk characteristics. Respondents were asked to judge flood on the following nine pre-defined, qualitative risk characteristics on a

5-point scale: *personal risk* (i.e. if flood is seen as a risk that threatens the respondent personally), the subjective estimated *probability to die* from an event, the degree of *scientific knowledge* about the hazard risk, the *familiarity* (old or new risk), if the risk is associated with emotions of *fear* ("Angst"), the perceived range of possibilities to *influence* the risk, the subjective estimated *frequency* of occurrence, the perceived *predictability* of events and if *future increase or decrease* is expected in terms of events and losses (for the risk characteristics, see also Brun 1992). In addition to analysis of variances, pair-wise non-parametric tests (U-Test) were employed to study differences between the three area samples.

Besides, a descriptive analysis of loss experiences in the three sample groups was performed to relate risk perception to experience. In addition to quantitative data analysis, the social history and recent local experience with flood were deduced from a content analysis of several documents, mainly newspapers and other accessible sources. The results are presented in Section 4.

3 Risk Perception from Flood and Loss Experience

As shown in Table 1, the ratings of perceived dangerousness (general risk assessment) from flood differs significantly between the respondents of the three flood areas (ANOVA, Duncan-Test on 5%-level). Against the expectations and against the finding, that persons with loss experience rated the dangerousness of flood higher (see Table 1), the respondents from often inundated areas in Cologne-Rodenkirchen (Rhine) and Passau (Danube) rated the general flood risk lower than respondents from the recently for the first time flooded Neustadt (Danube). This result could be explained with a familiarization effect to flooding of the often affected areas (Cologne, Passau).

With a closer look on the data, these findings could also be explained by loss experience from floods reported from the areas: the most loss experiences were reported by the respondents from Neustadt, the second most from those of Cologne, and the respondents from Passau reported the least loss experiences (see Table 2). The proportion of respondents reporting loss corresponds to the magnitude of the general risk assessment in the three areas. As reliable information on the amount of losses was not available, the relation between risk perception and extent of loss could not be investigated.

		General Risk Assessment of Flood *Median (scale 0-100)*	No. of respondents
(A)	Köln-Rodenkirchen	59	114
	Neustadt/D.	76	63
	Passau	44	96
	Total No. of respondents		*273*
(B)	All respondents from affected flood *without* loss	52	138
	All respondents from affected areas *with* loss	69	135
	Total No. of respondents		*273*

Table 1: General Risk Assessment (perceived dangerousness) of flood in three flood areas (A) and by persons with and without loss experience from flood (B).

	No. of respondents with loss	No. of respondents without loss	*Total No. of respondents*
Köln-Rodenkirchen	57	57	*114*
Neustadt/D.	55	9	*64*
Passau	23	72	*95*
Total No. of respondents	*135*	*183*	*273*

Table 2:   Reported experience of loss in the three survey areas.

However, there are further significant differences in the attribution of risk characteristics. Figure 1 shows the average perception of flood regarding nine risk characteristics for the three areas (Median). For five risk characteristics, significant differences between the respondents between the three area samples can be observed (ANOVA, 5 %- and 0.5 %-level): the degree of perceived personal threat, the likelihood of fatal consequences from flood, the familiarity (old or new risk), the degree of fear ("Angst") evoked by the flood risk, and the perceived possibilities to influence and control the risk. Again, the respondents from Neustadt differ clearly from the other two groups. One exception is the risk characteristic of the perceived personal threat which is also conceived as very high by the respondents from Cologne-Rodenkirchen. The observed results in the perception of risk characteristics correspond very well to the differences in general ratings in the perceived dangerousness for Neustadt and Passau, whereas the correspondence for Cologne-Rodenkirchen is somewhat less.

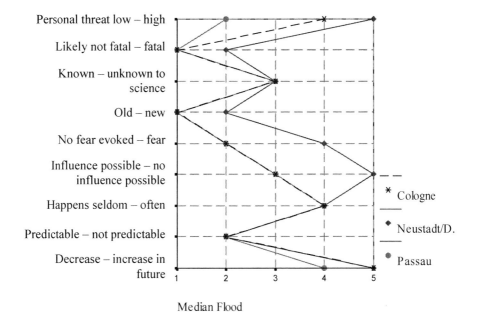

Figure 1: Median of flood risk characteristics in the three areas.

In sum, the respondents split into two groups: on the one side the respondents of Passau and Cologne-Rodenkirchen form the group of those with lower general risk perception, less fear evoked by the risk and the understanding of having at least some possibilities to influence their flood risk. Despite all similarities, the respondents from Cologne-Rodenkirchen perceive a higher personal risk from flood than those from Passau. On the other side, the respondents from Neustadt form the group with high general and personal risk ratings, rather intense fear and the understanding of having no possibilities to influence their flood risk.

4 Local Experience of Disaster Management: Neustadt/Donau versus Passau and Cologne-Rodenkirchen

For further understanding of these observed disparities between the groups, the past events and the respective social hazard history of the areas were studied in content analysis of reports from newspapers, a TV-documentation, and accessible reports of civil protection institutions.

4.1 Neustadt/Danube

Low lying parts of the community of Neustadt / Danube were inundated May 1999 when the dike broke due to overtopping (BLaWa 1999, DRV 1999). The dike was part of a floodplain management system constructed in the 1950/60ies (Neustadt 1987). Since that time the dam system had been able to withstand even the highest discharges of the river Danube in the Neustadt area. The floods stroke a completely unprepared community having a Pentecost folk festival.

From the view of the affected people, adequate reaction and response were impossible because of a bad information policy of the local emergency management: warnings and alert were issued too late. The failure of the dike could have been avoided if the responsible crisis management groups had reacted properly by defending the dike in time. In consequence, the disaster and the loss have to be attributed to the mismanagement and mistakes of the local authorities (BK 1999, Mittelbayerische Zeitung, SZ 1999, ZDF 1999).

In the perspective of the local disaster management team and the emergency organisations, the affected people reacted too late. Despite of the timely alert some of them continued to party at the folk festival which had to be broken up finally by the police. When the dike showed first signs of breaking, evacuation was started and more than 1.000 persons were evacuated (DRLG Bayern aktuell 6/99, BK 1999, SZ 1999, Mittelbayerische Zeitung, ZDF 1999).

Quickly a conflict about who to be blamed for the disaster and the losses arose. Only two days after the inundation, a stakeholders group of damaged persons started a court procedure for compensation of damages (BK 1999). The case was dismissed one year later. Thereafter individual law suits were filed. These lawsuits had not yet been decided by the time the survey was conducted.

4.2 Cologne-Rodenkirchen and Passau

The inhabitants of the flood area of Cologne-Rodenkirchen and Passau and the responsible local crisis management groups had to experience several floods with varying impacts in the last decades (Cologne in 1982, 1993, 1995, 1998; Passau in 1981, 1985, 1991, 1999, see München Rückversicherung 1999). Especially for Cologne-Rodenkirchen it is documented that these experiences were also taken as opportunities to improve emergency response (Pfeil 1999, 2000).

In Passau, people in the flood areas are quite well prepared by experience and the civil protection organisations respond to flood events effectively with routine (StMLU 2002, PNP 2002). Additionally, the fact that many students live in the flood-prone area of the old town, helps to keep the damage and loss in certain limits.

In Cologne-Rodenkirchen, on the one hand, there is a will of the local emergency centre to achieve a better cooperation with civil protection organisations and with affected people during the emergency. On the other hand, a citizen's initiative provides instructions for individual protective measurements and demands participation of affected people in the emergency response activities and the long-time flood risk management decisions (Pfeil 1999, 2000, Pegellatte 4/97+5/97).

Although the respondents from Cologne see at least some possibilities to influence the flood risk, their perception of personal threat is rather high compared to the respondents from Passau, likely due to higher proportion of loss experience. In opposite, the perception of personal threat of the respondents from Passau is rather low. This result may also be caused by the particularity of the Passau sample. The sample contains a large proportion of young persons, most of them students, and high education level and younger age corresponds with lower risk perception (Plapp 2004).

5 Summary and Conclusion

Quantitative data analysis show significant differences in the risk perception of the 273 respondents from flood regions: while the respondents living in areas flooded almost regularly (Passau, Köln-Rodenkirchen) regard the flood risk as quite low, the flood risk is rated rather high by those persons living in a small town being inundated the first time since the 1950ies (Neustadt/Donau). These results cannot be explained by reported losses, familiarisation to frequent flooding, emotional blunting effects or other psychological adjustment strategies alone. Also a false sense of security evoked by the dike (SwissRe 1998) in the Neustadt area could have an effect on the results.

Moreover, both the local history of flood events and the disaster management of these events have to be taken into account: while experience in the sense of damage experience seems to increase risk ratings, the experience of a functioning disaster management, at the same time, seems to put risk perception into another perspective. Hence, when concepts for local disaster management are designed to allow cooperation and participation with the affected persons, these concepts contribute to the process of building trust and confidence in the coping capabilities of the responsible institutions.

Regarding the psychometric paradigm as dominant approach in risk perception research, the results show that local characteristics play a role which cannot be caught using this approach. Whenever the psychometric approach is applied in the field of natural hazards it is therefore important to ensure that the underlying research design takes local features into account.

Acknowledgements

This project was part of the Interdisciplinary Postgraduate Programme "Natural Disasters". It was carried out at the Institute for Insurance, University of Karlsruhe and funded by the Deutsche Forschungsgemeinschaft DFG and the University of Karlsruhe. The mail survey was additionally supported by the Stiftung Umwelt und Schadenvorsorge, SV

Versicherungen, Stuttgart, Germany. I would like to thank all respondents for their time and patience to answer the questionnaire.

References

Barke, Richard P., Hank Jenkins-Smith, Paul Slovic: Risk Perceptions of Men and Women Scientists. *Social Science Quarterly,* 78: 167-176, 1997

BK 1999: Bürgernetz Kelheim e.V.: Das Jahrhundert-Hochwasser 1999 in Kelheim und Neustadt a. d. Donau, 1999. Available electronically at: http://www.keh.net/hw, 1999

BLaWa 1999: Bayerisches Landesamt für Wasserwirtschaft, Hochwassernachrichtendienst: Pfingsthochwasser. Available electronically at: www.bayern.de/lfz/hnd/ereignisse.htm, 1999

Brun, Wibecke: Cognitive Components in Risk Perception: Natural versus Manmade Risks. *Journal of Behavioral Decision Making,* 5: 117-132, 1992

Dake, Karl: Orienting Dispositions in the Perception of Risk. An Analysis of Contemporary Worldviews and Cultural Biases. *Journal of Cross-Cultural Psychology,* 22: 61-82, 1991

Douglas, Mary, Aaron Wildavsky: *Risk and Culture.* Berkeley et al., University of California Press, 1983

DRLG Bayern aktuell 6/99: Hochwasserkatastropheneinsatz in Bayern: Bewährungsprobe für die DRLG, in: DRLG Bayern aktuell, Offizielles Mitteilungsorgan des LV Bayern der Deutschen Lebensrettungsgesellschaft, Ausgabe 6/99

DRV 1999: Deutsche Rückversicherung AG: *Das Pfingsthochwasser im Mai 1999.* Düsseldorf, 1999

Lupton, Deborah: *Risk.* London/New York, Routledge, 1999

Mileti, Dennis S., Colleen Fitzpatrick: The Causal Sequence of Risk Communication in the Parkfield Earthquake Prediction Experiment. *Risk Analysis*, 12: 393-400, 1992

Mittelbayerische Zeitung. Online-Archiv

Münchener Rückversicherung: Naturkatastrophen in Deutschland. Schadenerfahrungen und Schadenpotentiale, München, 1999.

Neustadt 1987: Stadt Neustadt a. d. Donau, Landkreis Kelheim: Erläuterungen zum Flächennutzungsplan mit integriertem Landschaftsplan. 1987

Pegellatte 4/97 + 5/97, Bürgerinitiative Hochwasser, Altgemeinde Rodenkirchen e.V.

Peters, Ellen, Paul Slovic: The Role of Affect and Worldviews as Orienting Dispositions in the Perception and Acceptance of Nuclear Power. *Journal of Applied Social Psychology,* 26: 1427-1453, 1996

Pfeil, Jan: Maßnahmen des Katastrophenschutzes und Reaktionen der Bürger in Hochwassergebieten am Beispiel von Bonn und Köln. Diplomarbeit, Geographische Institute der Rheinischen Friedrich-Wilhelms-Universität Bonn, 1999

Pfeil, Jan: Maßnahmen des Katastrophenschutzes und Reaktionen der Bürger in Hochwassergebieten am Beispiel von Bonn und Köln. DKKV / German Committee for Disaster Reduction, Bonn, 2000

Plapp, Tina: *Wahrnehmung von Risiken aus Naturkatastrophen.* Karlsruher Reihe II Risikoforschung und Versicherungsmanagement, Bd. 2, Verlag für Versicherungswirtschaft, Karlsruhe, 2004

PNP 2002: Bericht der Freiwilligen Feuerwehr Ries/Passau: Hochwasser-Gefahr ist gebannt - jetzt geht's ans Aufräumen. Available electronically at: http://home.t-online.de/ home/markus.schwarz/berichte_pnp25030

Renn, Ortwin: Individual and Social Perception of Risk. In Fuhrer, Urs, editor: *Ökologisches Handeln als sozialer Prozess*, pp. 27-50, Basel, Birkhäuser, 1995

Siegrist, Michael: The Influence of Trust on Perceptions of Risk and Benefits on the Acceptance of Gene Technology. *Risk Analysis*, 20: 195-203, 2000

Sjöberg, Lennart: Factors in Risk Perception. *Risk Analysis*, 20: 1-11, 2000a

Sjöberg, Lennart: The Methodology of Risk Perception Research. *Quality and Quantity*, 34: 407-418, 2000b

Slovic, Paul: Perception of Risk. *Science,* 236: 280-285, 1987

Slovic, Paul: Reflections on the Psychometric Paradigm. In Krimsky, Sheldon, Dominic Golding, editors: *Social theories of risk,* pp. 117-152, Wesport: Praeger, 1992

Slovic, Paul: *The Perception of Risk*. London/Sterling, Earthscan, 2000

StMLU 2002: Bayerisches Staatsministerium für Landesentwicklung und Umweltfragen, Pressemitteilung vom 22. März 2002: Hochwasser an der Donau in Passau

Swiss Re: *Floods - an insurable risk?* Zurich, 1998

SZ 1999: „Saumäßiger Dusel". Helfer bei der Hochwasser-Katastrophe feiern und erinnern sich. In Süddeutsche Zeitung Nr. 174, 31.8.1999

Vaughan, Elaine, Brenda Nordenstam: The Perception of Environmental Risks Among Ethnically Diverse Groups. *Journal of Cross-Cultural Psychology,* 22: 29-60, 1991

ZDF, 1999: „Behörden schuld am Dammbruch?" *Frontal*, 7.9.1999

Project Man-Made Hazards (CEDIM) – First Results

Ute Werner[1], Christiane Lechtenbörger[2] Dietmar Borst[1]

[1] *Institute for Finance, Banking, and Insurance (FBV), Dept. for Insurance,*
Universität Karlsruhe (TH), D- 76128 Karlsruhe, Germany;
[2] *CEDIM, Head Office, Universität Karlsruhe (TH), D- 76128 Karlsruhe, Germany*

Abstract

The Risk Map Germany, as the first project of CEDIM, consists of 7 subprojects. Six groups are focusing on natural disasters whereas the research group "man-made hazards" is concentrating on human induced catastrophes, especially malicious actions. This is a very new research topic in Germany. Therefore, as a starting point, an inventory is being conducted of critical infrastructure in densely populated areas. To follow will come analyses of accidents, near-miss-events and former attacks. All information is stored in data bases and linked with a GIS for visualisation. Along with mapping the hazards one main goal of this group is to account for potential damages of events, as "unthinkable" as they might be. First results of the inventory and the analysis of former events are presented, as well as first concepts of scenario outlines.

Keywords: critical infrastructure, Man-Made Hazards, malicious actions, inventory, risk mapping, GIS, scenarios, CEDIM.

1 Introduction

The "Man-Made Hazards" research group is a multidisciplinary team with scientific backgrounds in business management (especially risk management and insurance), economical engineering, geography, remote sensing, and mathematics. It was constituted during the founding period of CEDIM[1] in order to contribute to the first common project of this institute, called "Risk Map Germany". Astonishingly, until end of 2002, no integrated mapping of natural and man made hazards in Germany existed or was publicly available.

Man-Made Hazards are induced by technical, human or organisational failures and malicious actions of persons or groups. They are mapped alongside hazards linked to earthquakes, floods, storms, space weather and network infrastructures. Mapping in this context means allocating areas that might be hit by extreme natural events, industrial accidents or intentional attacks such as sabotage or terrorist actions.

The mapping of Man-Made Hazards is a very new and distinct field of research in Germany. Since historical and statistical data on this type of hazard is scarce, approaches of data collection, analysis, as well as hazard and risk modelling have to be developed. The research group has, in fact, to start at the very beginning; this is explained in the paper.

---

[1] Center for Disaster Management and Risk Reduction Technology, situated in Karlsruhe and Potsdam, co-founded by University of Karlsruhe (TH) and GeoforschungZentrum Postdam (GFZ).

2 Goals of the Project

Risk Map Germany as the first project of CEDIM aims at providing information about hazards, i.e. their location, frequency and intensity. This information has to be integrated in order to model interdependencies of certain hazards, e.g. the conjunction of heavy storm and flood events, or the damage potential due to earthquakes hitting in densely industrialized regions. This hazard, of course, is co-created by humans,[2] whereas terrorist attacks on important infrastructure – just to mention one example - are completely man-made. The vision of *Risk Map Germany*, therefore, is to map the *risk* defined through hazard and vulnerability: An important part of the work is concentrated on modelling and estimating losses - especially monetary losses - due to catastrophes, and their consequences on human beings, infrastructure and nature.

Man-Made Hazards are difficult to model since they don't follow meteorological, hydrological or seismological patterns. Their characteristics vary in time, alongside with technological innovations or changes in human values and behavior. Moreover, malicious actions are patterned to be unpatternable – consequently it is hard to collect substantial pre-existing data about this kind of hazard. The statistical situation is somewhat better in the realm of accidents: Here, reliable aggregate data does exist for the past thirty years.[3] This observation period is, however, quite short, and since surrounding conditions such as risk management efforts are changing permanently, statistical analysis becomes very challenging.

The research group decided, therefore, to build up its own data base on critical infrastructure in Germany (cf. definition under point 3). Parallel to these efforts data about past accidents (chemical, nuclear etc.), malicious actions and so-called near-miss events[4] is being searched and analyzed. Events have scripts, initiating causes, participants with certain characteristics; they occur in space and time and produce effects that can be recorded along with data on intervening actions or other features of the event. This approach serves to identify major accident hazards and to deduce reference scenarios for accidents as well as scenarios for malicious actions (cf. points 4 and 5).

The risk map referring to Man-Made Hazards will show "hot spots", i.e. geographically coded sites where the potential of accidents or malicious attacks leading to disastrous effects is high. In a first step, effects considered are direct property losses and number of evacuated, injured or dead persons. In the long run it is intended to estimate the total loss potential in certain geographical areas, including damage propagation via interruption of businesses or life-lines such as energy supply (cf. Point6). Table 1 shows the different work packages.

3 Inventory of Critical Infrastructure

Critical Infrastructure can be defined as "…critical in that their incapacitation or destruction would have a debilitating effect on the nation's defense or economic security. These are: telecommunications, electrical power systems, gas and oil storage and transportation, banking and finance, transportation, water supply systems, emergency services, and continuity of government" (Haimes and Longstaff, 2002; cf. President's Commission on Critical Infrastructure Protection, 1997).

---

[2] Since we have collected enough knowledge so far to be able to distinguish between zones in Germany that are more or less prone to extreme natural events, we can and do shape the vulnerability of settlements, industry, environment (see Mileti, 1993).

[3] Compare publications of professional reinsurers such as Munich Re or Swiss Re.

[4] Failures or malicious actions not resulting in considerable damages.

Damages to this kind of infrastructure threaten to interrupt processes, systems or functions important to society, its institutions, and to individuals as well. It is therefore much more than defense and economic security that might become incapacitated: The performance of a broad variety of tasks can be restricted, depending on the kind, place, and duration of the potential failure.

Break-downs or initial disruptions due to *accidents* call for investigations into possible causes, whether technical, organizational or human. They tend to generate distrust in experts such as engineers, managers, or politicians, often leading to claims for changes in policy.

Steps	*Details*
I Inventory of Critical Infrastructure like	Nuclear power plants industrial plants life-lines like water and energy supply, traffic lines financial services government
II Reference Scenarios for Accidents	start with industries under review identification of major accident hazards
III Scenarios for Malicious Actions	Based on case studies, accident scenarios, and sketches of extreme events
IV Estimation of Immediate Loss Potential	Direct property losses (in €/area) personal damage (kind and number)
V Integration of Damage Propagation	via links to other subprojects via cooperation with external experts
VI Estimation of Total Loss Potential	exposure to direct and indirect losses of various magnitudes (incl. cumulative processes such as business interruption)

Table 1: Work packages of the project "Man-Made Hazards".

People with *malicious* motives are using weak spots, i.e. parts of infrastructure that can be reached easily and where even small, inconspicuous interventions will produce considerable damages of all sorts. Intention, therefore, alongside with prospective planning, is the distinguishing characteristic between disasters caused by extreme natural events or by malicious human acts.[5]

---

[5] In both cases, the vulnerability to disastrous consequences is furthered by humans, e.g. through settlements in regions subject to extreme natural events, or via interdependencies created by complex systems.

Terrorist acts, especially, aim at forcing/imposing/impressing political, economical or social changes that cannot be achieved otherwise.[6] RMS (2002) describes the problem as following: "The geography of terrorism hazard is defined not by coastlines or faults, but by the correlated dynamics of several factors. First, the 'terrorists' clear desire to maximize the 'utility'of their attacks, inflicting extreme damage and disruption in deliberately planned and executed attacks; second, their ability to obtain and deploy various classes of weapons; and third, their rational responses to, and anticipation of the moves and counter-moves of the security and intelligence services".

Keeping in mind these characteristics, an inventory of industrial sites - including chemical industrial sites – was set up for a pilot region, the Rhine-Neckar-Triangle. The documentation of "Financial Service Institutions" as another category of critical infrastructure is underway for the Frankfurt Area. As this type of infrastructure is very different from industrial plants, the parameters for the documentation had to be modified.

The data-files archived in the common data base consist of information about production processes, substances, emergency plans, number of employees etc. for industrial sites. Financial service institutions are covered through number of employees, building characteristics like height, branches inside. In addition to that work all nuclear power plants in Germany are archived using descriptive parameters such as type of reactor, emergency plans or nuclear charge (cf. fig. 1). A nationwide inventorying of football stadiums is also underway. It accounts for their size and type of usage (German football league).

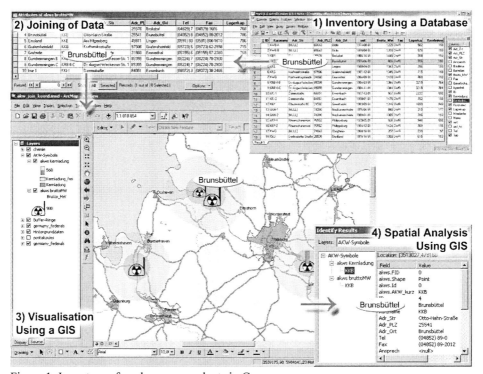

Figure 1: Inventory of nuclear power plants in Germany

---

[6] Worldwide reactions to the 9/11 attack in 2001 show that the states of mind of many people can be altered through terrorist acts, leading to extensive changes in policies and legislation.

These first examples show how the process of documentation is laid out: Different categories of critical infrastructure are selected independently of their spatial location in Germany. Another point of consideration are hazards situated in densely populated areas. This allows a) the creation of thematic layers within the data bank, and b) an arrangement of the archiving process based upon overall damage potentials.

Fig. 1 demonstrates the inventorying process starting with (1) archiving the data in a data base via (2) integrating geographical and thematical oriented data, then (3) visualization via GIS, and ending with spatial queries (4) in a GIS (ArcGIS).

This information serves as input for scenario development as well as for the assessment of loss potentials (steps IV to VI of the work packages).

4 Reference Scenarios for Accidents

One approach in estimating the probability of future accidents and their catastrophic outcomes is to analyse past events: considering causes and circumstances leading to injuries to human life, fatalities, as well as property damage.

Man-made hazards can be categorized according to the following scheme (Swiss Re, 2003):

- Fires, explosions and further industrial plant accidents,
- Accidents in air traffic,
- Accidents in shipping,
- Accidents in ground traffic (railway and road traffic),
- Mining accidents,
- Structural collapses,
- Others (including malicious actions).

Starting with *industrial* accidents (of major concern both for Germany as a heavily industrialized country and for our projects' pilot region Rhine-Neckar-Triangle) and *air traffic* events[7], data was collected on a global as well as on a national scale. For a detailed understanding of different subtypes of events, case studies were examined. Quantitative information was extracted from various data bases.

Both on a global and national scale, the amount of data available, and the quality of the records, increase for events dating back shorter periods of time. This is probably due to improved data exchange technologies and an enhanced commitment to systematic methods of record keeping in recent times. Concerning certain *types* of events like accidents in chemical plants[8], there hardly exist any publicly available databases containing information from the whole world. Databases on a global scale like EM-DAT[9] are nevertheless valuable for extracting general trends regarding frequency and characteristics of aggregated disaster types, e.g. for industrial accidents of all kinds versus traffic related disasters.

---

[7] thus illustrating the small distance between accidental hazards and those of malicious origin, i.e. September 11th, 2001
[8] of special interest for the project, since this kind of critical infrastructure is to be documented in the inventory
[9] EM-DAT: The OFDA/CRED International Disaster Database, www.em-dat.net

Specialized databases on a global scale do exist in the field of (commercial) air traffic[10], thus facilitating analyses of different courses of events as well as research of factors triggering accidents worldwide. Figure 2 gives an impression of the level of detail provided by the two databases mentioned above.

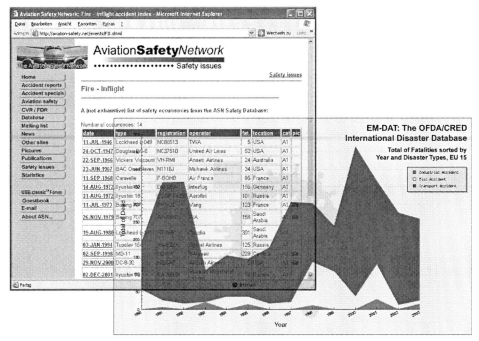

Figure 2: Databases containing worldwide information.

All sources of information on a global scale tend to focus on disastrous events while leaving out minor occurrences and near-miss events, although both can be of the same importance in estimating future risks. This may be the consequence of a lack of capacities at the service providers' level to handle and process their input data. Additionally, especially in case of near-miss events, the involved parties often do not disseminate their information on the incidents for various reasons. Therefore, it is practically impossible to analyse near-miss events on a global scale, since this would require a worldwide collection and aggregation of regional available information where there exist too many missing values.

On the European level as well as in Germany and many other countries (e.g. the USA), information on accidents is collected with the help of (federal) state authorities or by act of law. Therefore, more detailed data is available on this scale: the MARS database[11] contains all major events occurring in chemical plants or on their territory that have to be reported according to EU legislation. In Germany, records exist on incidents concerning nuclear plants[12], chemical plants[13] and air traffic[14]. This allows both statistical analysis and case

[10] The aviation safety network website (http://aviation-safety.net/index.shtml) is just one remarkable example.
[11] JRC Major Accident Reporting System (MARS), http://mahbsrv.jrc.it/mars/Default.html
[12] Bundesamt für Strahlenschutz, http://www.bfs.de/kerntechnik/ereignisse
[13] Umweltbundesamt, http://www.infosis.bam.de/zema/zema_main.php

study examinations and, therefore, facilitates the design of reference scenarios. Due to the high level of risk awareness in Germany regarding nuclear hazards, information on near-miss events in this area is collected by order of law. This is also true of air traffic incidents and events in chemical plants. For the latter, additional and systematically collected information can be found in a non state-driven database specialized on near-miss events.[15]

Together with information about locations of hazardous facilities stored in the GIS-based inventory of critical infrastructure, the newly gained reference scenarios will be used for estimating the risk due to man-made hazards in certain regions. Starting point and pilot area is the Rhine-Neckar-Triangle (step 6 of the work packages).

5 Scenarios for Malicious Actions

A much more challenging task in comparison to the search for data on accidents and its analytical, and synthetical, processing is the building of scenarios for events caused by malicious actions. In those cases, a clearly defined cause-and-effect chain is difficult to detect since the actions of various kinds of perpetrators depend on aims that are hard to understand and, therefore, even harder to predict.

Our strategy for coping with these difficulties consists of several parts. Together they make up a creative process of scenario planning that can be compared in some respects with the modus operandi of people committing malicious actions. Similar to the analysis of accidents, statistical research has been done on terrorist events of global significance[16]. This helps to gain knowledge about distinctive features of attacks such as technologies used, types of terrorist (groups), course of actions and reactions, as well as losses incurred. The most important goal of this investigation, however, has been to find out and categorize the motives and purposes of the assaults. At the same time, methodologies from futurology, psychology, and economics have been studied and used to create scenarios of malicious actions. This was done by involving students of different faculties during an interactive course at the University of Karlsruhe (TH). Additionally, technical expertise could be made available through cooperation with state authorities for the surveillance of critical infrastructures.[17] This will assist in our next step – the developing of plausible, credible terrorist attack scenarios, based on inventory data and models of terrorist behaviour.

6 Outlook

As the above illustrated work packages are in progress the next steps, estimation of immediate loss potential, integration of damage propagation and the estimation of total loss potential can be expected to start soon as overlapping procedures.

The basics of the estimation of immediate loss potential are conducted by the project-spanning research group *"Asset Estimation"* within the whole project *"Risk Map Germany"*. Members of the research group *"Man-Made Hazards"* are also part of this group. It's results

---

[14] Bundesstelle für Flugunfalluntersuchung, flight safety information,
http://www.bfu-web.de/flusiinfo/index.htm
[15] Dechema Fachsektion Sicherheitstechnik Ereignisdatenbank (event database),
http://fach-for.dechema.de/sicherheitstechnik/deutsch/fa/ahag2/schaden/index.html
[16] National terrorism in Germany is of minor importance in terms of the terrorist threat at present, whereas criminal acts are quite common but less severe in their consequences for living in society.
[17] Landesamt für Umweltschutz Baden-Württemberg, Referat 31, Bundesamt für Bevölkerungsschutz und Katastrophenvorsorge

(see "Estimation of Building Values as a Basis for a Comparative Risk Assessment" by Kleist et al. in this book) will be adapted to our research group's needs.

Via links to other subprojects and cooperation with external experts the Man-Made Hazards research team will try to integrate damage propagation into its approach. The main challenge will be to estimate the total loss potential with the help of scenarios and simulation as this research work cannot be restricted to modelling of historical data.

The upcoming relevance of this research field in Germany is documented through current activities of public authorities like the "Bundesamt für Bevölkerungsschutz und Katastrophenhilfe", project "KRITIS" (critical infrastructure) or the next state of deNIS, deNIS II (deutsches Notfallvorsorge-Informationssystem) (Werner and Lechtenbörger, 2004). Both institutions contacted the research group "Man-Made Hazards" for further cooperation.

References

Haimes, Yacov Y. and Longstaff, Thomas: The role of risk analysis in the protection of critical infrastructure against terrorism. *Risk Analysis*, 22 (4): 439-444, 2002.

Mileti, Dennis: *Disasters by Design. A Reassessment of Natural Hazards in the United States*, Washington D.C.: Joseph Henry Press, 1999.

President's Commission on Critical Infrastructure Protection, (ed.): *Critical Foundations: Protecting America's Infrastructures: The Report of the President's Commission on Critical Infrastructure Protection*. Washington D.C.: U.S. Government Printing Office, 1997.

RMS, (ed.): *Understanding and Managing Terrorism Risk*. Available electronically at http://www.rms.com, 2002.

Swiss Re: *sigma no. 2/2003*, Zürich, 2003.

Werner, Ute and Lechtenbörger, Christiane: Phänomen Terrorismus. Die institutionelle Bearbeitung dieser Bedrohung in Deutschland. *Homeland Security*, 1 (2): 16-22, 2004.

Panel Discussion

Summary of the Panel Discussion

Dörthe Malzahn[1]**, Max Wyss**[2]
[1] *Institute for Mathematical Stochastics, University of Karlsruhe (TH), D- 76128 Karlsruhe, Germany;*
[2] *World Agency of Planetary Monitoring and Earthquake Risk Reduction, 36A Route de Malagnou, CH-1208 Geneva, Switzerland*

1 Introduction

The conference "Disasters and Society – From Hazard Assessment to Risk Reduction" 2004 Karlsruhe was closed by a panel discussion. The panelists were (in alphabetical order):

- **Walter Ammann**, Eidgenössisches Institut für Schnee- und Lawinenforschung, SLF Davos
- **Janos J. Bogardi,** United Nations University, Institute for Environment and Human Security, Bonn
- **Louise K. Comfort**, University of Pittsburgh, Pittsburgh
- **Alok Goyal**, Indian Institute of Technology, Civil Engineering Department, Bombay
- **Lothar Stempniewski**, Center for Disaster Management and Risk Reduction Technology (CEDIM), Universität Karlsruhe and GeoForschungsZentrum Potsdam
- **Friedemann Wenzel**, Sonderforschungsbereich "Starkbeben", Universität Karlsruhe
- **Max Wyss**, World Agency for Planetary Monitoring and Earthquake Risk Reduction (WAPMERR), Geneva

This is a summary of some of the ideas put forward during the dialogue between members of the panel and the audience. The discussion was moderated by Max Wyss. Dörthe Malzahn took the minutes of the panel discussion.

2 Definitions and Communication

There was some concern that not all technical expressions were used consistently and that some were not well enough defined. For example, *vulnerability* was used by some speakers to characterize the resistance of buildings to shaking, as well as the inability of a society to recover from a disaster. In contrast, it was pointed out that in the US the properties of buildings are referred to as *fragility*, whereas *vulnerability* is reserved for the condition of society. Other poorly defined words included *resistance, exposure* and *resilience*. There was general agreement that it would be desirable to adhere to specific definitions of words in technical communications. It was also argued that some terms, like *risk*, should be defined quantitatively by means of an equation. However, it was not clear how generally accepted definitions could be implemented. It was pointed out that the UN based International Strategy for Disaster Reduction (ISDR) is preparing a dictionary of scientific terms.

Some of the difficulty in sticking to specific definitions of terms may come from the interdisciplinary nature (science, engineering, sociology) of work addressing hazard analysis and risk reduction. In addition, it was stressed that researchers and experts must be able to communicate effectively with journalists, community leaders, decision makers and risk managers, if their work is to benefit society. This requires a vocabulary and a mode of phrasing that is understandable to non-specialists.

3 Information Technology, Training and Disaster Management

The need for effective real-time flow of information in a disaster situation was recognized. The bowtie-model, in which information flows from many directions to and from a center, was viewed as a good solution to the problem, provided that measures are in place to prevent collapse of the center. It was stressed that specific skills are required to manage such a center, because different organizations and individuals with different backgrounds interact with the center and with others through the center. It was pointed out that in Karlsruhe, and in other universities of Germany, courses in disaster management are now being offered or planned.

The question was discussed whether or not there are problems specific to interdisciplinary collaboration and which they may be. The ideal disaster manager would manage the whole circle from disaster prevention to disaster management. However, it was agreed that this is too complex to be done by one person. It was remarked that humans are limited in their problem solving capacity and that their ability to absorb and process information drops under stress. Thus, it was suggested that we should use information technology to increase the efficiency of human decision makers. For promoting the development from hazard oriented prevention techniques to the reduction of risk by organizational measures, tools for rapid and successful information exchange are central.

Finally, it was remarked that case studies are needed for all disciplines to learn about mistakes.

4 The Human Dimension

It was pointed out that risk managing measures are only as good as they can be implemented in the praxis during disasters. Therefore, the interaction with the public should be given more attention. The opinion was offered that the need to involve the public in dealing with disasters is underrepresented in current research. Misunderstandings in the communication between experts and affected people may be hampering mitigation measures. It was emphasized that a "technocratic" approach alienates people and thus reduces the effectiveness of the help intended.

5 Urbanization and Disasters

It was generally recognized that risk is increasing exponentially with time. One reason for this development is the growth of megacities in developing countries at an unprecedented rate. Activities to mitigate this risk are increasing at a very modest rate, or not at all. However, the unmitigated risk is not only a serious problem in megacities. If unprepared, smaller communities are hit by disasters, it can be so major that the community is unable to recover. The fact that the population of megacities alone is smaller than that of the entire population can be used to argue that it is not only these cities that need attention. On the other hand, they influence the economy and public life of a country in critical ways. Therefore, their protection from disasters is of special importance. To better plan the development of cities is becoming a key issue. Finally, it was stressed that on all levels the population should be involved in augmenting preparedness and disaster response.

Acknowledgements

During the panel discussion, the following were among the participants who contributed ideas: the seven panel members, D. Sakulski, U. Werner, R. Sinha, E. Plate, J. Birkmann, M. Bostenaru Dan, M. Zwick, J. Calmet and T. Mitchell. The names of some people who contributed to the discussion may be missing in this list.

Author Index